Fire Safety Design and Concrete

Concrete Design and Construction Series

SERIES EDITORS

PROFESSOR F. K. KONG
Nanyang Technological University, Singapore

EMERITUS PROFESSOR R. H. EVANS CBE
University of Leeds

OTHER TITLES IN THE SERIES

Concrete Radiation Shielding: Nuclear Physics,
Concrete Properties, Design and Construction
by M. F. Kaplan

Quality in Precast Concrete: Design — Production —
Supervision *by John G. Richardson*

Reinforced and Prestressed Masonry
edited by Arnold W. Hendry

Concrete Structures: Materials, Maintenance and Repair
by Denison Campbell-Allen and Harold Roper

Concrete Bridges: Design and Construction
by A. C. Liebenberg

Assessment and Renovation of Concrete Structures
by T. Kay

Concrete Structures in Earthquake Regions: Design and Analysis
edited by Edmund Booth

Concrete Structures: Eurocode EC2 and BS8110 Compared
edited by R. S. Narayanan

Fire Safety Design
and Concrete

T. Z. Harmathy

Longman
Scientific &
Technical

Longman Scientific & Technical
Longman Group UK Limited
Longman House, Burnt Mill, Harlow,
Essex, CM20 2JE, England
and Associated Companies throughout the world.

Copublished in the United States with
John Wiley & Sons, Inc., 695 Third Avenue, New York, NY 10158

© Longman Group UK Limited 1993
First published 1993

British Library Cataloguing in Publication Data
A catalogue record for this book is available from the British Library

ISBN 0582 07687 0

Library of Congress Cataloging-in-Publication Data
A catalogue record for this book is available from the Library of Congress

ISBN 0470 20005 7 (USA only)

Set by 4 in Compugraphic 10 pt Times
Printed and Bound in Great Britain at Bookcraft (Bath) Ltd.

Contents

Preface

About Fire Safety and this Book

Fire safety has to do with the probability of fire occurrence and the gravity of consequences if fire does occur. A wood shack may be regarded as a fire-safe place if it is occupied by people who are aware of the consequences of their actions. Conversely, a building equipped with an assortment of fire safety facilities may not be a safe place if it is inhabited by people who constantly probe society's tolerance. Analyzing fire losses in international dimensions, Banks and Rardin (1982) commented that virtually all indicators of lack of social cohesion, such as crime rate, rate of divorce, and illegitimate birth rate, positively correlate with per capita fire losses in buildings. Unflattering fire statistics often suggest some problems in the social fabric of a country. Solving the problems that cause social degradation would seem to be the most effective way of improving fire safety. Unfortunately, the tools to achieve that goal are in the hands of politicians who are rarely inclined to touch 'delicate' issues. All the building designer can do is to try to mitigate the consequences of fires. It would be naive to believe, however, that fire safety can be achieved solely by the application of fire science.

The seriousness of property losses and human losses is related to the extent of fire spread. Fire spread and destruction are associated ideas. Destruction is usually regarded as the cause, and fire spread as the result. Traditionally, spread is visualized as the advance of fire due to the structural failure of, or heat conduction through, building elements that separate the various compartments. To address this concern, the concept of fire resistant compartmentation has been developed. This concept regards the use of compartment boundaries resistant to structural or conductive failure as the basic tool in fire protection.

With modern construction practices, the probability of spread by destruction is very small, and the probability of spread by heat conduction is practically nonexistent (see Chapter 15). Spread takes place mostly by convection, i.e. by the advance of flames and hot gases through doors left open by fleeing occupants, through windows broken by the fire, or through other openings between compartments. The duration of the fully developed period of a fire in any one compartment in nonindustrial, nonmercantile buildings is, as a rule, between

20 min and 1 h. Fiercely burning fires that last longer than 1 h usually indicate intercompartmental spread by convection.

Although nowadays more and more attention is paid in building regulations to the possibility of convective spread, fire resistant compartmentation is still the backbone of the coercive protection measures. It is hoped, however, that as science keeps expanding the frontiers of knowledge, the stereotyped safety measures will be gradually abandoned, and avenues will be opened for the designer to find fire safety solutions commensurate with the prevailing problems.

The prescriptive fire safety measures contained in various building regulations will not be discussed in detail in this book. The purpose of the book is to provide the fire safety practitioner and the student of fire safety science with a knowledge base necessary for recognizing some vital safety problems and devising solutions. Since the book is being published in a series dealing with various aspects of the design of concrete structures, the discussion is slanted toward safety problems encountered in the design of concrete structures. It should be realized, however, that fire safety is the resultant of many factors, and cannot be appreciated if viewed from a narrow angle.

Fire safety science is an interdisciplinary field. It is a mishmash of subjects drawn from engineering, natural science, social science, and mathematics. Because of the great diversity of the areas to be covered, the author of a book on applied fire science has to be selective as to the topics to be included and to be left out. It is quite natural that the author's own expertise is the decisive factor in the selection. The reader will find that a few important topics, such as detection and suppression of fires, and evacuation of fire-stricken buildings, are not adequately covered in this book. More detailed discussions on these topics can be found in the references.

The SI system is used throughout the book. It offers great convenience in the field of fire science, which is a meeting ground for many disciplines. The only deviation from the SI system is the occasional use of °C instead of K, and h and min instead of s.

It is rather unfortunate that, although the SI system is still struggling for adoption, the adulteration of the system has already started. Many authors use in their publications the prefixed versions of the SI units (such as kW, MPa, GPa, mm, etc.) or compound units (such as $N\,mm^{-2}$) as base units (instead of W, Pa, m, etc.). Of course, the use of the prefixed units is perfectly all right in listing information, and presenting the results of tests or calculations. However, their use may lead to annoying conversion problems in areas where various disciplines overlap. For the sake of consistency, most equations in this book are written in terms of the the base SI units. The prefix M (mega) is used occasionally in tables or in the text, in an effort to avoid using powers larger than 6 in scientific notations.

T. Z. Harmathy
Ottawa, May 1992

Acknowledgements

We are indebted to the following for permission to reproduce copyright material:

American Concrete Institute for Figs 7.3 (Tovey, 1986), 10.5−10.7 (Harmathy, 1983a), 12.9 (Abrams & Gustaferro, 1969), 13.15 (Gustaferro *et al.*, 1971a), 13.16 (ACI Committee, 1989), 16.1 (Harmathy, 1968); the managing editor, *American Journal of Science* for Figs 6.13, 6.14 & Table 6.9 (Birch & Clark, 1940); American Society for Testing & Materials (ASTM) for Figs 4.5, 4.6 (Harmathy, 1965a), 6.2−6.5, 6.7, 6.8, 6.19 (Harmathy, 1970a), 7.5−7.8 (Harmathy & Stanzak, 1970), 7.11 (Harmathy, 1967d), 12.1, 12.6, 12.7 (Harmathy, 1979a), 13.8 (Harmathy, 1961), 14.7 (Issen *et al.*, 1970) & Tables 4.2 (ASTM, 1988a), 6.1, 6.2, (Harmathy, 1970a), 6.11 (Harmathy, 1970a), 7.2, 7.3 (Harmathy & Stanzak, 1970), 12.1 (ASTM, 1988a), 12.2 (Harmathy, 1970b) Copyright ASTM. Reprinted with permission; Elsevier Applied Science Publishers Ltd. for Figs 10.2 (Harmathy, 1980/81), 12.11 (Harmathy, 1977/78); Elsevier Sequoia S.A. and the author, H. Gross for Fig. 6.17 (Gross, 1975); Energy, Mines & Resources Canada for Fig. 6.11 & Table 6.6 (Geller *et al.*, 1962); the author, T.T. Lie for Fig. 3.5 (Lie, 1972); National Fire Protection Association for Figs 4.4 (Harmathy & Lie, 1971); 10.3 (Harmathy, 1972), 10.8, (Lie, 1974), 11.4, 11.5 (Harmathy, 1976), 12.4 (Harmathy, 1965b), 15.1 (National Fire Protection Association, 1986) & Table 1.4 (Clarke & Ottoson, 1976) Copyright © 1965, 1971, 1972, 1974, 1976 & 1986 National Fire Protection Association; National Research Council of Canada for Figs 6.9, 8.2−8.4 (Harmathy, 1983b), 13.5 (Harmathy, 1963), 14.9 (Lie *et al.*, 1984) & Tables 4.1 (Harmathy, 1967a), 6.12, 12.3−12.5, 14.1 (Associate Committee on the National Building Code, 1990), 13.2 (Harmathy, 1963); Portland Cement Association for Figs 12.8 & 13.14 (Abrams & Gustaferro, 1968); Society of Fire Protection Engineers for Table 9.4 (Purser, 1988) Copyright © 1988 Society of Fire Protection Engineers; John Wiley & Sons Ltd. for Figs 3.7 (Harmathy & Mehaffey, 1982), 10.4 (Harmathy, 1979a), 11.2, 11.3 (Harmathy & Mehaffey, 1984) & Tables 3.1 (Harmathy & Mehaffey, 1982), 9.1 (Harmathy, 1984), 15.2−15.4 (Harmathy *et al.*, 1989) Copyright © 1979, 1982, 1984 & 1989 John Wiley & Sons Ltd.; John Wiley & Sons Inc. for Table 8.3 (Kroschwitz *et al.*, 1985−1989) Copyright © 1985−1989 John Wiley & Sons Inc.; the editor, *Wood and Fiber Science* for

Fig. 9.1 (Harmathy, 1977); the authors, D. Yung & V.R. Beck for Table 15.1 (Yung & Beck, 1989).

Whilst every effort has been made to trace the owners of copyright material, in a few cases this has proved impossible and we take this oportunity to offer our apologies to any copyright holders whose rights we may have unwittingly infringed.

1 Fire in the Light of Statistics

The value people put on their personal safety depends on their perception of the level of risk. In an experiment conducted at Dundee University under contract to the Fire Research Station of the Building Research Establishment, UK (Green and Brown 1978), 70 students were asked to assess how much riskier, in their opinion, are a number of activities in relation to a hypothetical 'absolutely safe' activity. Table 1.1 shows the results. Also shown in the table are some FAFR (Fatal Accident Frequency Rate per 10^8 h of exposure) values, as reported by Kletz (1971). From a comparison of the two sets of values it appears that (1) people tend to consider mainly lethal accidents in judging the riskiness of various activities, and (2) intuition is a fair guide in ranking personal risks, but it is a poor guide in assessing the magnitude of these risks. In general, there is a tendency to underestimate the riskiness of voluntary activities, especially if they are undertaken with intent to derive some benefits.

It is clear from the first item in Table 1.1 that people do not think much of the possibility of dying from fire in their homes. Their apathy comes as no surprise in the light of some factual information on mortality in accidents (US Bureau of Census 1990), which shows the death rate by fire as approximately equal to that by drowning (Table 1.2).

The individual's perception of safety is, however, quite different from society's perception. For the most part, society is more concerned with the possibility of multiple deaths than with (more frequent) single deaths. If a single person is killed, society tends to blame the person for his fate. If, however, many people are killed at the same time, the blame is often shifted to society, for not recognizing and acting on some potentially dangerous situations.

It is, of course, inaccurate to judge the ills of fire mainly on the basis of mortality statistics. As Table 1.3 shows (Wilmot 1979, 1989; Perry 1986; Tudhope 1989), direct property losses alone amount to 0.07−0.4 per cent of the gross national product (GNP). The associated costs (including indirect fire losses, the cost of fire-fighting organizations, and insurance administration costs) add up to 0.31−0.47 per cent of the GNP (Wilmot 1979, 1989; World Fire Statistics Centre 1983), and the cost of fire protection of buildings is estimated to amount to

Table 1.1 Perceived Relative Riskiness (RR) and Fatal Accident Frequency Rate (FAFR) values for some common activities and situations (Green and Brown 1978; Kletz 1971; Institution of Engineers of Australia 1989; for a hypothetical no-risk activity RR = 1.0)

Activity	RR	FAFR per 10^8 h exposure
Staying at home	1.63	3
Swimming in a swimming pool	2.12	
Staying in a hotel	2.62	
Traveling by train	2.72	5
Crossing the road	3.61	
Traveling by bus	3.68	3
Traveling by car	4.06	57
Skiing	5.12	
Traveling by plane	5.25	240
Riding a motorcycle	8.08	660
Rock climbing	10.56	4000

Table 1.2 Annual mortality in accidents for 1986 (US Bureau of Census 1990)

Accident	Annual mortality per million people
Motor vehicle	199
Falls	47
Poisoning	24
Fires, flames	20
Drowning	20
Inhalation or ingestion of objects	15
Medical procedures	13
Transportation (other than motor vehicle)	12
Firearms	6
Electrical	4
All other	35

0.14−0.7 per cent (Beck *et al.* 1989). All in all, somewhere between 0.5 and 1.5 per cent of the GNP goes to waste because of fire.

The information available on fire losses is by no means reliable. (In fact, some of the data in Table 1.3 are clearly inaccurate.) In North America the majority of data are based on a sampling of the fire departments' records across the country. No allowance is made for unreported fires or losses, for fires attended solely by private fire brigades, or for those extinguished by fixed suppression systems without participation by the fire department (Karter 1988). The claims processed by insurance companies could, in principle, provide a better basis for preparing national fire statistics. Unfortunately, many property owners do not carry fire insurance, or under-insure their properties. Furthermore, insurance companies do not keep accurate records of the payments made against fire-related claims

Table 1.3 Direct losses of property as a percentage of the GNP, and fire fatalities per million people (Wilmot 1979, 1989; Perry 1986; Tudhope 1989)

Country	Property loss (% GNP)	Year(s)	Fire fatalities per million people	Year(s)
Australia	0.27	1981–1984	11	1983
Austria	0.18	1986	8	1986
Canada	0.19	1986	22	1986
Denmark	0.25	1986	11	1986
Finland	0.12	1986	19	1986
France	0.29	1981–1982	16	1984
Italy	0.15	1970–1975	8	1970–1975
Japan	0.19	1984–1985	16	1984–1985
Netherlands	0.12	1986	4	1986
Norway	0.40	1986	19	1986
South Africa	0.16	1986	29	1986
Sweden	0.23	1986	12	1986
Switzerland	0.07	1986	4	1986
United Kingdom	0.12	1986	17	1986
United States	0.16	1986	21	1986
West Germany	0.22	1986	8	1983

to homeowners with multiple-peril insurance policies. Clearly, information on fire fatalities is the most reliable part of the fire statistics.

Several countries also publish information on fire-related injuries. For 1987, the National Fire Protection Association, USA (NFPA) records (Karter 1988) show 5810 deaths and 28 215 injuries. Strictly speaking, these 'human losses' should also be taken into account when determining national fire losses.

The value of human life and the cost of injury are very controversial subjects. There are many who question the morality of putting a price tag on human life and suffering. The economist, however, cannot be sensitive about moral issues. It is customary to choose US$500 000 as the value of life, and $20 000 as the cost of injury (Ruegg and Fuller 1984). With these figures the human losses in the USA would amount to about 0.08 per cent of the GNP. Clearly, even though the deaths and injuries receive much of the attention, fire is primarily a problem that concerns society's material wealth.

From a survey of a number of publications on national (Magnusson 1978; Bryan 1979; Berl and Halpin 1980; Labour Canada 1988; Karter 1988; Fire Defense Agency 1988; Beck et al. 1989; Home Office 1990), and international (Banks and Rardin 1982; Beck et al. 1989; Taday 1989, 1990) fire statistics, several interesting conclusions can be drawn. Some of the most noteworthy conclusions are as follows:

• Residential buildings account for the largest percentage of fires, usually 50 per cent or more. A disproportionately high percentage of deaths occur in one- or two-family dwellings, mobile homes, apartments, hotels, motels, and boarding homes.

- Roughly 50 per cent of all fires are restricted to the object ignited, and 80 per cent are confined to the area of origin.
- Fires are most frequent during the heating season, especially in winter and early spring. Well over half of the fire fatalities occur at night.
- Incendiarism, careless use of smokers' materials, and accidents related to cooking and heating are the main causes of fire.
- In residential buildings the kitchen, the living room, and the bedroom are the three most prominent areas of fire origin. Robes, housecoats, upholstered furniture, nightgowns, pajamas, and dresses are usually the items first ignited.
- Since the building's structural components and the interior finish are seldom the items first ignited, the building contents contribute far more to fire initiation and development than the construction elements.
- More than one-third of fire-related casualties occur in the area of fire origin, and more than 50 per cent occur on the floor of origin. The victims are often asleep or impaired in some way at the time of exposure to lethal conditions.
- Smoke inhalation is the most frequent cause of death when the fire or its effects spread beyond the area of origin. The immediate cause of death is usually carbon monoxide intake.

After surveying the available information, Clarke and Ottoson (1976) identified and ranked the most frequent scenarios for fire death in the United States. Their findings are shown in Table 1.4. The scenarios highlight the areas where education of the public rather than the application of technology may prove fruitful.

From Table 1.3 it appears that for countries with populations of similar cultural and ethnic background the fire losses depend, to some degree, on the country's climate, probably on account of the prominence of heating among the causes of fire.

Banks and Rardin (1982) studied the fire loss experience in various developed countries. They found very significant, positive correlations between human or property losses in fire and various economic, technological, sociological, and cultural factors. However, they shied away from drawing any conclusion concerning the causalities. Their cautiousness is characteristic of the present age in which it is considered uncivil to talk of differences between identifiable groups of persons or nations.

Fires are caused, more often than not, by people of deficient sense of social responsibility. Although there are relatively few of them, they are responsible for most of the troubles that plague today's society. Short-sighted administrators constantly clamor for newer and stricter fire safety regulations, which drive up the cost of housing, and thereby penalize the whole of the community, the irresponsible and the conscientious alike.

It is doubtful if problems that have a strong social component can be solved by regulations.

Table 1.4 The most important fire death scenarios (Clarke and Ottoson 1976). Reprinted with permission from *Fire Journal* ®, (Vol. 70, No. 3), copyright © 1976, National Fire Protection Association, Quincy, MA 02269. *Fire Journal* ® is a registered trademark of the National Fire Protection Association Inc., Quincy, MA 02269

Rank	Occupancy	Item ignited	Ignition source	Percentage of US fire deaths
1	Residential	Furnishings	Smoking	27
2	Residential	Furnishings	Open flame	5
3	Transportation	Flammable liquids	Several	4
3	Various	Apparel	Heating, cooking equipment	4
3	Residential	Furnishings	Heating, cooking equipment	4
4	Various	Apparel, flammable liquids	Several	3
4	Residential	Flammable liquids	Heating, cooking equipment	3
4	Residential	Flammable liquids	Open flame	3
4	Various	Apparel	Open flame	3
5	Residential	Interior finish	Heating, cooking equipment	2
5	Residential	Interior finish	Electrical equipment	2
5	Various	Apparel	Smoking	2
5	Residential	Structural components	Electrical equipment	2
5	Residential	Trash	Smoking	2
Others, all less than 2 per cent of total				34
				100

2 Building Regulations

Control over safety and public health in buildings is achieved through the following three devices (Garnham Wright 1983):

(1) through lawful coercion;
(2) through the device of finance, used either as constraint or inducement by financing organs (e.g. by a government that has responsibilities over certain types of buildings, or by mortgage and insurance companies) toward achieving certain ends (better housing, increased safety, energy conservation, protection of financial interests); and
(3) through the device of privilege exercised by the public at large (in selecting certain types of person to perform tasks that require expertise and/or experience), and by the suppliers of services (water, electricity, gas) who make rules for installations and enforce them by withholding the supply from those who do not obey.

The need for using legal instruments when dealing with the problems of public safety was recognized as early as the 16th century BC, when Hammurapi, King of Babylon, ordered that the death penalty be imposed on any builders whose work was so seriously faulty as to cause the collapse of the building and the death of its occupant.

Efforts to deal with the problem of public safety usually gain momentum after major disasters. Thus, following the conflagration of Rome in 65 AD, Emperor Nero implemented regulations concerning the minimum widths of streets and materials from which houses could be built. The office of *praefectus vigilum* was established; his duty was to investigate the cause of fires and to impose monetary or corporal punishments, or even the death penalty, on those found guilty of starting the fire.

The great fire of London in 1666 gave rise to the London Building Act. The Lord Mayor and the City Council were empowered to appoint surveyors to enforce the requirements.

The Building Research Station (BRS) in the United Kingdom conducted studies on the structure of building control in several countries in Europe. Mainly on the basis of BRS study (Atkinson 1971, 1974; Cibula 1971), Garnham Wright (1983) reviewed the international scene.

The structure of building control depends on

(1) how much the central government participates in the control,
(2) which (public or private) organ is in charge of the instruments of control, and
(3) how the control is administered and enforced.

The participation by the central government can be

(a) practically nonexistent (as in France and Switzerland),
(b) rather weak, consisting mainly of passing legislation in specific cases of national importance (as in the United States),
(c) more substantial, consisting of passing primary acts that identify what the controls should contain and how they should work, and (in the interest of uniformity of safety provisions) of preparing or arranging for the preparation of model documents that give the essentials of regulations (as in Canada, Germany, and the Netherlands),
(d) decisive, consisting of producing building laws and national building regulations (as in the United Kingdom and the Scandinavian countries).

The organ in charge of the instruments of control (i.e. in charge of bringing in regulations, or writing rules and criteria with which the buildings must conform) may be

(a) the central government, perhaps leaning to a degree on lower levels of government (as in the United Kingdom and the Scandinavian countries),
(b) state (provincial) governments which may share their law-making functions with, or pass them on entirely to, the local governments (as in some states in the United States, and in Canada and Germany),
(c) local governments (as in some other states of the United States),
(d) local governments, making use of the model documents prepared under the auspices of the central government (as in the Netherlands),
(e) insurers who indemnify the building designers (as required by law) and set their premiums according to standards approved by public bodies (as in France),
(f) designers operating under professional codes of practice, and liable to their clients under private contract for their actions (as in Switzerland).

The administration of the regulations and the tasks for their enforcement are delegated to the local authorities everywhere except France and Switzerland. In these two countries the courts of law are the ultimate enforcing bodies. However, in France the insurance companies exercise a substantial amount of control through on-site inspections, and by requiring that the construction materials and products be certified by testing under an arrangement known as the Agrément system.

As mentioned earlier, in Canada, Germany, and the Netherlands the central government has some role in the preparation of model documents which form the basis of building regulations. In the United States the model documents are prepared by nonprofit, nongovernment organizations.

Building regulations make frequent reference to standards and codes of practice, which are documents giving detailed prescriptions or advice with respect to performance of materials and products, and methods of manufacture or construction. The standards represent consensus by those with some interest in the industry, and do not originate as legislation. The bodies charged with the preparation of standards may be

(a) for the most part professional and research-oriented organizations (as in the Scandinavian countries, Switzerland, and France),
(b) widely based organizations, consisting of representatives from the industry, consumer groups, and the academic and research community (as in most other countries).

The standards-writing bodies may (as in most countries) or may not be involved in the certification and listing of materials and products that conform to the requirements of the various standards.

The French method of building control has evolved from the sound principle that the central government should not interfere in a major way in the everyday operation of society. Unfortunately, the rule of the free market system is all but defeated by the constitutional edict that the designers and contractors are liable for any major defects in the building for a period of ten years, and by the legal requirement that the designers and contractors protect themselves with insurance. Consequently, the insurers, who provide their indemnity, have become involved not only with framing the technical requirements, but also with the processes of control and inspection.

Controls that rely on public authorities to set the rules and enforce compliance are thought to be superior to the control exercised by insurance companies. It is believed that the elected representatives will always have the true interests of their electorate in sight, whereas the insurers are influenced mainly by considerations of economics. Those who advocate this view usually have a blind faith in democratic institutions. In fact, democratically constituted committees are known to be inefficient and to turn often into battlegrounds of opposing interests. There are ample signs that committees in which highly qualified professionals have the upper hand (such as those in the Scandinavian countries and Germany) are capable of keeping abreast with the advances in science and technology much better than their democratically constituted counterparts.

The lack of ability to adjust to the rapidly changing world of science and technology is a serious drawback in any field of engineering which is under substantial controls. To shorten the protracted delays in putting the results of research to use, there is a move in some countries toward the adoption of the performance concept. The essence of the concept is to specify materials, products, and technical solutions in terms of what is meant to be achieved, rather than in terms of particular kinds, brands, values, and arrangements. England and Wales have already taken a big step in the right direction by introducing in 1985 a new set of building regulations, with a format changed from prescriptive to functional requirements (Fisher 1985). The writers of the new regulations have hoped that this new format will encourage the use of improved materials and solutions founded on scientific knowledge (rather than on conventionalism), and result in more cost-effective construction.

The opponents of the performance concept wasted no time to point out that the advantages of adopting the new format are more than offset by some serious drawbacks. Although the performance concept does allow the architect and engineer to incorporate in the design the latest materials and solutions founded on scientific knowledge, the cost of proving the worthiness of using state-of-the-art technology may be prohibitive, and the time lag between the design and implementation may extend to several years. Schulman (1988) claims that in New York City the approval process may drag on for two years.

Unfortunately, approval does not always ensure successful use. Failure means lawsuits and insurance problems. According to Schulman, the liability insurance for firms whose yearly gross is less than $500 000 has risen over the past decade from 2.3 to 5.7 per cent. In a profession where the profit margin is traditionally about 6 per cent, the cost of liability insurance for small and medium-sized practices now equals their profit. As a result, increasing numbers of firms choose to practice without insurance and to accept liability for their designs. Thus, state-of-the-art solutions, innovation, imagination, and all the other exciting possibilities once promised by the advocates of performance regulations are fading in the glare of lawsuits and exorbitant insurance premiums.

Disturbed by colorful reports on disastrous fires, and stirred up by consumers' advocate groups, the public clamors for more and more regulations. It appears that once the need for regulations is accepted by society, the natural tendency for the regulated items is to grow year by year. A paradoxical situation has arisen in some countries: even measures aimed at allaying the harmful effects of the conventionalism built into the regulations must be implemented by further regulations. Researchers, instead of fighting for the partial dismantling of the regulatory maze, try to gain respectability for their research findings by having them written into the regulations.

New agencies have been organized to deal with the problems associated with the introduction of new materials and products. The Agrément Board in the United Kingdom can make 'authoritative assessments of new building products' and issue certificates valid for three years. The Canadian Construction Materials Centre offers the service of evaluating new and innovative products and products where standards exist but a certification program has not yet been established.

As to the application of nonconventional designs, Yung and Beck (1989) have suggested a technique for assessing, in terms of two parameters (expected risk to life, and fire cost expectation), whether an innovative design is equivalent to the requirements specified in the building regulations. Harmathy *et al.* (1989), relying on statistical data and Delphi decisions (see Chapter 15), also developed a technique by which the equivalence of any two sets of fire safety measures (one may be based on conventionalism, the other on scientific findings) can be determined. The big question is, of course, whether the insurance companies are willing to accept nonconventional solutions (in which there is always some element of unpredictability) in lieu of so-called 'time-tested' solutions incorporated in the existing building regulations, without raising the insurance premium for the building designer.

What seems to be rather paradoxical about those time-tested solutions is that many of them were adopted on historical grounds, and their soundness has never been examined, let alone proved. But even if they had been verified, creating rules for the future from past experience has serious limitations (Wilson 1973). Many years may be required to produce a statistically significant record of events for analysis. Factors that gave rise to past events may no longer be present, and therefore the accumulated experience may not be suitable to establish probabilities of failure of current buildings, not to mention buildings now designed for future use. Clearly, those (enforcing agencies or insurance companies) who decide on the acceptability of equivalent solutions must, to a great extent, rely on the reputation of the designer.

It appears that the Swiss system of building control, which places the responsibility for the proper functioning of buildings squarely on the building designer and contractor and uses the least amount of lawful coercion, is the best system from the public's point of view. A drastic deregulation of the building industry in all the other developed countries would perhaps result in a temporary increase of fire deaths and property losses, but would no doubt lead to better and less expensive housing, and thereby serve the long-term interests of the public. Statistical data (see Chapter 1) indicate that fire is not the most menacing danger that a person has to face in everyday life. Even a tenfold increase in the number of fire deaths would not match the number of deaths due to transportation accidents.

As mentioned in the Preface, the law-enforced safety measures will not be discussed in detail in this book. The subject of fire safety will be considered mainly from the point of view of technical soundness.

3   Fire Test Standards

Since building regulations, usually rather voluminous tomes, cannot cover all necessary details, they frequently refer to other documents. For example, the National Building Code of Canada, a model code prepared under the auspices of the federal government, cites, directly or indirectly, almost 200 other references, mainly standards which, as mentioned earlier, are documents prepared by professional organizations or widely based standards-writing committees. Among them, the fire test standards are especially important, since they provide the foundation for classifying the fire performance of materials, products, and building components.

There are numerous fire-related standards upon which building regulations lean. In the United States, for example, 26 standards are under the care of ASTM (American Society for Testing and Materials) Committee E-5 which is responsible for most fire test standards of principal concern to the building designer. Some 21 other committees are involved in preparing and updating standards relevant to fire safety.

Many of these test standards were originally written in response to some urgent need for quality control in the industry in an early stage of scientific and technical development, and contain cookbook-type prescriptions. The results of these standard tests are, in a strict scientific sense, of limited applicability at best, or utterly misleading at worst. Efforts to update them or to scrap them altogether are, however, fiercely resisted by one or another segment of the industry for fear of losing their competitive edge.

On the international scene, efforts by the International Standards Organization (ISO) to harmonize various national standards run into yet another kind of difficulty: rivalry among the leading industrial countries.

A succinct review of the most common American fire test standards has been presented in Volume 7 (pp. 196−205) of the *Encyclopedia of Polymer Science and Engineering* (Kroschwitz *et al.* 1985−1989). A paper by Malhotra (1980) has examined the history of fire testing in the United Kingdom and outlined the role of the ISO in the development of fire tests in the European Economic Community countries, in the Scandinavian countries, and in Eastern Europe.

Some of the numerous fire test standards will be discussed later in the appropriate chapters of this book. Two groups of standards, those of foremost interest to the building designer, will be reviewed in this chapter: the so-called 'reaction to fire' tests, and the fire resistance tests.

The results of tests in the first group are looked upon primarily as information on materials or products with respect to their behavior prior to 'flashover' (to be discussed in Chapter 9). That information is not any fundamental material property, but some kind of measure which is influenced partly by all those properties that have some bearing on the test specimen's ignitability, burning behavior, and smoke-producing propensity, and partly by the conditions imposed on the specimen in the test.

In the United States ASTM E 84 (American Society for Testing and Materials 1989a) is the standard test commonly used for characterizing the 'surface burning behavior' of building materials, particularly lining materials applied to walls or ceilings. The test equipment is shown in Fig. 3.1. The E 84 test ('tunnel test') calls for a specimen of 7.3 m by 0.5 m which, when installed for testing, will form the upper surface of a tunnel-like, horizontal furnace chamber of 0.5 m by 0.3 m inside cross-sectional dimensions and 7.6 m in length. In the test, the specimen is exposed to a controlled air flow and to flames delivered at one end of the furnace by two methane burners. The rates of air and gas flow are adjusted in a calibration procedure to cause the fire to spread in 5.5 min along the entire length of a specimen made from select grade red oak. The outlet end of the furnace is equipped with a photometer system consisting of a lamp and a photoelectric cell. Recorded in the test are the position of the flame front (based on observations through windows on one side of the furnace) and the photoelectric cell output, as functions of time. The test is aimed at the evaluation of the *flame spread index* (FSI) and the *smoke index*. These indices are calculated by comparing the areas under the flame spread distance versus time and the photocell output versus time curves, respectively, with those for red oak.

In Europe, a three-test package has been suggested for use as a *reaction to fire* test. These are (1) the BS 476:Part 7 test (British Standards Institution 1981), (2) the NF P 92-501 test (Norme française 1985), and (3) the DIN 4102 Part 1 'Brandschacht' (flue) test (Deutsches Institut für Normung 1981). The results yielded by the BS 476:Part 7 test are probably least dependent on the conditions imposed on the material or product in the test.

The specimens required for these three tests are much smaller than those used in the ASTM E 84 test. In the BS 476:Part 7 and NF P 92-501 tests, the specimen is exposed to a radiation source of defined intensity, and the burning of the specimen is initiated by pilot flame(s). In the DIN 4102 Part 1 'Brandschacht' test, the burning is initiated (without imposed radiation) by flame(s) touching the lower edge of vertically placed specimen(s). The materials or products are classified on the basis of the distance and rate of flame spread, the effects of heat evolved from the burning specimen(s), and from other records and observations taken during the tests.

Fig. 3.1 Test equipment in the laboratory of the National Research Council of Canada, for performing ASTM E 84 'tunnel' tests

To the designer of concrete structures, the fire resistance test standards are of primary interest. Fire resistance testing began in Germany in the 1880s and the United States and England in the 1890s (Babrauskas and Williamson 1978). Efforts to standardize fire testing started in the early 1900s (Shoub 1961).

The various national standards, such as

- AS 1530.4 Part 4 in Australia,
- NBN 713020 in Belgium,
- CAN 4-S101 in Canada,
- DS 1051 in Denmark,
- DIN 4102 Part 4 in Germany,
- BS 476:Parts 20, 21, 22, and 23 in the United Kingdom,
- ASTM E 119 in the United States

have evolved on slightly divergent paths and have undergone significant changes during the years. Consequently, the results yielded by them are not necessarily interchangeable.

Because the fire resistance test standard under the care of the International Standards Organization (ISO) contains all the essential features of similar standards, and because the ISO standard is expected to gain acceptance eventually by all nations (with the possible exception of the United States[1]), only the ISO 834 standard (International Standards Organization 1990) will be discussed here in detail.

The ISO 834 standard consists of three parts. The general requirements are presented in the first part, whereas the second part deals with special requirements for various building elements. The third part is a commentary on the test method and on the use of the test results.

To be able to test all kinds of building elements, a laboratory may need four test furnaces:

(1) a wall furnace for heating vertical separating elements[2] (walls, partitions) on one side,
(2) a floor furnace for heating horizontal separating elements (floors, roofs) on the underside,
(3) a beam furnace for heating flexural elements (girders, beams, trusses) on three or four sides, and,
(4) a column furnace for heating columns or walls from all sides.

Floor furnaces may also be used for performing tests on flexural elements. Figures 3.2, 3.3, and 3.4 show the wall-, floor-, and column-furnaces of the National Research Council of Canada.

In the case of wall-, floor-, and beam-furnaces, five sides of the furnace chamber are formed by the furnace itself, and one side by the test specimen. A column-furnace completely surrounds the specimen. Each furnace is equipped with burners, flue, and devices for measuring and controlling the temperature and pressure of the furnace gases.

In the fire resistance test, a specimen of the building element to be examined is exposed on one side (separating elements), three sides (flexural members), or all sides (columns, and possibly some walls) to the hot gases in the test furnace, the average temperature of which is controlled to vary with time in a prescribed way. The temperature–time curves specified in the various national standards are shown in Fig. 3.5 (Lie 1972). The upper curve in Fig. 3.6 shows the temperature–time curve specified in ISO 834.

The temperature of the furnace gases is measured by five to eight bare-wire thermocouples,[3] uniformly distributed in the furnace at specified distances from the specimen's surface. The 'furnace temperature' is interpreted as the average of the temperatures at the points of measurement. That temperature is monitored in the test, and controlled to follow the relation

$$T_F = 293 + 345 \log_{10}(1 + 0.133t) \qquad [3.1]$$

---

1   The relation between the ISO 834 and ASTM E 119 test methods and the results of the two kinds of test have been discussed by Harmathy *et al.* (1987) and Harmathy and Sultan (1988).
2   See discussion on *key elements* and *separating elements* in Chapter 12.
3   The ASTM E 119 standard (American Society for Testing and Materials 1988a) specifies nine thermocouples inserted into and shielded by steel or Inconel pipes.

Fig. 3.2 Test equipment in the laboratory of the National Research Council of Canada, for performing fire resistance tests on vertical separating elements

Fig. 3.3 Test equipment in the laboratory of the National Research Council of Canada, for performing fire resistance tests on horizontal separating elements (floors, roofs) and flexural members

Fig. 3.4 Test equipment in the laboratory of the National Research Council of Canada, for performing fire resistance tests on columns

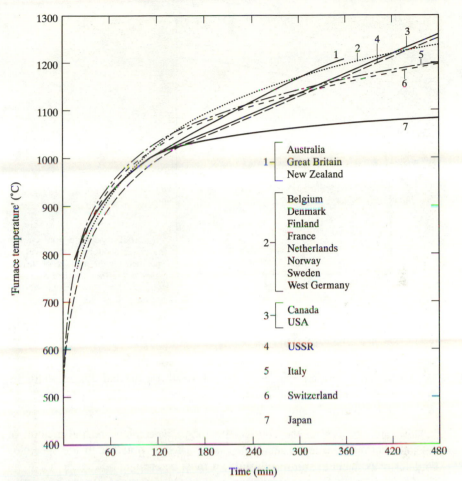

Fig. 3.5 Temperature–time curves specified for the average of furnace gases in various national fire resistance test standards (Lie 1972)

where T_F = space-averaged temperature of furnace gases,[4] K;
     t = time, s.

The pressure is measured at two points. In general, it is required that the pressure near the top of the furnace be kept at about 20 Pa above the pressure of the ambient atmosphere.

---

4   In some calculations related to the prediction of fire resistance of building elements, use will be made of the time-averaged 'furnace temperature', $(T_F)_{av}$ (K), defined as

$$(T_F)_{av} = \frac{1}{\tau} \int_0^\tau T_F dt \qquad [3.2]$$

where τ (s) is the duration of fire test (i.e. duration of fire exposure of the test specimen). The $(T_F)_{av}$ versus τ curve is also shown in Fig. 3.6.

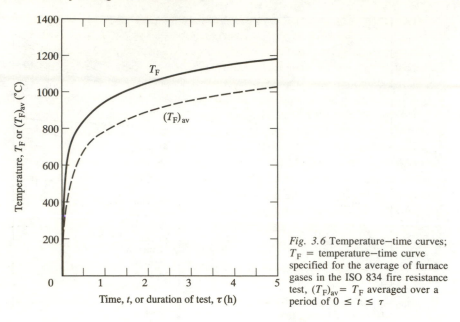

Fig. 3.6 Temperature–time curves; T_F = temperature–time curve specified for the average of furnace gases in the ISO 834 fire resistance test, $(T_F)_{av}$ = T_F averaged over a period of $0 \leq t \leq \tau$

The test specimens are required to be of the following minimum dimensions:

- vertical separating elements: 3 m in height and width,
- horizontal separating elements: 4 m in length (span) and 3 m in width,
- flexural elements: 4 m in length,
- columns: 3 m in height.

The materials used in the construction of the specimen and the method of construction must be representative of those used in practice. If the specimen contains or is liable to absorb moisture, it must be brought before the test to an air-dry condition, defined as equilibrium with an atmosphere of 50 per cent relative humidity at 23 °C (296 K).

When testing specimens of vertical and horizontal separating elements, the temperature of the specimen's surface away from the furnace (the 'unexposed surface' temperature) must also be measured at five points (near the center and near the centers of the quarter-sections) in a manner specified in the standard.

The test specimen has to be mounted in a special support or restraint frame in such a way as to simulate the boundary (edge or end) conditions expected to arise in practice. These conditions may mean (partial) restraint against thermal expansion and edge (end) rotation (see Chapter 12), or they may mean freedom for deformation to occur. The specimen may be tested with one or other of these boundary conditions applied to all or only some of its edges. The choice is made on the basis of a careful analysis of the circumstances. If the boundary conditions are uncertain or undefinable, the specimen must be tested with the most adverse edge (end) conditions that may arise in service.

Specimens of load-bearing building elements must be loaded during the fire resistance test. The imposed load is determined on the basis of the actual or characteristic properties of the components that carry the principal stresses, according to a design method specified in a recognized structural code. Alternatively, it can be determined on the basis of the service load specified in a code of practice.

With specimens of load-bearing walls, the load is to be imposed vertically, and the vertical edges are to be left unrestrained. With floors, roofs, and flexural elements, if some of the specimen's dimensions are smaller than those of the actual building element, it is important that the nature and magnitude of the stresses arising in the actual element be reproduced. With columns, the load on the specimen must be calculated in such a way that both the compressive stresses and bending moments in the actual column are reproduced. The way the ends are held (fixed, hinged, restrained) is also an important consideration.

The fire resistance test is usually terminated shortly after the 'failure' of the test specimen. Failure may occur in three ways:

- with specimens representing load-bearing elements: loss of load-bearing capacity,
- with specimens representing any kind of vertical or horizontal separating element: loss of integrity,
- with specimens representing any kind of vertical or horizontal separating element: loss of insulating capability.

The loss of load-bearing capacity is evaluated

(1) with horizontal separating elements and flexural elements: from the deflection and rate of deflection of the specimen at midspan,
(2) with axially loaded elements (walls, columns): from the axial deformation and rate of deformation.

Loss of integrity is judged to arise if sufficiently large gaps or openings form in the specimen. Whether the gaps or openings are large enough to cause failure is decided from measurements specified in the standard.

The insulating capability of the specimen of a separating element is deemed to have been lost when

(1) the average temperature of its unexposed surface reaches a limit 140 K above the initial temperature, or
(2) the temperature at any point of its unexposed surface exceeds 180 K above the initial temperature.

The failure of the test specimen is usually deemed to occur at the time when the first of the applicable failure criteria is observed. This time is referred to as the *fire resistance* of the building element represented by the specimen.[5]

Some errors in the measured fire resistances may occur on account of the testing

---

5   However, ISO 834 allows that the three kinds of failure (namely, by loss of load-bearing capacity, by loss of integrity, and by loss of insulating capability) be separately reported.

laboratory not being able to follow accurately the temperature–time relation prescribed for the furnace in the test specification (Eq. 3.1). There is in the ASTM E 119 standard provision for correcting the test result if substantial deviations from the prescribed and actual temperature–time relations have occurred during the test. The ASTM correction formula is

$$\frac{\Delta \tau}{\tau} = \frac{2}{3} \frac{A - A_s}{A_s + L} \qquad [3.3]$$

where $\Delta \tau$ = fire resistance correction (positive if $A > A_s$, negative if $A < A_s$), s (min);[6]

τ = measured fire resistance time (time to failure in test), s (min);

A = area under the actual furnace temperature versus time curve (above the 293 K level) for the first three-quarters of τ, s K (°C min);

A_s = area under the furnace temperature versus time curve (above the 293 K level) specified in the standard, for the first three-quarters of τ, s K (°C min);

L = area associated with the lag of furnace thermocouples during the initial period of test, $\simeq 0.108 \times 10^6$ s K (1800 °C min).

Harmathy (1985a) examined the validity of Eq. 3.3, and concluded that the correction yielded by the equation is usually insufficient. He suggested that the accurate way of correcting for inadvertent deviations from the prescribed temperature–time curve is by comparing the normalized heat load (to be discussed) on the test specimen in the flawed test with that in a flawless test.

The fire resistance is by no means an easily reproducible quantity. A study conducted at the National Research Council of Canada (Sultan *et al.* 1986) indicated that the results of fire tests may depend on

- the volume of the furnace,
- the ratio of the exposed surface of specimen to the surface of furnace boundaries,
- the thermal properties of the furnace boundaries and of the test specimen, and
- the absorption (emission) coefficient of the furnace gases which, in turn, depends on the nature of the fuel and the excess air for combustion.

Perfect reproducibility of the test results could be achieved only if all test furnaces (serving the same purpose) were built according to the same plan, heated with the same fuel, and operated at the same fuel-to-air ratio. However, furnaces of different designs that meet the calibration requirements described in the cited paper could be expected to yield reasonably uniform results for a wide variety of building elements.

Those who developed the first fire resistance test standards many decades ago did so with the intent of offering the building designer the convenience of viewing

6   In practice, the fire resistance of building elements is expressed in minutes or hours.

the merits and demerits of various building elements on a 'fire resistive quality' scale. That the performance of these elements in standard tests could ever be related to their performance in real-world fires probably did not enter their minds. It was not until the early 1980s that the relation between the requirements for the boundary elements of a compartment[7] to withstand a real-world fire and the fire resistance values assigned by standard tests to these elements was understood (Harmathy 1981).

The destructive potential of a fire with respect to a given building element is determined by the maximum temperature reached at a depth where important load-bearing components are located. It was shown (Harmathy 1980a, 1983a; Harmathy and Mehaffey 1982) that the maximum temperature at such depths can be expressed in terms of the *normalized heat load*, defined as

$$H = \frac{1}{\sqrt{k\rho c}} \int_0^\tau q \mathrm{d}t \qquad \text{[3.4]}$$

where H = normalized heat load, $s^{1/2}$ K;

$\sqrt{k\rho c}$ = thermal absorptivity[8] of the building element on the fire-exposed side (k is thermal conductivity, ρ is density, and c is specific heat), $J\,m^{-2}\,s^{-1/2}\,K^{-1}$;

q = heat flux penetrating the element's surface, $W\,m^{-2}$;

t = time, s;

τ = duration of fire exposure, s.

In words: the normalized heat load is the total heat absorbed by unit area of a building element during fire exposure, divided by the thermal absorptivity of the element on the exposed side.

Since the normalized heat load depends on the total heat absorbed by the building element, but is (approximately) independent of the temperature history of the fire that imposes the 'heat load', it can be used as a convenient descriptor of the destructive potential of compartment fires, test fires as well as real-world fires.

Figure 3.7 shows curves of normalized heat load (H) versus duration of standard fire exposure (τ) for test specimens made from a variety of materials, of which the thermal properties are listed in Table 3.1.[9] Clearly, in the case of standard fire tests (for which the fire's temperature history is uniquely defined), the H versus τ relations lie, for the most common types of nonmetallic construction

---

7   A compartment is defined as a building space (enclosed by walls, floor, and ceiling) which is usually separated from other building spaces by closed door(s).

8   Some authors refer to the group $k\rho c$ as thermal inertia. See a discussion on this subject by Harmathy (1985b).

9   The first is a hypothetical material (perfect insulator) with zero thermal absorptivity. Three (mineral wool, insulating fire brick, and fire clay brick) are materials used mainly for lining test furnaces. Two (lightweight concrete and normal-weight concrete) are common construction materials. The last two (steel and aluminum) are metals used in the construction industry. Finally, two (quartz and silicon carbide) are materials selected to fill the gap between the insulators and the conductors.

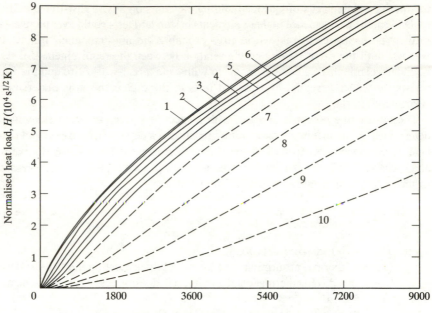

Length of exposure to standard fire test, τ (s)

Fig. 3.7 The H versus τ relation for standard fire tests performed in a high-efficiency furnace. Curves: 1, hypothetical material (perfect insulator); 2, mineral wool (fiberfrax); 3, insulating fire brick; 4, lightweight concrete; 5, fire clay brick; 6, normal-weight concrete; 7, quartz; 8, silicon carbide; 9, steel; 10, aluminum (Harmathy and Mehaffey 1982)

Table 3.1 Thermal properties of the materials (for the appropriate temperature intervals) used in preparing Fig. 3.7 (Harmathy and Mehaffey 1982)

Material	Thermal conductivity (W m^{-1} K^{-1})	Density (kg m^{-3})	Specific heat (J kg^{-1} K^{-1})	Thermal absorptivity (J m^{-2} s$^{-1/2}$ K^{-1})
1 Hypothetical				0
2 Mineral wool (fiberfrax)	0.04	160	1000	80
3 Insulating fire brick	0.25	750	900	410
4 Lightweight concrete	0.50	1450	1000	850
5 Fire clay brick	1.20	2300	900	1580
6 Normal-weight concrete	1.70	2250	1200	2140
7 Quartz	4.6	2640	1050	3570
8 Silicon carbide	17.0	2950	1000	7100
9 Steel	40	7800	540	13000
10 Aluminum	225	2700	1000	24600

materials, in a fairly close cluster, and therefore may be replaced by a single-valued function (see Fig. 11.2).

The convenience offered by the normalized heat load concept in fire safety design will be discussed in some detail in Chapters 10 and 11. At this time it may suffice to point out that the results of standard fire resistance tests, unless they relate to building elements made from or lined with steel or aluminum (or some other metallic material), do have significance in fire safety design.

Nomenclature

A	area under the actual furnace temperature versus time curve (above the 293 K level) for the first three-quarters of τ, s K (°C min)
A_s	area under the furnace temperature versus time curve (above the 293 K level) specified in the standard for the first three-quarters of τ, s K (°C min)
H	normalized heat load, $s^{1/2}$ K
$\sqrt{k\rho c}$	thermal absorptivity of the material on the fire-exposed side of the building element (k is thermal conductivity, ρ is density, and c is specific heat), $J\ m^{-2}\ s^{-1/2}\ K^{-1}$
L	area associated with the lag of furnace thermocouples during the initial period of test, $\simeq 0.108 \times 10^6$ s K (1800 °C min)
q	heat flux that penetrates surface of the building element, $W\ m^{-2}$
t	time, s (min, h)
T_F	space-averaged temperature of furnace gases, K (°C)
$(T_F)_{av}$	T_F averaged over a period of $0 \leq t \leq \tau$, K (°C)
τ	measured fire resistance time (time to failure in test); duration of fire exposure, s (min)
$\Delta\tau$	fire resistance correction, s (min)

4   Moisture in Building Materials

What is loosely referred to as *fire science* encompasses nearly all branches of natural science. It has been conventional to divide fire science into two major areas: one concerned mainly with material behavior, and the other mainly with fire-related processes and phenomena. It may seem appropriate to refer to the former as *input-end fire science*, and to the latter as *output-end fire science*. Subjects that fall within the area of input-end fire science will be treated in Chapters 5 to 8, and those related to output-end fire science in Chapters 9 and 10. The application of fire science to fire safety design will be discussed in Chapters 11 to 15.

4.1   Porosity and Related Properties

What is commonly referred to as a solid is actually all the material within the bulk volume of a solid object. Clearly, if the solid is porous — and most building materials are — it consists of at least two phases: a solid-phase matrix and a gaseous phase (air) in the solid matrix.

The *porosity*, P ($m^3 \, m^{-3}$), of a porous solid can be calculated as[1]

$$P = \frac{\rho_t - \rho}{\rho_t} \qquad\qquad [4.1]$$

where ρ = (bulk) density of the solid, kg m^{-3};
$\qquad \rho_t$ = true density of the (poreless) solid, kg m^{-3}.

For solids that do not exist in poreless condition, ρ_t is a theoretical value to be determined from the crystal parameters, or estimated using some standard test procedure, e.g. ASTM C 135 (American Society for Testing and Materials 1986).[2]

---

1   To be more exact, $P = (\rho_t - \rho)/(\rho_t - \rho_a)$, where ρ_a (kg m^{-3}) is the density of air. However, since usually $\rho_a/\rho_t < 0.001$, ρ_a in the denominator can be neglected without noticeable loss of accuracy.

2   Some confusion seems to exist in connection with the interpretation of density. It has been caused partly by the presence of non-interconnected pores in most solids, and partly by the practice of regarding the density as a property related to the weight (rather than the mass) of the solid. It

The pores may be *interconnected* or *non-interconnected*. Fluids can move through the pores only if at least some of the pores are interconnected. The interconnected pore space is the *effective* pore space (Muskat and Wyckoff 1946; Collins 1961).

The pores are of various sizes, and hardly ever cylindrical in shape. Yet, tacitly assuming cylindrical shape, the size of pores is always stated in terms of pore radius or pore diameter. The pore size distribution is usually determined by the mercury injection technique (Washburn 1921; Ritter and Drake 1945; Scheidegger 1960).

Another important characteristic of the pore structure is the *specific surface*, S $(\text{m}^2\,\text{m}^{-3})$ or $\bar{S}$ $(\text{m}^2\,\text{kg}^{-1})$: surface area of the pores per unit volume or unit mass of the solid. The relation between S and $\bar{S}$ is

$$S = \rho\bar{S} \qquad\qquad [4.2]$$

In the case of concrete (cement paste, to be exact), $\bar{S}$ may be estimated from certain characteristics of the concrete mixture and from the age of the material (see Chapter 6). Unfortunately, for concrete the meaning of pore surface is rather ambiguous (Ramachandran *et al.* 1981).

It is customary to express the average pore radius, r_{av} (m), as

$$r_{av} = \frac{2P}{S} \qquad\qquad [4.3]$$

The *permeability* of a solid is the measure of the ease with which fluids can pass through its pores. The movement of fluids is governed by Darcy's law. For an isotropic material,

$$v = -\frac{\kappa}{\mu}\frac{dp}{dl} \qquad\qquad [4.4]$$

where v = velocity of fluid flow, m s^{-1};
p = pressure, Pa;
l = length, m;
κ = coefficient of permeability, m^2;
μ = viscosity of the fluid, $\text{kg m}^{-1}\,\text{s}^{-1}$.

Researchers in the field of concrete technology usually conduct permeability tests according to the original experiments of Darcy (Scheidegger 1960), and express the pressure gradient in terms of the drop in water head through the sample.

---

has been common to define the different kinds of density in terms of the technique of measurement. The terminology to be used in this book is as follows. *True density* is the mass (kg) of the unit volume (m^3) of the solid in a (usually hypothetical) poreless condition. *Density* or *bulk density* is the mass of a unit volume of the real (porous) solid. The expression *unit mass, loose* will be used to denote the mass of a unit volume of materials consisting of grains or lumps. With such materials, the unit volume is interpreted as consisting of the true volume of the solid matrix, the pore volume within the solid matrix, and the space between the particles.

The dimension of the coefficient of permeability arrived at from these tests, to be denoted by $\bar{\kappa}$, is $m\,s^{-1}$. The relation between κ and $\bar{\kappa}$ (provided the permeability of the solid is sufficiently low) is

$$\kappa = \left(\frac{\mu_w}{\rho_w g}\right)\bar{\kappa} \qquad\qquad [4.5]$$

where μ_w = viscosity of water = $1.002 \times 10^{-3}\,kg\,m^{-1}\,s^{-1}$ at 20°C;
$\qquad \rho_w$ = density of water = $1000\,kg\,m^{-3}$;
$\qquad g$ = gravitational acceleration $\simeq 9.8\,m\,s^{-2}$.

Thus $\kappa/\bar{\kappa} \simeq 1.02 \times 10^{-7}\,m\,s$.

According to the Kozeny equation (see Scheidegger 1960),

$$\kappa = \eta\,\frac{p^3}{s^2} \qquad\qquad [4.6]$$

where η (dimensionless) is an empirical factor whose value, according to theory, is about 0.55. Numerous modifications to the Kozeny equation have been proposed. Since none of the variables that appear in Eq. 4.6 can be measured with high accuracy, refining the Kozeny equation makes little practical sense. But even in its original form, the equation is a valuable guide in sorting out the often-contradictory values reported in the literature concerning pore dimensions, permeabilities, and specific surfaces.

4.2   Sorption Characteristics

Under normal atmospheric conditions, the pores of a solid contain moisture, held by adsorption (or by other forces of the van der Waals type) on the pore surfaces, or by capillary condensation in the pores. *Sorption* is the general term for the propensity of solids to hold moisture (or any gas or vapor in condensed state) in their pores. The *sorption isotherm* is a plot of

$\qquad m$ against p/p_o

where m = moisture content of the porous solid (referred to a specified moistureless condition, to be discussed) in equilibrium with the ambient atmosphere, $kg\,kg^{-1}$;
$\qquad p$ = partial pressure of water vapor in the atmosphere at a temperature T (K), Pa;
$\qquad p_o$ = vapor pressure of water at the same temperature, Pa.

The ratio p/p_o is referred to as the relative humidity of the atmosphere.

The sorption isotherm has two branches: the *adsorption branch*, obtained by monotonically increasing the relative humidity of the ambient atmosphere from 0 to 100 per cent in very small, equilibrium steps; and the *desorption branch*, obtained by monotonically decreasing the relative humidity from 100 to 0 per

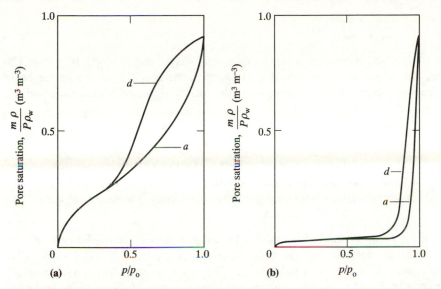

Fig. 4.1 Typical sorption isotherms: (a) hygroscopic solid (Solid A); (b) non-hygroscopic solid (Solid B); a, adsorption branch; d, desorption branch

cent. The actual moisture content of a porous material is, as a rule, represented by a point somewhere between the two branches.

Figure 4.1(a) shows the typical shape of the two branches of the sorption isotherm[3] of a so-called hygroscopic material. Up to a relative humidity of about 0.4, the water molecules are held on the internal surfaces of the host material (adsorbent) by true adsorption, i.e. by the action of intermolecular forces. The two branches of the sorption isotherm usually run the same course, and the amount of adsorbed vapors can be described by the Brunauer–Emmett–Teller equation (Brunauer *et al.* 1938; Brunauer 1945).

$$\frac{m}{m_1} = \frac{Cp/p_0}{(1-p/p_0)(1-p/p_0+Cp/p_0)} \qquad [4.7]$$

where m_1 = mass of water held by a unit mass of the host material in a layer one molecule deep on its internal surfaces, kg kg^{-1};

C = constant for a given host material, characteristic of the heat of adsorption of one layer of water vapor on the pore surfaces, dimensionless.

Equation 4.7 contains two material constants, m and C. Plotting the results of a series of experiments in the form

---

3   The ordinate in the figure, $m\rho/P\rho_w$, is the fractional saturation of the pores: m^3 moisture per m^3 pore volume.

$$\frac{1}{m} \frac{p/p_o}{1 - p/p_o} \text{ versus } p/p_o$$

a straight line is obtained. From its intercept, $1/m_1 C$, and its slope, $(C-1)/m_1 C$, m_1 and C can be evaluated. If m_1 is known, the specific surface of the host material can also be calculated, using the following relation between m and $\bar{S}$:

$$m_1 = \frac{M_w}{N} \frac{\bar{S}}{s_m} \qquad [4.8]$$

where M_w = molecular mass of water = $18.016 \text{ kg kmol}^{-1}$;
$\quad\quad N$ = Avogadro's number = $6.023 \times 10^{26} \text{ molecules kmol}^{-1}$;
$\quad\quad s_m$ = surface area occupied by one water molecule $\simeq 0.114 \times 10^{-18} \text{ m}^2 \text{ molecule}^{-1}$.

If the pores are interconnected, the porosity, P, is a measure of the maximum amount of moisture that the host material is capable of holding when fully saturated.[4] If, on the other hand, $p/p_o < 0.4$, the sorption isotherm is determined jointly by $\bar{S}$ and C. Thus the upper and lower regions of the sorption isotherm can be approximately mapped out if the three material constants, P, $\bar{S}$ and C, are known.

In Table 4.1, typical values of these three constants are presented for a number of materials (Harmathy 1967a). The sorption isotherms of solids consisting of several components can be estimated using the simple mixture rule (Eq. 5.1 with $m = 1$).

Assessing the course of the sorption isotherm over the $0.4 \leq p/p_o \leq 1$ interval may be somewhat problematic. Once the pore surfaces are covered with one layer of water molecules, further moisture will settle in the pores mainly by capillary condensation. Capillary condensation is subject to hysteresis, which is responsible for the existence of the two branches of the sorption isotherm and, in general, for the dependence of the moisture content on the previous history of the porous system.

Capillary condensation is brought about by the lowering of vapor pressure over the curved surfaces of water in the pores. The equilibrium vapor pressure over a curved surface is given by the Kelvin equation. Applied to capillary water, the equation takes the form

$$\ln \frac{p}{p_o} = - \frac{\sigma_w M_w}{\rho_w R T} \frac{2}{r} \qquad [4.9]$$

where p = vapor pressure over curved water surfaces in the pores, Pa;
$\quad\quad p_o$ = vapor pressure over flat water surface, Pa;
$\quad\quad r$ = radius of mean normal curvature of water surface, m;

---

4   However, complete saturation can rarely be achieved merely by exposing the material to an atmosphere of 100 per cent relative humidity.

Table 4.1 Typical values of the constants characterizing the moisture sorption of various materials [collected by Harmathy (1967a) from sources listed in his report]

Material	$\bar{S}$ (10^3 m² kg⁻¹)	P (10^2 m³ m⁻³)	C	ρ_t (10^3 kg m⁻³)
Cement paste				
normal	150–250	28–55	17–92	2.40–2.56
autoclaved	5–10		24	2.4–2.5
Gypsum products	8–55	60	40	2.32
Aggregates				
quartz	<0.3	<1		2.65
siliceous sand	2.5–9	<2.5		2.7
quartzite	14–22	2.5–6.0	3–10	2.7
limestone	0–16	1.7–4.5	6–50	2.7
dolomite	4.3	3.6	10	2.8–2.9
expanded shale	18–30	30–45	11–13	2.6–2.8
expanded slag	35–59	25–45	6–9	2.0–2.5
expanded vermiculite	17–26	70–90	5–6	2.2–2.7
expanded perlite	14–20	70–90	14–18	2.6–2.8
diatomite	40–90	35	6–21	2.0–2.3
pumice	0.4	30–40		2.6–2.7
Brick				
clay brick	0.3–17	15–35	5–66	2.6–2.8
sand–lime brick	8	30	47	2.6–2.7
Miscellaneous building materials				
wood	240–260	50–80	6–8	1.5
paper	125–250		16–26	
asbestos fiber	30		5.5	2.2
glass wool	5		75	2.4–2.6
unhydrated cement	0.3–1.1			3.15
Soils	17–83		8–30	
Industrial adsorbents				
silica gel	300–600	50–65	4.5–8	2.1–2.3
activated alumina	145–260	50	23–49	3.25–3.35
charcoal	95–185	35–60	4–20	1.75–2.10

σ_w = surface tension of water = 0.073 kg s⁻² at 20 °C;
ρ_w = density of water = 1000 kg m⁻³;
R = gas constant = 8315 J kmol⁻¹ K⁻¹;
T = temperature, K.

Since r is comparable in magnitude with some characteristic dimension of the pores, it is usually referred to as equivalent pore radius.

Clearly, if the porous solid is in equilibrium with its environment, the vapor pressure in its pores, p, must be equal to the partial pressure of water vapor in the atmosphere. Hence, p/p_0 on the left-hand side of Eq. 4.9 is identical with the relative humidity of the ambient atmosphere.

With some manipulation, the sorption isotherm over the interval of capillary condensation (i.e. over $0.4 \leq p/p_o \leq 1$) can be converted into a kind of pore size distribution function. Multiplying m by $\rho/P\rho_w$ (where ρ and P are the bulk density and the porosity of the solid, respectively), the resulting group of variables, $m\rho/P\rho_w$, is an expression of the fractional saturation of the pores (m^3 water per m^3 pore volume). Expressing from Eq. 4.9 the equivalent pore radius, and plotting

$$\frac{m\rho}{P\rho_w} \text{ against } r$$

a curve is obtained which describes the cumulative (fractional) volume of the pores with radii smaller than a given r.

Unfortunately, such an estimate of pore size distribution cannot be given full credence. One reason is that the properties of water are affected by capillary and adsorption forces. Sorption hysteresis is another reason. Even if all the pores were of cylindrical shape, the pores would not empty (on desorption) and fill (on adsorption) at the same relative humidity (de Boer 1958). The relation between the relative humidities pertaining to the emptying and filling of cylindrical pores is approximately

$$\left(\frac{p}{p_o}\right)_d = \left(\frac{p}{p_o}\right)_a^2 \qquad\qquad [4.10]$$

where the subscripts d and a designate desorption and adsorption, respectively. However, depending on the dominant pore shape, $(p/p_o)_d$ may be larger or smaller than $(p/p_o)_a^2$.

De Boer (1958) distinguished 15 groups of pore shapes, and analyzed the types of hysteresis loops associated with them. Everett (1958) argued that the pore dimensions cannot be adequately determined from adsorption or desorption measurements, unless the type of pore structure is known from independent evidence. Nevertheless, the general appearance of the sorption isotherm does give a rough indication of the nature of the solid's porosity. Figures 4.1(a) and 4.1(b) show the isotherms of two hypothetical solids; one, Solid A, is hygroscopic, whilst the other, Solid B, is non-hygroscopic. At 50 per cent pore saturation, p/p_o is equal to about 0.6 for Solid A, and about 0.95 for Solid B. On the strength of Eq. 4.9, one may reason that the pores of Solid B are roughly an order of magnitude larger than those of Solid A.

It is important to note that the sorption isotherm of concrete (or rather, cement paste in concrete) does not follow the usual pattern, in that the two branches of the isotherm do not join at low relative humidities. This kind of sorption behavior is relatively rare, and can be found only with some graphites, clay minerals, cellulose-type materials, and hydrated portland cement. According to Feldman and Sereda (1969), the sorption hysteresis of cement paste at low relative humidities is attributable to irreversible intercalation of water in the layered structure of the paste. The typical shape of the sorption isotherm for cement paste is shown in Fig. 4.2.

Table 4.2 The desorption branch of an 'all-purpose' isotherm for portland cement paste (American Society for Testing and Materials 1988a)

$\left(m\,\dfrac{\rho}{\rho_w}\right)_{pt}$ $(\mathrm{m^3\,m^{-3}})$	$\dfrac{p}{p_0}$	$\left(m\,\dfrac{\rho}{\rho_w}\right)_{pt}$ $(\mathrm{m^3\,m^{-3}})$	$\dfrac{p}{p_0}$
0.15	0.4	0.225	0.7
0.16	0.45	0.24	0.75
0.175	0.5	0.255	0.8
0.185	0.55	0.275	0.85
0.195	0.6	0.30	0.9
0.21	0.65		

To allow for the adjustment of the results of fire resistance tests performed on specimens with moisture contents in excess of a specified limit, the ASTM fire test standard (American Society for Testing and Materials 1988a) tabulates the desorption branch of an 'all-purpose' isotherm for portland cement paste. Since these values (see Table 4.2) are used solely for adjusting the test results for different moisture levels, it is unlikely that their use may lead to major errors. It should be borne in mind, however, that they are by no means of general applicability. The sorption isotherms of cement pastes depend on a number of factors, principally on the water−cement ratio in the concrete mixture and on the age of the paste.

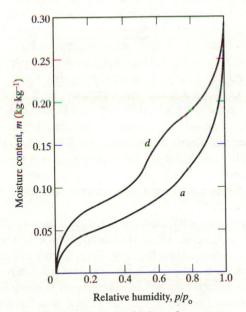

Fig. 4.2 Typical sorption isotherm for cement paste: *a*, adsorption branch; *d*, desorption branch

4.3   Drying

Since the presence of moisture may have significant and often unpredictable effects on the properties of materials at any temperature below 100 °C, it is imperative to have all material property tests conducted on specimens brought prior to the test to a moistureless 'reference condition'. The reference condition is usually defined as that attained by heating the specimen in an oven at 105 °C until its weight shows no change. A few building materials, however, among them all gypsum products, may undergo irreversible chemical changes when held at that temperature for an extended period. Specimens of these materials are usually heated in a vacuum oven at some lower temperature (at 40 °C in the case of gypsum products).

The drying of concrete may also present some problems, owing to the complex role of water in the material's microstructure (Feldman and Sereda 1969). The water molecules are held in the cement paste by forces that vary almost continuously in strength from those characteristic of simple capillary condensation to those representative of strong chemical (ionic or covalent) bonds. Thus, no clear-cut line can be drawn between sorbed moisture and water of hydration. It is more appropriate, therefore, to refer to the water held in the paste by capillary condensation and van der Waals bonds as *evaporable water*, and that held by stronger bonds as *nonevaporable water*.

In precise scientific work, the demarcation line between the evaporable and nonevaporable water content of the cement paste is drawn by drying techniques that do not expose the test specimen to conditions that may lead to irreversible changes in the microstructure of the material. Two such techniques have gained recognition: (i) P-drying: drying at room temperature over a mixture of $Mg(ClO_4)_2 \cdot 2H_2O$ and $Mg(ClO_4)_2 \cdot 4H_2O$, at a vapor pressure of 6×10^{-4} mm Hg; and (ii) D-drying: drying over dry ice at 195 K (sublimation temperature of CO_2), at a water vapor pressure of 5×10^{-4} mm Hg.

The drying of large specimens to be exposed to full-scale fire resistance tests is another area of interest to fire safety practitioners. The test standards [e.g. ISO 834 (International Standards Organization 1990) and ASTM E 119 (American Society for Testing and Materials 1988a)] require that the test specimens be in a 'reference condition' which, with respect to moisture content, is usually interpreted as that in equilibrium with an atmosphere of 50 per cent relative humidity at about 20 °C.

Early work by Keen (1914), Fisher (1923, 1927), Sherwood (1929, 1930, 1932), and Sherwood and Comings (1933) indicated that the drying of a moist, porous solid proceeds at a constant rate for some time. During this time, the evaporation of moisture takes place at the solid's surface, kept constantly wet by water drawn from the interior by capillary forces. The rate of evaporation is approximately equal to that from a free water surface. If heat transfer from the surrounding air to the drying surface takes place predominantly by convection, the surface temperature of the solid will be equal to the wet-bulb temperature of the atmosphere,[5] and the rate of loss of moisture by the solid during the *constant rate period of drying* can be expressed as

$$W \frac{dm}{dt} = \frac{hA}{\Delta h_w} (T_a - T_{wb})$$ [4.11]

where W = mass of the (dry) solid, kg;
   A = surface area of the solid, m^2;
   h = coefficient of convective heat transfer, $W\,m^{-2}\,K^{-1}$;
   Δh_w = heat of vaporization of water = $2.44 \times 10^6\,J\,kg^{-1}$ at room temperature;
   T_a = temperature of the ambient atmosphere, K;
   T_{wb} = wet-bulb temperature of the atmosphere, K;
   t = time, s.

The coefficient of heat transfer for air flow parallel to plane surfaces can be assessed from the following equation (Shepherd *et al.* 1938):

$$h = 16.67v^{0.8}$$ [4.12]

where v ($m\,s^{-1}$) is the velocity of air, and 16.67 is a dimensional constant.

Figure 4.3(a) shows the variation of the average moisture content and average temperature of a solid (pictured as an infinite slab) during the drying process. Figure 4.3(b) depicts the moisture distribution in the solid, based mainly on information published by Kamei (1937), Kamei and Shiomi (1937), Corben and Newitt (1955), and Johnson (1961).

The moisture distribution appears to be fairly uniform during the constant rate drying. At the *upper critical point* (point C_u in the figure), dry patches begin to show up on the surface, indicating that the rate of flow of moisture by capillary action is no longer capable of matching the rate of moisture loss from the surface. The *falling rate period of drying* sets in. The onset of this period is independent of the solid's dimensions; it depends solely on its pore structure.

At C_u the temperature of the solid begins to rise, and keeps on rising at an increasing rate. Meanwhile the moisture is depleted in a layer below the surface, and the thickness of that layer increases steadily. The *lower critical point* (C_l in the figure) is soon reached, at which the last wet patches disappear from the surface, and the rate of temperature rise begins to decelerate. At this point the migration of moisture in the liquid state comes to an end. From this point on, the movement of moisture in the pores takes place mainly in the gaseous phase, by an evaporation–condensation mechanism. This mechanism of drying was discussed by Harmathy (1967b, 1969), Huang *et al.* (1979), Huang (1979), and Dayan and Gluekler (1982).

---

5   In general, the contribution of radiant heat transfer to the overall heat transmission process can only be kept at a negligible level if forced convection is applied. If the contribution of radiant heat transfer is not negligible, the temperature of the surface will attain a value somewhere between the wet-bulb temperature and the temperature of the atmosphere.

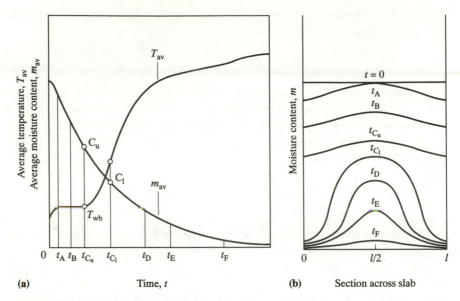

Fig. 4.3 The drying process: (a) average moisture content and average temperature; (b) moisture distribution (l is slab thickness; the times t_A, t_B, ... t_F are reference points between the two figures) (Harmathy 1967b)

At some point before the end of the drying process, the temperature of certain kinds of solids tends to level off a few degrees below the temperature of the atmosphere. Nissan *et al.* (1959) referred to this temperature as 'pseudo-wet-bulb temperature'.

The fire resistance test standards require that at the time of testing the test specimens be in equilibrium with an atmosphere of no higher than 50 per cent relative humidity at about 20 °C. Among the common building materials, only concrete (more accurately: the cement paste in concrete) and wood, because of their complex microstructures, are capable of holding moisture in amounts large enough to be taken into account in fire performance assessments.

In the case of concrete specimens, reducing the moisture content to the level required by the standard is a lengthy and expensive procedure. To comply with the requirements, some testing laboratories subject their test specimens to accelerated drying a few weeks after their construction. This is a rather unfortunate practice. If the relative humidity in the pores of the cement paste drops below 80 per cent, the hydration process stops, and the mechanical and some thermal properties of the concrete may become adversely affected.

Abrams and Orals (1965) studied the problems associated with accelerated drying, using seven-day-old concrete slab specimens, 152 mm thick. They found that the time required to lower the moisture content of the slabs to the required level could be reduced significantly by kiln-drying at 93 °C. However, in the subsequent fire tests the kiln-dried specimens yielded much poorer performance than the companion specimens naturally dried to the same moisture level.[6]

4.4   Correction of Fire Test Results

Testing laboratories and clients seeking fire resistance ratings are advised to have fire resistance tests conducted on specimens cured in a fog chamber for at least six months, and then conditioned at moderately elevated temperatures. When testing separating elements (see Chapters 3 and 12), they may also consider having the tests performed while the moisture content of the specimen is still above the specified level, and taking advantage of the correction formula that allows for adjusting the test results.

Since the reference condition for fire resistance test specimens is defined in terms of equilibrium relative humidity in the pores of the material, the testing laboratories usually monitor the progress of drying by taking readings from humidity gauges installed in the specimen.[7] However, the effect of moisture on the fire test result is not related directly to the equilibrium relative humidity in the pores. It is the heat absorbed in the desorption of moisture that affects the fire test result, and that heat depends on the amount of water actually held in the pores in liquid (or liquid-like) state. In the case of concrete specimens, Table 4.2 may be used to convert the relative humidity values to moisture concentration in the cement paste, expressed in volume fraction, $(m \, \rho/\rho_w)_{pt}$.

The amount of moisture held by the aggregates is rarely significant enough to have a noticeable effect on the results of fire tests. Hence, the moisture content of concrete is, for all intents, equal to the moisture content of the cement paste, the volume fraction of which can be determined from information on the concrete mixture.[8]

An experimental study conducted by Harmathy (1967c) indicated that the effect

---

6   The reason for the poor fire performance of concretes whose hydration is not allowed to proceed to completion will be understood after having gained an insight into the effect of dehydration reactions on heat flow (Chapters 5 and 6).

7   These humidity gauges are often referred to as Menzel gauges (Menzel 1955).

8   A publication by the Portland Cement Association, entitled *Design and Control of Concrete Mixtures* (Portland Cement Association 1968), tabulates suggested trial mixtures in pounds per cubic yard for air-entrained and non-air-entrained concretes of medium consistency. If any of the suggested mixtures has been used, calculate the mass fractions of the components (water, cement, fine aggregate, and coarse aggregate), then, using Eq. 5.2, convert them to volume fractions. The densities are: water, 1000 kg m^{-3}; cement, 3150 kg m^{-3}; fine and coarse aggregates, approximately 2700 kg m^{-3}. The sum of the volume fractions of water and cement is equal to the volume fraction of the cement paste. Note that the 'effective' amount of water may be less or more than the amount added to the mixture. If the aggregates are dry, they absorb some water, and the absorbed amount has to be deducted from the total amount of water. If, on the other hand, the aggregate are moist, the amount of surface moisture brought into the mixture may be as much as 2—5 per cent of the mass of aggregates.

The volume fraction of the paste usually amounts to 0.26—0.43 (m^3 per m^3 concrete) in non-air-entrained concretes, and 0.24—0.40 in air-entrained concretes.

of moisture on the results of fire resistance tests of vertical or horizontal separating elements (see Chapter 3) can be described by the following equation:[9]

$$\tau_d^2 + \tau_d\left[4b\left(m\frac{\rho}{\rho_w}\right) + 4 - \tau_m\right] - 4\tau_m = 0 \qquad [4.13]$$

where τ_d = fire resistance of the element in oven-dry condition, h;
 τ_m = fire resistance of the element at a moisture content of m, h;
 b = material constant, related to the material's permeability, h.

As mentioned earlier, m is moisture content referred to oven-dry mass (kg kg^{-1}), and ρ and ρ_w (kg m^{-3}) are the (bulk) density of the material and the density of water, respectively.

Typical values of b are given in Table 4.3. The reason for the increase of this constant with increasing permeabilities is that in materials with larger effective pore space the moisture can more easily retreat toward the unexposed (cool) surface, ahead of the heat flow. As will be discussed soon, the farther the location of the moist zone from the fire-exposed surface, the more beneficial its effect on fire resistance.

Equation 4.13 can be solved for either τ_d or τ_m.

$$\tau_d = \frac{1}{2}\sqrt{\left[4b\left(m\frac{\rho}{\rho_w}\right) + 4 - \tau_m\right]^2 + 16\tau_m} - \frac{1}{2}\left[4b\left(m\frac{\rho}{\rho_w}\right) + 4 - \tau_m\right]$$

$$[4.13a]$$

$$\tau_m = \frac{\tau_d^2 + 4\tau_d\left[1 + b\left(m\frac{\rho}{\rho_w}\right)\right]}{\tau_d + 4} \qquad [4.13b]$$

The use of these equations is illustrated through a worked example in Section 4.6.

Equations 4.13a and 4.13b are applicable only if the distribution of moisture is uniform or symmetrical about the specimen's central plane, as shown in Figs 4.4(a) and 4.4(b). If the moisture distribution is asymmetric [Figs 4.4(c) and 4.4(d)], the gain in fire resistance will depend not only on the average moisture content, but also on the moment of moisture about the fire-exposed surface. More exactly: the gain will be proportional to the ratio of the moisture moment at asymmetric distribution to that at symmetrical distribution (Harmathy and Lie 1971). With the notations shown in Fig. 4.4, this ratio is $2d/l$, where d (m) is the distance of the center of gravity of moisture from the fire-exposed surface, and l (m) is the thickness of the specimen.

The asymmetric distribution of moisture can be taken into account by replacing b in Eq. 4.13 (and in Eqs 4.13a and 4.13b) by $\bar{b}$, defined as

---

9   In a strict sense, this equation is applicable only if the separating element has failed, or is expected to fail, in the fire test by loss of insulating capability (see Chapter 3).

Table 4.3 Typical values of *b* for use in Eq. 4.13 (Harmathy 1967c; American Society for Testing and Materials 1988a)

Material	Oven-dry density (kg m^{-3})	*b* (h)
Normal-weight and gun-applied concrete, brick	>2100	5.5
Lightweight concrete	1360–1840	8.0
Insulating concrete	<800	10.0

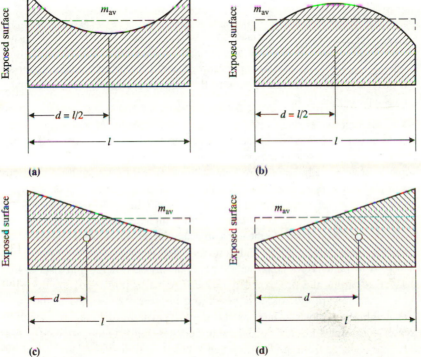

Fig. 4.4 Illustration of moisture moment: (a), (b) symmetrical moisture distributions; (c), (d) asymmetrical moisture distributions (Harmathy and Lie 1971). Reprinted with permission from *Fire Technology* (Issue 7), copyright © 1971 National Fire Protection Association, Quincy, MA 02269

$$\bar{b} = b\,\frac{2d}{l} \qquad [4.14]$$

and by interpreting *m* in the equation as average moisture concentration.

Equation 4.13 is not applicable to columns, which in a standard fire test are heated on all four sides. According to Lie (1989), the effect of moisture on the fire resistance of reinforced concrete columns is, within reasonable limits, negligible (see also Chapter 14).

4.5   Spalling

The presence of moisture is beneficial for fire resistance only if its concentration is not so high as to cause *spalling*. Concrete structures, when exposed to fire, are especially susceptible to this phenomenon.

Spalling may occur on account of a number of factors, such as high moisture content, low permeability, high local stresses in the concrete, stresses due to differences in the thermal expansion of the cement paste and the aggregates, stresses due to differences in the thermal expansion of the concrete and the reinforcing steel, and general deterioration of concrete at high temperatures.

According to Kordina and Meyer-Ottens (1981) and Dougill (1983), the principal kinds of spalling are:

(1) local spalling, such as surface spalling, aggregate splitting, and corner separation, caused by physical or chemical changes at elevated temperatures,
(2) 'sloughing off', i.e. partial separation of small layers of surface material, a process that may continue slowly throughout the fire or fire test, and
(3) explosive spalling.

Whilst the first two kinds of spalling affect the performance of building elements only in a minor way, the third kind presents very serious problems, and it may lead to the complete disintegration of the element at an early stage of fire exposure.

Some lightweight aggregates are known to cause explosive spalling. Various kinds of thermal stresses may also become contributing factors.[10] It is generally recognized, however, that explosive spalling is not likely to occur unless the moisture in the structure is at an unacceptably high level.

Harmathy (1965a) referred to the explosive spalling resulting from high moisture level as *moisture clog spalling*, and described its mechanism as follows. As the heat begins to penetrate a concrete element, desorption of moisture starts in a thin layer adjoining the fire-exposed surface. A major portion of the desorbed vapors moves toward the colder regions, and becomes readsorbed in the pores of the neighboring layer. Thus, as the thickness of the dry region gradually increases, a saturated layer of considerable thickness (called moisture clog) builds up at some distance from the fire-exposed surface. A little while later, a sharply defined front forms between the dry and saturated regions (Fig. 4.5). Further vaporization of moisture will take place from this frontal area, indicated by line C–D in the figure.

In the meantime, as the temperature of the fire-exposed surface keeps rising, a steep temperature gradient develops across the dry region, resulting in a high rate of heat flow and intense moisture vaporization at the C–D plane. Having

---

10 The suggestion by Saito (1966), that high compressive stresses near the fire-exposed surface, brought about by markedly nonlinear temperature distribution in the concrete element, are the principal causes of explosive spalling, is hardly tenable. Since the tensile stresses that concrete is capable of carrying are insignificant, the interior of the element will undergo extensive cracking on uneven heating, and thereby the high compressive stresses are relieved.

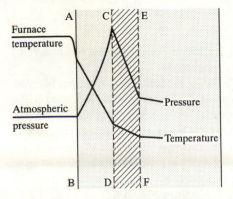

Fig. 4.5 Illustration of moisture clog formation (Harmathy 1965a)

hardly any outlet toward the colder regions, the vapors are compelled to move toward the exposed surface through the A−B−C−D layer which, being strongly compressed, does not offer easy passage in the form of cracks. Moreover, the vapors heat up, expand, and meet increasing resistance along the flow path.

With further steepening of the temperature gradient, there will be a rapid pressure buildup at the C−D plane. If the permeability of the material is relatively high (i.e. the resistance of the pores to moisture flow is relatively low), the entire moisture clog (the region C−D−E−F) will be eased, under the effect of a high-pressure gradient across the region, toward the unexposed side of the element, and the pressure buildup will level off. If, on the other hand, the permeability is low, the pressure at plane C−D will continue to grow, and will eventually exceed the tensile strength of the material.[11] When this point is reached, a layer of a thickness approximately equal to that of the dry layer (usually about 25 mm) is blown off with a bang from the exposed surface.

In standard fire tests, the first spalling usually occurs at 10−25 min into the test. The disintegration of the test specimen is not likely to stop at this point. A new surface layer of high moisture concentration becomes suddenly exposed to the temperature of the furnace gases, already well over 700 °C. The buildup of moisture clog in a subsurface region starts again, and spalling usually continues at increasing frequency.

Based on the above outlined model, Harmathy (1965a) developed a criterion for explosive spalling, in which the fractional pore saturation, $m\rho/P\rho_w$(m^3 moisture per m porous material), is related to a number of variables characterizing the material as well as the process. In a simplified form,

$$\frac{m\rho}{P\rho_w} > \left[1 + \frac{2.1 \times 10^{-10}}{P\kappa\sigma_u}\left(\frac{k}{k+3.6}\right)\right]^{-1} \qquad [4.15]$$

11 Studies by Waubke and Schneider (1974) and Dayan (1982) have indicated that conditions characteristic of fire exposure can, in fact, give rise to excessive tensile stresses that may lead to spalling.

where σ_u = ultimate tensile strength of concrete (cement paste), Pa;
  k = thermal conductivity of concrete, $W\,m^{-1}\,K^{-1}$.

and the constants, 2.1×10^{-10} (N) and 3.6 ($W\,m^{-1}\,K^{-1}$), are representative of the heat and mass transfer process that leads to spalling.

Figure 4.6 shows the susceptibility of a particular concrete to spalling, in terms of fractional pore saturation and permeability. The curve was calculated using Eq. 4.15 and the following material properties: $P = 0.3\,m^3\,m^{-3}$, $\sigma_u = 1.7\,MPa$, and $k = 1.0\,W\,m^{-1}\,K^{-1}$ (which may be regarded as typical of concrete under conditions prevailing in a fire test). If the desorption branch of the sorption isotherm of the material is known, the fractional pore saturation can be converted to equilibrium relative humidity. The desorption curve of the isotherm shown in Fig. 4.2 was used to perform the conversion in Fig. 4.6.

Copier (1979), who regarded the aging of concrete as a straight drying process, ventured the opinion that, since in two years the moisture content of concrete usually decreases to below 7.5 per cent by volume, spalling is unlikely to occur with old building elements. In reality, old concrete elements, especially those prepared with low water—cement ratios, are the best candidates for explosive spalling in fires. As concrete ages, its effective pore space, and with it its permeability, diminishes on account of the segmentation of its capillary pores.[12] Figure 4.6 shows that an order of magnitude reduction in its permeability may

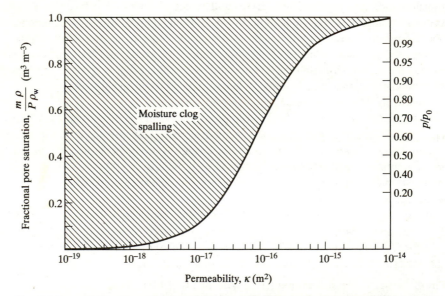

Fig. 4.6 Susceptibility of a particular concrete to spalling, in terms of fractional pore saturation (or equilibrium relative humidity) and permeability (Harmathy 1965a)

---

12 The segmentation of capillary pores is discussed in Chapter 6.

make a building element prone to spalling, even though its evaporable moisture content may have decreased with the passing of time.

In a large series of experiments, Meyer-Ottens (1974) investigated the spalling of specimens of building elements exposed to fire on two or three sides. He found that, with the particular concretes and conditioning techniques used in his studies, spalling did not occur if the moisture content of the specimen was less than 2 per cent by mass (i.e. about 4.4 per cent by volume). In the case of specimens with higher moisture content, the susceptibility to spalling increased with the applied load and with the rate of heating, and decreased with the cross-sectional dimensions of the specimen.

4.6 Worked Example

Effect of Moisture on Fire Resistance

Given

A normal-weight concrete slab, when tested at a moisture content in equilibrium with an atmosphere of 75 per cent equilibrium relative humidity, yielded 1.5 h fire resistance. The volume fraction of cement paste in the concrete was $0.30 \, \text{m}^3 \, \text{m}^{-3}$. Calculate what the fire resistance would have been, had the test been conducted, as required by the standard, at a moisture content in equilibrium with an atmosphere of 50 per cent relative humidity.

Input Information

Moisture content of the cement paste at 75 per cent and 50 per cent relative humidities (Table 4.2):

$$\left(m \frac{\rho}{\rho_w} \right)_{pt} = 0.240 \, \text{m}^3 \, \text{m}^{-3}$$

$$\left(m \frac{\rho}{\rho_w} \right)_{pt} = 0.175 \, \text{m}^3 \, \text{m}^{-3}$$

Moisture content of the concrete at the same relative humidities:

$$m \frac{\rho}{\rho_w} = 0.3 \times 0.240 = 0.0720 \, \text{m}^3 \, \text{m}^{-3}$$

$$m \frac{\rho}{\rho_w} = 0.3 \times 0.175 = 0.0525 \, \text{m}^3 \, \text{m}^{-3}$$

b for normal-weight concrete (Table 4.3):

$$b = 5.5$$

Fire resistance of the slab in oven-dry condition (Eq. 4.13a):

$$\tau_d = \frac{1}{2} \sqrt{(4 + 4 \times 5.5 \times 0.0720 - 1.5)^2 + 16 \times 1.5} - \frac{1}{2}(4 + 4 \times 5.5 \times 0.0720 - 1.5)$$

$$= 1.147 \, \text{h}$$

Fire resistance of the slab in standard condition (i.e. at 50 per cent relative humidity; Eq. 4.13b):

$$\tau_m = \frac{1.147^2 + 4 \times 1.147 \times (1 + 5.5 \times 0.0525)}{1.147 + 4} = 1.404\,\text{h}$$

Conclusion

In standard condition, the fire resistance of the slab would have been 6 min (0.1 h) less.

Alternatively, a nomogram given in ASTM E 119 can be used for the assessment of the fire resistance in standard condition.

Nomenclature

A	surface area of solid, m^2
b	material constant, related to permeability, h
$\bar{b}$	b modified to take account of asymmetric moisture distribution, h
C	constant for a given adsorbent, dimensionless
d	distance of the center of gravity of moisture from the fire-exposed surface, m
g	gravitational acceleration $\simeq 9.8\,\text{m}\,\text{s}^{-2}$
h	coefficient of heat transfer, $W\,m^{-2}\,K^{-1}$
Δh	heat of vaporization, $J\,kg^{-1}$
k	thermal conductivity, $W\,m^{-1}\,K^{-1}$
l	length; thickness of specimen, m
m	moisture content; moisture concentration, $kg\,kg^{-1}$
m_1	mass of water held by a unit mass of the host material in a layer one molecule deep on its internal surfaces, $kg\,kg^{-1}$
M	molecular mass, $kg\,kmol^{-1}$
N	Avogadro's number $= 6.023 \times 10^{26}$ molecules $kmol^{-1}$
p	pressure; partial pressure of water; vapor pressure over curved surfaces, Pa
p_o	vapor pressure of water (over flat surfaces), Pa
P	porosity, $m^3\,m^{-3}$
r	pore radius; radius of mean normal curvature, m
R	gas constant $= 8315\,J\,kmol^{-1}\,K^{-1}$
s_m	surface area occupied by one water molecule $\simeq 0.114 \times 10^{-18}$ m^2 molecule^{-1}
S	specific surface, $m^2\,m^{-3}$
$\bar{S}$	specific surface, $m^2\,kg^{-1}$
t	time, s
T	temperature, K
v	velocity of fluid flow; velocity of air, $m\,s^{-1}$
W	(dry) mass of solid, kg
η	empirical factor $\simeq 0.55$ according to theory, dimensionless
κ	coefficient of permeability, m^2
$\bar{\kappa}$	coefficient of permeability, $m\,s^{-1}$
μ	viscosity, $kg\,m^{-1}\,s^{-1}$
ρ	density, $kg\,m^{-3}$
σ_u	ultimate tensile strength, Pa
σ	surface tension, $kg\,s^{-2}$
τ	fire resistance, h

Subscripts

a	of air; of ambient atmosphere; at adsorption
av	average
d	at desorption; in oven-dry condition
m	at a moisture content m
pt	of the cement paste
t	true
w	of water
wb	wet-bulb

5   Rudiments of Materials Science

Although fire safety is discussed in this book with particular reference to concrete construction, the reader will no doubt realize that the question of safety cannot be adequately addressed with a single material in mind. Clearly, the fire safety practitioner must have at least some rudimentary knowledge of the behavior of other materials that may be used in conjunction with concrete, or may significantly affect the building's performance in fire.

5.1   Terminology

The materials and products built into or used in buildings are manufactured by many different techniques. The three most widely employed techniques are:

(1) solidifying the material from a melt and (usually) forming the product by hot or cold work,
(2) molding a wet, plastic mass of powders and firing it in a kiln, and
(3) mixing finely ground powders (and aggregates) with water, placing the mixture in forms, and letting it harden by hydration.

In general, materials formed by solidification from a melt show the highest degree of homogeneity. The other two techniques usually yield porous solids with complex microstructures.

Homogeneous solids have the same composition and the same properties throughout their volume. Most construction materials are *heterogeneous*, yet their heterogeneity is often glossed over when dealing with practical problems. Furthermore, most of them are treated as *isotropic* materials, as though they possessed the same properties in all directions.

Among the material properties, those that are unambiguously defined by composition and phase are referred to as *structure-insensitive*. Some others depend on the microstructure of the solid and (indirectly) on its previous history. These properties are *structure-sensitive*.

The majority of solids are crystalline, characterized by an orderly arrangement of atoms. Yet, this orderly array is interrupted

(a) on a submicroscopic scale: by atomic dislocations and vacancies, and by foreign atoms, and

(b) on a microscopic scale: by *grain boundaries* (which separate regions of crystals of differently oriented axes) and (usually) by pores.

Solids composed of grains are referred to as *polycrystalline* solids, as opposed to single crystals, which are rarely found in nature in sizes larger than a few centimeters. The grains are, in general, equiaxed and randomly oriented. Mechanical shaping, known as cold work, results in elongated, distorted grains.

Noncrystalline solids are called *amorphous* materials. *Gels* and *glasses* are amorphous materials. Gels are formed by the coagulation of a colloidal solution. Glasses (vitreous materials) are solids with a liquid-like submicroscopic structure. For lack of long-range crystalline order in the arrangement of their atoms, amorphous materials do not form grains and, therefore, are not subdivided by grain boundaries. Although glasses are regarded as fully homogeneous materials, they do exhibit some heterogeneity: they may be phase-separated on a submicroscopic scale, and may contain small bubbles and crystalline imperfections on a microscopic scale.

Most inorganic, nonmetallic building materials are mixtures of two or more crystalline or amorphous phases, forming a solid matrix which is interspersed with pore spaces.

Synthetic polymers (plastics) are made up of long macromolecules created by polymerization from smaller repeating units: monomers. These molecular chains are present in an entangled mass, like spaghetti in a bowl. At some locations, however, the adjacent chains tend, on account of weak (van der Waals) binding forces, to become aligned and produce regions of ordered, crystalline structure.

In *thermoplastics*, the mobility of the molecular chains increases on heating. The material softens, as glasses do. In some other types of plastics, polymerization also produces cross-bonds between the chains. These materials, called *thermosetting plastics*, are permanently hard and do not soften on heating.

Whether *nonburnable* or *burnable*, a building material is of primary interest to the fire safety practitioner with regard to its thermal and structural behavior in fire. In the case of burnable materials, however, the interest is twofold. The positive role assigned to these materials by design (i.e. performing as building components or insulation) may change in fire into a negative role: becoming fuel and adding to the severity of fire. This latter role will be discussed in some detail in Chapter 9.

From the point of view of their performance in fire, building materials may be divided into the following groups:

● Group L (load-bearing) materials: materials capable of carrying high stresses, usually in tension. With these materials, the mechanical properties related to behavior in tension are of principal interest.

● Group L/I (load-bearing/insulating) materials: materials capable of carrying moderate stresses and, in fire, providing thermal protection for Group L materials. With Group L/I materials, the mechanical properties (related mainly to behavior in compression) and the thermal properties are of equal interest.

- Group I (insulating) materials: materials not designed to carry load. From the point of view of fire safety, their role is to resist heat transmission through building elements and/or to provide insulation to Group L or Group L/I materials. With Group I materials, only the thermal properties are of interest.
- Group L/I/F (load-bearing/insulating/fuel) materials: L/I type materials that may become fuel in fire.
- Group I/F (insulating/fuel) materials: I type materials that may become fuel in fire.

Building materials are not expected to function normally in a fire environment. They may (and most of them do) become unstable and undergo physicochemical changes ('reactions' in a generalized sense[1]) on heating. Information on their elevated-temperature behavior is not easy to come by, and relying on generic data may lead to grossly erroneous conclusions. It is imperative, therefore, that the fire safety practitioner know how to extend, based on a priori considerations, the utility of the scanty data that can be gathered from the technical literature.

In this chapter, the rudiments of materials science are outlined. Many excellent books and reports have been written on the subjects to be discussed here. The reader may find the books by Kingery (1959), Kingery et al. (1976), and Keyser (1974), and reports by Hidnert and Souder (1950), and Lynch et al. (1966) particularly useful. Dozens of other publications will be mentioned at the appropriate places in this chapter and in the next three chapters.

5.2   Mixture Rules

Some properties of materials of mixed composition or mixed phase can be calculated by simple rules if the properties of the constituents are known. The simplest mixture rule (Bruggeman 1936) is

$$\pi^m = \sum_i v_i \pi_i^m \qquad\qquad [5.1]$$

where π is a property of the composite material, π_i is that of the composite's ith constituent, $v_i (m^3 m^{-3})$ is the volume fraction of that constituent, and m (dimensionless) is a constant with a value between -1 and $+1$.

The constituents of a composite are often quantified in terms of mass fraction (weight fraction). The relations between the volume fractions and the mass fractions are

---

1   The term 'reaction' covers any change in the structure, phase, or composition of the material which is accompanied by measurable absorption or evolution of heat. Thus, phase changes (melting, evaporation, sublimation), desorption of moisture, decomposition, vitrification, polymorphic inversion, order–disorder transformations, magnetic and ferromagnetic transformations, and, naturally, chemical reactions are 'reactions' (physicochemical changes).

$$v_i = \frac{\dfrac{\omega_i}{\rho_i}}{\sum_i \dfrac{\omega_i}{\rho_i}}$$

[5.2]

and

$$\omega_i = \frac{v_i \rho_i}{\sum_i v_i \rho_i}$$

[5.3]

where ω_i (kg kg^{-1}) is the mass fraction and ρ_i (kg m^{-3}) the density of the ith constituent of the composite material. Naturally

$$\sum_i v_i = 1 \quad \text{and} \quad \sum_i \omega_i = 1$$

[5.4]

Hamilton and Crosser (1962) recommended the following rather versatile formula for two-phase solids:

$$\pi = \frac{v_1 \pi_1 + \beta v_2 \pi_2}{v_1 + \beta v_2}$$

[5.5]

where

$$\beta = \frac{n \pi_1}{(n - 1)\pi_1 + \pi_2}$$

[5.6]

Phase 1 must always be the principal continuous phase. The value of n depends on the phase distribution geometry and, in general, has to be determined experimentally. With $n \to \infty$ and $n = 1$, Eqs 5.5 and 5.6 convert into Eq. 5.1 with $m = 1$ and $m = -1$, respectively. With $n = 3$, a relation is obtained for a system in which the discontinuous phase consists of spherical particles (Maxwell 1904). Applied repeatedly, Eqs 5.5 and 5.6 may be used to calculate the properties of systems consisting of more than two constituents, e.g. those of a moist, porous solid which consists of three essentially continuous constituents: a solid matrix, and air and moisture within that matrix (Harmathy 1970a).

5.3   Dilatometric Curve, Thermal Expansion

The dilatometric curve is a record of the fractional change of a linear dimension of a solid at steadily increasing or decreasing temperature. With mathematical symbolism, the dilatometric curve is a plot of

$$\frac{\Delta l}{l_o} \text{ against } T$$

where $\Delta l = l - l_o$, l and l_o (m) are the changed and original dimensions of the solid (the latter usually taken at room temperature), respectively, and T (K) is temperature. The difference Δl reflects not only the linear expansion or shrinkage of the solid, but also dimensional effects caused by physicochemical changes (reactions).

The heating of a sample of the solid usually takes place at an agreed rate, 5 °C per minute as a rule. Since the physicochemical changes proceed at a finite rate and some of them are irreversible (or not immediately reversible), dilatometric curves obtained at various rates of heating may look different, and the branches obtained on heating and cooling rarely coincide. Sluggish reactions may bring about (positive or negative) changes in the slope of the curve, whereas very fast reactions may show up as discontinuities in the slope. Heating rates higher than 5 °C min^{-1} usually cause the reactions to shift to higher temperatures and to develop in a narrower temperature interval.

Within regions of physicochemical stability, a crystalline solid expands on heating, because of increasing mean separation between the constituent atoms on account of enlarged vibrational amplitudes. Since both the thermal expansion and the heat capacity of the solid depend on the same factor, namely the amplitude of thermal vibrations, the ratio of specific heat to coefficient of thermal expansion is nearly constant at all temperatures.

The coefficient of thermal expansion is defined as

$$\alpha = \frac{1}{l} \frac{dl}{dT} \qquad [5.7]$$

or

$$\bar{\alpha} = \frac{1}{v} \frac{dv}{dT} \qquad [5.8]$$

where α = coefficient of linear thermal expansion, $m\,m^{-1}\,K^{-1}$;
$\quad\;\; \bar{\alpha}$ = coefficient of volume expansion, $m^3\,m^{-3}\,K^{-1}$;
$\quad\;\; v$ = volume at a temperature T, m^3.

Since $l \simeq l_o$, the coefficient of linear thermal expansion is, for all intents, the tangent to the dilatometric curve.

The coefficient of volume expansion for crystals is roughly equal to the sum of the coefficients of linear expansion along the crystallographic axes. This value is also applicable to polycrystalline materials. For solids that are isotropic in a macroscopic sense, the relation between the volume and linear expansions is approximately

$$\bar{\alpha} \simeq 3\alpha \qquad [5.9]$$

Materials with strong crystal bonds show lower thermal expansion than those with weak bonds. For this reason, there seems to be a correlation between the expansion coefficients and the melting points. Carnelley (1879) and Lémeray (1900) showed (see also report by Hidnert and Souder 1950) that for the elements the coefficient of linear expansion, α, is inversely proportional to melting point, T_m (K), and Wiebe (1906) found that the product of $\bar{\alpha}$ and T_m is approximately constant and equal to 0.06. Similar correlations exist for certain other groups of materials, e.g. some oxides and halides (Kingery 1959).

Turner (1946) suggested the following equation for the coefficient of thermal expansion of solids consisting of isotropic phases:

$$\alpha = \frac{\sum_i \dfrac{\alpha_i K_i \omega_i}{\rho_i}}{\sum_i \dfrac{K_i \omega_i}{\rho_i}} \qquad [5.10]$$

where α_i, (Pa), ω_i (kg kg^{-1}) and ρ_i (kg m^{-3}) are, respectively, coefficient of linear thermal expansion, bulk modulus (see Eq. 5.16), mass fraction, and density of the ith phase.

If the bulk moduli of the phases are approximately equal, on account of Eq. 5.2

$$\alpha \simeq \sum_i v_i \alpha_i \qquad [5.11]$$

in other words, the simple mixing rule applies with $m = 1$.

In the case of multiphase solids, grain-boundary stresses may affect the observed expansion and cause intergranular fracture.

5.4   Thermogravimetric Curve, Density, Porosity

The thermogravimetric curve shows the fractional variation of the mass of a solid at steadily increasing or decreasing temperature. With mathematical symbolism, a thermogravimetric curve is a plot of

$$\frac{W}{W_o} \text{ against T}$$

where W and W_o (kg) are the changed and original masses of the solid, respectively, the latter usually taken at room temperature. If the curve is obtained by heating the solid, the rate of heating is again 5 °C min^{-1}, as a rule.

Thermogravimetric curves reveal only those reactions which are accompanied by loss or gain of mass, e.g. desorption of moisture, dehydration, pyrolysis, and oxidation.

Again, an increase in the rate of heating usually causes the reactions to shift to higher temperatures and to develop in narrower temperature intervals.

Assuming that the material is isotropic with respect to its dilatometric behavior, the (bulk) density of the solid, ρ (kg m^{-3}), can be calculated at any temperature from the thermogravimetric and dilatometric curves as

$$\rho = \rho_0 \frac{\left(\frac{W}{W_0}\right)_T}{\left[1 + \left(\frac{\Delta l}{l_0}\right)_T\right]^3}$$

[5.12]

where ρ_0 (kg m^{-3}) is the density of the solid at the reference temperature (usually room temperature), and the T subscripts indicate that the values pertain to a temperature T in the thermogravimetric and dilatometric records.

The density of composite solids at normal temperatures can be calculated by means of the mixture rule in its simplest form (Eq. 5.1 with $m = 1$):

$$\rho = \sum_i v_i \rho_i$$

[5.13]

where the i subscripts relate to information on the ith component. At elevated temperatures, the expansion of the components is subject to constraints (Eq. 5.10), and therefore the mixture rule is not strictly applicable. It may be used, however, as a crude approximation.

The relation between density and porosity was discussed earlier (see Eq. 4.1). The overall porosity of a composite solid consisting of several porous constituents is

$$P = \sum_i v_i P_i$$

[5.14]

where again the i subscripts relate to information on the ith component.

5.5   Elasticity, Plasticity, Strength, Creep

The elastic behavior of solids is characterized by their elastic moduli: Young's modulus (commonly referred to as modulus of elasticity), E (Pa); modulus of rigidity (or shear modulus), G (Pa); and bulk modulus (or modulus of volume elasticity), K (Pa). The latter two are related to Young's modulus through Poisson's ratio, μ (dimensionless), as follows:

$$G = \frac{E}{2(1 + \mu)}$$

[5.15]

$$K = \frac{E}{3(1 - 2\mu)}$$

[5.16]

Poisson's ratio is about 0.29 for steel. For concrete, it usually lies between 0.15 and 0.2, but it may be as high as 0.3. A value of 0.2 is recommended as a reasonable approximation for materials for which experimental information is not available.

Among the elastic moduli, Young's modulus is of principal significance. It is expected to be the same in tension and compression.

The value of Young's modulus is insensitive to the grain structure and to minor changes in the material's composition, but it is dependent on its porosity. The following relation has been proposed (Weil 1964):

$$\frac{E}{E_o} = \frac{1 - P}{1 + aP} \qquad [5.17]$$

where E (Pa) is Young's modulus for the (perhaps hypothetical) nonporous solid, and a is a constant (dimensionless) whose value is determined by the size, shape, and distribution of pores. The value of a is probably somewhere between 1 and 3.

For multiphase solids, Young's modulus may be estimated using the mixture rule in its simplest form

$$E = \sum_i v_i E_i \qquad [5.18]$$

where the i subscripts refer to information on the ith phase. It has been found, however, that the values yielded by Eq. 5.18 are often higher than the experimental values, and the disagreement increases with greater differences in the elastic moduli of the phases (Hashin 1962).

The mechanical properties of solids are usually derived from conventional tensile or compressive tests. Figure 5.1(a) shows the typical variation of the *stress*, σ (Pa), with increasing *strain*, ϵ (m m^{-1}) (i.e. deformation), in the tensile test of a metal. In practice, σ is interpreted as force divided by the original (rather than the momentary) cross-sectional area of the test specimen, and ϵ as $(l - l_o)/l_o$, where l and l_o (m) are the changed and original specimen lengths.

Section $0-e$ of the stress–strain curve represents the elastic deformation of the specimen, which is instantaneous and reversible. The slope of this straight-line section is equal to the modulus of elasticity (Young's modulus), E. Between points e and u the deformation is plastic, essentially nonrecoverable, and quasi-instantaneous. The yield strength of the material, σ_y (Pa) (measured at 0.2 per cent permanent deformation), and the ultimate strength, σ_u (Pa), are important design parameters. After some necking (localized reduction of cross-sectional area), the specimen ruptures at point r.

According to Weil (1964), the dependence of strength on porosity can be described with an equation similar to Eq. 5.17:

$$\frac{\sigma}{\sigma_o} = \frac{1 - P}{1 + aP} \qquad [5.19]$$

where σ and σ_o (Pa) are the (ultimate or yield) strengths of the porous and nonporous solids, respectively, and a is a constant (dimensionless) characteristic of the pore geometry.

Plastic deformation involves the sliding of flat, parallel layers of atoms within the crystals along certain preferred crystallographic planes. Assuming a perfect crystal, slip could occur only if the bonds between all atoms on either side of the slip plane break simultaneously. Calculations based on this mechanism yield

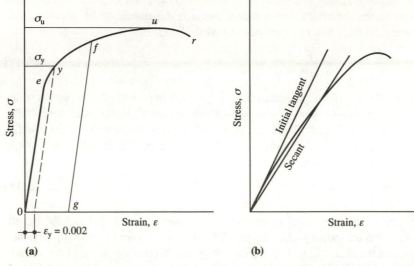

Fig. 5.1 Stress–strain curves: (a) test of a metallic material in tension; (b) test of a nonmetallic material in compression

theoretical strengths as much as 1000 times higher than the observed strengths. However, if atomic dislocations are also taken into account, the calculated and observed strengths are in good agreement. In the presence of a dislocation, slip can occur by the movement of the dislocation across the slip plane, severing only one atomic bond at a time.

Clearly, any obstruction that tends to reduce the mobility of dislocations will help increase the strength of the solid. Foreign atoms in the crystal structure, in the form of either impurities or dispersed particles introduced purposely (as in alloys), may hamper the movement of dislocations, and thereby improve the strength. Grain boundaries also act as blockages. Studies have indicated that the fracture strength is roughly proportional to the inverse square root of the mean grain diameter.

If the load is removed at some point f during plastic deformation, the test specimen will respond by an elastic contraction, as depicted by the f–g line in Fig. 5.1(a). If the load is later reapplied, the new stress–strain curve will retrace the g–f line. Thus, a new elastic limit, greater than the original, will be observed. Such an increase in the elastic strength is referred to as *strain hardening*. It results from the disappearance of dislocations during plastic deformation.

Deformation of a solid at temperatures much lower than the melting temperature, referred to as *cold work*, brings about a distorted grain structure (elongated in the direction of the deformation) and an increase in the yield strength. *Hot work* does not cause strain hardening. The grain distortion produced by hot work is only temporary; soon new, undistorted grains form which absorb the distorted grains.

The effects of cold work can be eliminated by *annealing* at an appropriate elevated temperature. In the annealing process, the distorted grains are replaced by equiaxed grains which, however, are finer than the grains of the original material. Thus, although the yield strength of the material drops, its ultimate strength remains fairly high on account of the finer grain structure.

The area under the stress–strain curve is a measure of the ability of the material to absorb energy up to the point of fracture. If this area is large, the material is *tough*; if it is small, the material is *brittle*. Nonmetallic materials, such as ceramics and concrete, are, on the whole, brittle. They do not exhibit appreciable plastic deformation, and their main role in structural design is to carry compressive load.

Figure 5.1(b) shows the stress-strain curve for a nonmetallic material in compression. For many nonmetallic materials, the straight-line section of the curve, characteristic of elastic behavior, is short or practically nonexistent. In a strict sense, Young's modulus should be interpreted as the tangent to the curve at the origin (known as *initial tangent modulus*). In engineering practice, however, the modulus of elasticity is usually interpreted as the *secant modulus*, the slope of a line connecting the origin with a point on the stress–strain curve, usually at 45 per cent of the ultimate strength.

At uniform temperatures, test specimens of brittle materials fracture in compression by shearing along oblique planes. Under transient conditions, however, thermal stresses may cause crushing in thin layers at the solid's surface. Unless dislodged by spalling (see Chapter 4), the crushed material is capable of carrying a reduced amount of compressive stress. In theoretical studies concerned with the response of solid bodies to heat penetration, it is customary to extend, more or less arbitrarily, the stress–strain curves beyond the point of fracture, in a way shown in Fig. 14.9.

Hardness tests, mostly Brinell tests, are often used for a preliminary assessment of the mechanical properties of the material. The Brinell hardness, HB ($kgf\,mm^{-2}$), is a measure of the strength of a solid, and its lack of ductility.

In room-temperature tensile or compressive tests, the specimen is deformed at a more or less constant rate (i.e. at constant crosshead speed), which is usually of the order of $1\,mm\,min^{-1}$. The stress–strain curves can be reproduced within a reasonable tolerance, and the shapes of the curves do not depend significantly on the crosshead speed. This finding is attributable to the relatively low mobility of the dislocations, and to the fact that the plastic deformation is accompanied by the disappearance of dislocations. At elevated temperatures, however, strain hardening is counteracted by viscous flow at the grain boundaries, and by diffusional phenomena that enhance the mobility of dislocations and the formation of new dislocations. Hence, the material will undergo plastic deformation even at constant stress, and the $e–r$ section of the stress–strain curve [Fig. 5.1(a)] will become markedly dependent on the crosshead speed.

The time-dependent plastic deformation of the material is referred to as creep strain and denoted by ϵ_t ($m\,m^{-1}$). In a creep test, the variation of ϵ_t is recorded

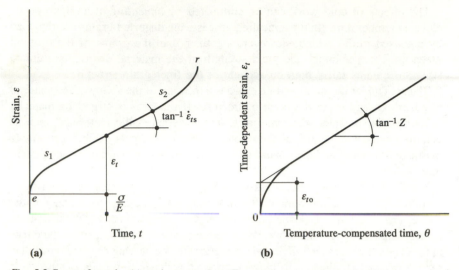

Fig. 5.2 Creep of metals: (a) strain versus time (T = constant, σ = constant); (b) time-dependent strain versus temperature-compensated time (σ = constant)

against time,[2] t (h), at constant stress, σ (Pa) (or rather, constant load), and at constant temperature. A typical strain–time curve is shown in Fig. 5.2(a). The total strain, ϵ (m m^{-1}), is

$$\epsilon = \frac{\sigma}{E} + \epsilon_t \qquad [5.20]$$

The $0-e$ section of the curve represents the instantaneous elastic (and recoverable) part of the deformation; the rest is creep, which is essentially nonrecoverable. The creep is relatively fast at first (primary creep, section $e-s_1$ in the figure), then proceeds for a long time at an approximately constant rate (secondary creep, section s_1-s_2), and finally accelerates until rupture occurs (section s_2-r). As illustrated in Fig. 5.3, the strain–time curves become steeper if the test is conducted at higher temperatures or at higher loads (stresses).

A concept advanced by Dorn (1954) is particularly suitable for dealing with deformation processes that develop at variable temperatures. Dorn eliminated the temperature as a separate variable by the introduction of the 'temperature-compensated time', θ (h), defined as

$$\theta = \int_0^t e^{-\Delta H_c/RT} \, dt \qquad [5.21]$$

where ΔH_c = activation energy of creep, J kmol^{-1};
 R = gas constant = 8315 J kmol^{-1} K^{-1}.

---

2   In engineering practice, the time of creep is measured in hours.

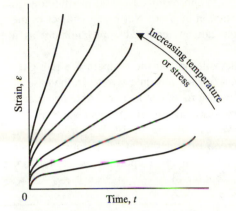

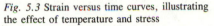

Fig. 5.3 Strain versus time curves, illustrating the effect of temperature and stress

The activation energy for creep has been found to be approximately equal to the activation energy of self-diffusion, and insensitive to (1) the microstructure of the material, (2) the temperature level, provided it is higher than one-half of the melting temperature, and (3) the stress level, unless it is extremely high.

From a practical point of view, only the primary and secondary creeps are of interest. Harmathy (1967d,e) has shown that the primary and secondary creep deformation can be satisfactorily described by the following equation:

$$\epsilon_t = \frac{\epsilon_{to}}{\ln 2} \cosh^{-1}(2^{Z\theta/\epsilon_{to}}) \qquad \text{if } \sigma = \text{constant} \qquad [5.22]$$

or approximated by the formula:

$$\epsilon_t = \epsilon_{to} + Z\theta \qquad \text{if } \sigma = \text{constant} \qquad [5.23]$$

ϵ_{to} (m m^{-1}) and Z (m m^{-1} h^{-1}) are two creep parameters whose meanings are explained in Fig. 5.2(b). Z is referred to as the Zener–Hollomon parameter, and defined (Zener and Hollomon 1944) as

$$Z = \dot{\epsilon}_{ts} \, e^{\Delta H_c/RT} \qquad [5.24]$$

where $\dot{\epsilon}_{ts}$ (m m^{-1} h^{-1}) is the rate of secondary creep at a given temperature, T, and stress, σ [see Fig. 5.2(a)]. The two creep parameters, ϵ_{to} and Z, have been found (Harmathy 1967e) to depend on the applied stress only (i.e. they are roughly independent of the temperature).

The σ = constant restriction imposed on Eqns 5.22 and 5.23 is an indication that the two creep parameters are, in fact, functions of not only the applied (constant) stress but also of the stress history of the material. Of the two, σ_{to} seems to be much more sensitive than Z to the way the load is applied.

At an advanced stage of creep deformation, i.e. when $\epsilon \gg \epsilon_{to}$, instead of Eq. 5.23, the following equation is applicable:

$$\epsilon_t \simeq Z\theta \qquad \text{if } \epsilon_t \gg \epsilon_{to} \qquad [5.25]$$

Harmathy (1967e) claimed that Eq. 5.25 is approximately valid in situations in which the stress is a slowly varying function of time. Not being subject to the σ = constant restriction, this equation can be regarded approximately as a mechanical equation of state.

For most solids, creep becomes noticeable only if the temperature is higher than about one-third of the melting point (on the absolute scale). In the case of concrete, creep may be quite significant even at room temperature. At temperatures below 105 °C, it is due mainly to the presence of evaporable water in the microstructure of the cement paste (Powers 1968; Neville 1971; Ramachandran *et al*. 1981). The mechanism of the creep of concrete at temperatures above 105 °C, where the cement paste no longer contains evaporable water, has not been satisfactorily explained (see also Chapter 6).

5.6   Thermal Conductivity

Conduction of heat in solids may take place by two mechanisms: coupling between the atomic vibrations in the crystal lattice, and movement of electrons. Debye (1914) postulated that in the former mechanism the 'mean free path of thermoelastic waves' plays a dominant role. Since it is proportional to $1/T$, the thermal conductivity of crystalline materials is expected to decrease as the temperature increases. However, even small departures from the perfect crystalline structure may bring about substantial reductions in the mean free path.

For glasses and crystalline materials with highly disordered structure (e.g. alloys and multiphase, polycrystalline solids), the mean free path becomes practically independent of the temperature. These materials exhibit relatively low conductivities at room temperature, and slowly increasing conductivities as their temperature rises.

In pure metals, the second kind of conduction mechanism (i.e. movement of electrons in the crystal lattice) is predominant. Consequently, there is a close relation between their thermal and electrical conductivities, as expressed by the Lorentz ratio

$$\frac{k}{\gamma T} = 2.45 \times 10^{-8} \, \mathrm{W\,\Omega\,K^{-2}} \qquad [5.26]$$

where k = thermal conductivity, $\mathrm{W\,m^{-1}\,K^{-1}}$;
$\quad\;\;\gamma$ = electric conductivity, $\mathrm{\Omega^{-1}\,m^{-1}}$.

Experimental values of the constant usually fall between 2.1 and 2.6.

If the material is transparent, radiation will also contribute to the transmission of heat. This contribution is proportional to the third power of temperature. For glasses, it becomes significant at temperatures above 500 °C (Kingery 1959).

The thermal conductivity of layered, multiphase, solid mixtures depends on whether the phases lie in the direction of the heat flow, or perpendicular to it.

The simple mixture rule is applicable, with $m = 1$ in the former case, and $m = -1$ in the latter. Thus, in these two limiting cases

$$k = \sum_i v_i k_i \qquad [5.27]$$

and

$$\frac{1}{k} = \sum_i \frac{v_i}{k_i} \qquad [5.28]$$

where k and k_i ($\text{W m}^{-1}\text{K}^{-1}$) are the thermal conductivities of the mixture and of its ith constituent, respectively.

With two-phase systems, if $k_1 \geq k_2$, $k \simeq v_1 k_1$ from Eq. 5.27, and $k \simeq k_2/v_2$ from Eq. 5.28. Expressing this in words: if the heat flow is parallel to the layers, the conduction is dominated by the better conductor; if it is perpendicular to the layers, the conduction is dominated by the poorer conductor.

The formula recommended for two-component systems by Hamilton and Crosser (1962) offers a much wider choice with respect to phase configurations than Eqs 5.27 and 5.28. Applied to thermal conductivity, the Hamilton–Crosser formula takes the form

$$k = \frac{v_1 k_1 + \beta v_2 k_2}{v_1 + \beta v_2} \qquad [5.29]$$

where

$$\beta = \frac{n k_1}{(n-1)k_1 + k_2} \qquad [5.30]$$

and subscript 1 designates the principal continuous phase.

There is substantial evidence (Hamilton and Crosser 1962; De Vries 1952) that Eqs 5.29 and 5.30, with $n = 3$, are applicable to any two-phase system consisting of a continuous and a discontinuous phase, irrespective of the geometry of the latter phase. When both phases are essentially continuous, as with most porous materials (air being the second phase), a lower value of n seems to be applicable: $n \simeq 1.5$ (Harmathy 1970a).

Since the conductivity of air ($\simeq 0.026 \text{ W m}^{-1}\text{K}^{-1}$ at room temperature, $0.046 \text{ W m}^{-1}\text{K}^{-1}$ at 300 °C) is usually negligible in comparison with that of the solid, the application of Eqs 5.29 and 5.30 to porous solids will result in an equation similar to Eqs 5.17 and 5.19:

$$\frac{k}{k_o} = \frac{1 - P}{1 + aP} \qquad [5.31]$$

where k_o ($\text{W m}^{-1}\text{K}^{-1}$) is the thermal conductivity of the (possibly hypothetical) nonporous solid. If the pores are discontinuous, the appropriate value for n in Eq. 5.30 is 3, and Eq. 5.29 yields $a = 0.5$. If, on the other hand, the pores are fully interconnected, $n = 1.5$ and $a = 2$ result. These values reflect the increased importance of the poorer conductor (air) as the pores assume the semblance of layers perpendicular to the heat flow.

At high temperatures, because of radiative heat transfer through the pores, the contribution of the pores to the thermal conductivity of the solid must not be disregarded.[3] The overall thermal conductivity of a porous solid can be expressed as

$$k = k_o \frac{1 - P}{1 + aP} + P(k_a + 4\sigma\epsilon b\delta T^3)$$ [5.32]

where a = constant (probably between 0.5 and 2), characteristic of pore geometry, dimensionless;

b = constant (probably between 0.6 and 1), characteristic of the pore geometry, dimensionless;

k_a = thermal conductivity of air, $W\,m^{-1}\,K^{-1}$;

σ = Stefan−Boltzmann constant = $5.67 \times 10^{-8}\,W\,m^{-2}\,K^{-4}$;

ϵ = emissivity of pores (probably between 0.7 and 1), dimensionless;

δ = characteristic pore size, m.

The fire safety practitioner need not be overly concerned about the empirical constants in Eq. 5.32. Thermal conductivity is a structure-sensitive property; its exact value is difficult to assess. The measured values depend to some extent on the method of measurement.[4] Unfortunately, no scanning technique is available for the determination of a continuous thermal conductivity versus temperature curve from a single test. Such a curve must be estimated from the results of tests conducted at discrete temperature levels.

Special problems arise in temperature intervals of physicochemical instability. The test methods require that prior to the test either a steady-state temperature distribution or a temperature level (and thereby a certain microstructural pattern in the material) be established. The test results can, therefore, be viewed as points on a continuous thermal conductivity versus temperature curve obtained by an imaginary scanning technique performed at an extremely slow scanning rate. Since each point pertains to a more or less stabilized microstructural pattern, there is no way of knowing how the thermal conductivity would vary during the course of physicochemical changes developing at finite rates and, consequently, at transient microstructures.

If the solid is not oven-dry, the temperature gradients induce migration of moisture, either in the liquid phase or by an evaporation−condensation mechanism (see Chapter 4). The movement of moisture is usually (but not necessarily) in the direction of the heat flow, and manifests itself as an apparent increase in the

---

3   It may be added that, unless the pore size is larger than, say, 5 mm, the contribution of convective heat transmission to the thermal conductivity is negligible.

4   Numerous steady-state and variable-state methods are available for the measurement of thermal conductivity. See, for example, Kingery et al. (1954), Kingery (1959), and Harmathy (1964).

thermal conductivity. Furthermore, if the solid undergoes some decomposition (e.g. dehydration) reaction at elevated temperatures, convective heat transfer by the gaseous decomposition products moving in the pores adds to the complexity of the heat flow process. At present, there is no way of satisfactorily handling these mass transfer processes in studies of heat flow under fire conditions.

5.7   Enthalpy, Specific Heat

The branch of physics that deals with phenomena in which temperature changes play an important role is called *thermodynamics*. A *system* is taken to mean any material or collection of materials under consideration, and the change that the system undergoes is called *process*. The processes to be discussed in this chapter are physicochemical changes, 'reactions' in a generalized sense (see Footnote 1). A composite system may consist solely of *stable materials(s)* [material(s) that do(es) not undergo any physicochemical change], or solely of *unstable material(s)* which in the reaction zone(s) change(s) from *reactant(s)* to *product(s)*, or (in general) of both stable and unstable materials.

When dealing with processes involving physicochemical changes, it is convenient to describe the system and its constituents in terms of molar ('per unit mole') properties, rather than 'per unit mass' properties. The molar properties will be denoted by capital letters, the 'per unit mass' properties by lower-case letters. To obtain 'per unit mass' properties, divide the molar properties by the molecular mass ('molecular weight'), M (kg kmol^{-1}), of the component or of the system.

In the case of a composite system, the relations between mass fractions and mole fractions are

$$x_i = \frac{\dfrac{\omega_i}{M_i}}{\sum_i \dfrac{\omega_i}{M_i}} \qquad [5.33]$$

and

$$\omega_i = \frac{x_i M_i}{\sum_i x_i M_i} \qquad [5.34]$$

where x_i (kmol kmol^{-1}) and M_i (kg kmol^{-1}) are the mole fraction and the molecular mass of the ith component of the system, respectively. Obviously,

$$\sum_i x_i = 1 \qquad \text{and} \qquad \sum_i \omega_i = 1 \qquad [5.35]$$

The molecular mass of the composite system is

$$M = \sum_i x_i M_i \qquad [5.36]$$

From the point of view of the subject to be discussed, two properties of the material (or system) are of particular interest: the *internal energy*, U (J kmol^{-1}), and the *enthalpy* (or heat content[5]), H (J kmol^{-1}). The relation between them is

$$H = U + pV \qquad [5.37]$$

where p = pressure, Pa;
$\quad\ V$ = specific (molar) volume, m^3 kmol^{-1}.

The internal energy is the energy that a unit mole of a material (or system) possesses on account of the configuration and motion of its atoms or molecules. The product pV in Eq. 5.37 is the 'external energy', which may be regarded as the energy a unit mole of the material (system) possesses by virtue of the space it occupies.

According to the principle of conservation of energy, the heat, dQ (J kmol^{-1}), received by the system during a time dt is

$$dQ = dU + pdV \qquad [5.38]$$

Hence, for a physicochemically stable system at constant volume (V = constant),

$$\left(\frac{dQ}{dT}\right)_V = \left(\frac{dU}{dT}\right)_V = C_V \qquad [5.39]$$

and at constant pressure (p = constant),

$$\left(\frac{dQ}{dT}\right)_p = \left(\frac{dH}{dT}\right)_p = C_p \qquad [5.40]$$

where C_V = heat capacity (molar specific heat) at constant volume, J kmol^{-1} K^{-1};
$\quad\ C_p$ = heat capacity (molar specific heat) at constant pressure, J kmol^{-1}K^{-1}.

In words: C_V and C_p are the amounts of heat required to raise the temperature of a unit mole of the system by one degree, while its volume or the pressure, respectively, is unchanged. For solids, the difference between C_V and C_p is small at room temperature, but it may become appreciable at higher temperatures, especially in the case of solids with high coefficients of thermal expansion.

The variation of C_p with temperature is usually described by an equation of the form

$$C_p = e + fT - \frac{g}{T^2} \qquad [5.41]$$

where e, f and g are empirical constants of various dimensions.

---

5   According to modern usage, heat is energy in the process of transfer between a system and its surroundings as a result of temperature differences. The use of heat in the sense of accumulated energy is discouraged.

The most comprehensive sources of information on the heat capacities of inorganic and organic solids at elevated temperatures are probably two books: one by Perry (1950) and the other by Eitel (1952). The data presented in these books have been assembled mainly from a series of publications by Kelley (1934, 1936, 1949, 1960).

The heat capacity of inorganic solids not listed in handbooks may be estimated using a modified version of the Neumann–Kopp rule (Eitel 1952). It has been shown that the heat capacity of inorganic compounds at room temperature can be estimated by the summation of empirically developed 'atomic heats' $(J\,kmol^{-1}\,K^{-1})$:

Silicon	15.90×10^3	Oxygen	16.74×10^3
Boron	11.30×10^3	Hydrogen	9.62×10^3
Germanium	23.01×10^3	Phosphorus	22.59×10^3
Sulfur	22.59×10^3	Carbon	7.53×10^3

All the other elements 26.78×10^3

For example, the heat capacity of orthoclase (potash feldspar), $KAlSi_3O_8$ can be calculated as

$$C_p = 10^3[1 \times 26.78(K) + 1 \times 26.78(Al) + 3 \times 15.90(Si) + 8 \times 16.74(O)]$$
$$= 0.235 \times 10^6\,J\,kmol^{-1}\,K^{-1}$$

The measured value is $0.203 \times 10^6\,J\,kmol^{-1}\,K^{-1}$.

Harmathy (1970a) suggested that the heat capacity of silicates could be assessed by the summation of the heat capacities of the constituent oxides, weighted according to their chemical formulas.[6] Since the heat capacity versus temperature relation for numerous oxides is available from the literature, Harmathy's method makes it possible to estimate for uncommon silicates not only the value of the heat capacity but also its variation with temperature. (The method will be further discussed in Chapter 6.)

Among some other summation techniques, the one proposed by Rihani and Doraiswamy (1965) may prove useful. It allows the calculation of the heat capacity of organic compounds at room temperature and at elevated temperatures from 'group contributions'.

The specific heat of solids is not sensitive to the microstructure of the material and to minor changes in its composition. In the case of physicochemically stable materials (such as ceramic materials and metals designed for high temperature use), the 'drop method' (or method of mixtures; Kingery 1959) is probably the

---

6   Clearly, the heat capacity for a compound, in which the oxides are tied together by chemical bonds, cannot be the same as that for a mixture of the same oxides, and yet the technique suggested by Harmathy seems to hold remarkably well. The error resulting from its use is less than 15 per cent, and usually consists of an overestimation of the true heat capacity of the compound.

simplest and most accurate technique for the experimental assessment of specific heat. This method consists of heating in a furnace a sample of the solid to the target temperature, dropping it from the hot furnace into a calorimeter held at a reference temperature [usually at 25 °C (298 K)] and measuring the heat released by the sample. This heat is numerically equal to the enthalpy gain by the sample on heating from the reference temperature to the target temperature. Having measured the enthalpy gain for a series of elevated temperatures, one can determine the c_p (J kg^{-1} K^{-1}) versus T relation by graphic differentiation of the $(h - h_o)$ (J kg^{-1}) versus T curve (where h_o is enthalpy at a reference temperature).

In the case of building materials, the temperature ranges of stability are often limited or practically nonexistent. If physicochemical changes (either nonreversible or not immediately reversible) occur, the heat released by the sample on cooling is not equal to the gain of enthalpy on heating.

For a long time, adiabatic calorimetry was the most widely used method for studying the h versus T relation for unstable materials. With this technique, a sample of the material is heated electrically in a spherical furnace through the temperature interval of interest. The heat input into the sample and the sample's temperature are continuously recorded. Heat exchange between the sample and the furnace is eliminated by equalizing (with the aid of scores of thermocouples) the temperature of the sample's surface and the furnace interior. The enthalpy versus temperature curve is evaluated from the heat input and temperature records.

Nowadays differential scanning calorimetry (DSC) is the most commonly used method for mapping out the h versus T curve in a single sweep at the desired rate of heating. The test sample is heated together with a sample of a chemically inert reference material. By controlling the heat inputs into the two samples, their temperatures are made to rise evenly. The enthalpy—temperature relation for the test material is calculated from the known h versus T relation for the reference material and from differences between the amounts of heat supplied to the two samples.

Differential thermal analysis (DTA) is an older and simpler version of DSC. Whilst the apparatus for adiabatic and differential scanning calorimetry are very expensive, DTA tests can be performed with easy-to-build laboratory equipment. Again, the test sample is heated together with a sample of an inert (reference) material in a furnace, the temperature of which is raised at a preselected rate. With this technique, however, the temperature of the test sample is allowed slightly to run ahead or lag behind that of the reference sample. The difference between the temperatures of the two samples plotted against the furnace temperature (or the temperature of the reference sample) is the DTA curve. The peaks of the curve (maxima or minima) are indicative of the nature and rate of the reactions that the test sample undergoes. Although DTA is used mainly for material identification, it may provide valuable information (for example on reaction temperature, rate of reaction) for those who wish to evaluate the h versus T relation on theoretical grounds.

In the temperature interval of a physicochemical change, a system is not completely defined by the thermodynamic properties mentioned so far. An additional variable, called the reaction progress variable, ξ (dimensionless), is needed for describing the momentary state of the system.

Since the (molar) enthalpy, H, is a function of p, T, and ξ, the heat to be communicated to the system to cause an incremental rise of temperature,[7] dT, at constant (atmospheric) pressure is

$$\left(\frac{dQ}{dT}\right)_p = \left(\frac{dH}{dT}\right)_{p,\xi} + \left(\frac{dH}{d\xi}\right)_p \left(\frac{d\xi}{dT}\right)_{p,T} \qquad [5.42]$$

Here $(dQ/dT)_p$ will be referred to as *apparent heat capacity* (or *apparent molar specific heat*) and denoted by C_p (J kmol^{-1} K^{-1}); $(dH/dT)_{p,\xi}$, denoted by $\overline{C}_p$ (J kmol^{-1} K^{-1}), is the sensible heat contribution to C_p, and represents the heat capacity of a mixture of materials consisting, in general, of stable material(s) and unstable material(s) [reactant(s) and product(s)] that are present in the system at a given value of the reaction progress variable, ξ (i.e. at a given temperature, since ξ is assumed to be function of T). $(dH/d\xi)_p$ is the (latent) heat of reaction at constant pressure, i.e. the amount of heat absorbed (or evolved) by the system per unit reaction. It will be designated as ΔH_r (J kmol^{-1}). With these symbols, Eq. 5.42 can be recast into the following form:

$$C_p = \overline{C}_p + \Delta H_r \left(\frac{d\xi}{dT}\right)_{p,T} \qquad [5.43]$$

Clearly, for a physicochemically stable system, or for an unstable system over temperature intervals of physicochemical stability, $C_p \equiv \overline{C}_p$.

Those who do not have access to the expensive facilities needed for the experimental determination of the C_p versus T curves for unstable building materials may assess the curve on theoretical grounds. The following information is needed:

- The approximate composition of the material.
- The nature and temperature of change(s) [reaction(s)] that the unstable compound(s) is (are) expected to undergo on heating. This information may be obtained from the literature, calculated from the thermodynamic properties of the reactant(s) and product(s), or developed by standard laboratory techniques. DTA records may prove very useful. They reveal not only the temperature interval(s) of reaction(s), but also give some hint of the heat(s) and rate(s) of reaction(s).
- The heat capacity versus temperature relation for all components of the system: stable compound(s), unstable compound(s) [i.e. reactant(s), and product(s) of reaction(s)].
- The latent heat(s) of physicochemical change(s) [reaction(s)]. This information

---

7 Although there are physicochemical changes that occur strictly at constant temperature, it is convenient to assume that all changes develop within a temperature interval (reaction zone), and that the reaction progress variable, ξ, is a function of temperature.

may also be obtained from the literature. If not available, the heat(s) of reaction can be determined by taking the difference between the enthalpies of the product(s) and reactant(s) at the reaction temperature(s), provided the reference enthalpy level for all the compounds participating in the reaction(s) is chosen as their heats of formation,[8] ΔH_f (J kmol^{-1}).

● The progress variable(s) for the reactions (to be discussed).

The calculation procedure will be illustrated using the example of a hypothetical material, the composition of which is:

Stable compound: I   Mass fraction ω_I Molecular mass M_I
Unstable compound: J   Mass fraction ω_J Molecular mass M_J

Compound J is known to decompose at a (nominal) reaction temperature of T_r (K) according to the following scheme:

$$J \rightarrow kK + lL + mM(g)\uparrow \qquad\qquad [5.44]$$

where k, l and m are stoichiometric coefficients (dimensionless). M is a gaseous compound that leaves the system.

Assume that the heat capacity versus temperature relation (see Eq. 5.41)[9] is known for all compounds: I, J, K, L, and M.

$$C_I = e_I + f_I T - \frac{g_I}{T^2} \qquad\qquad [5.45]$$

$$C_J = e_J + f_J T - \frac{g_J}{T^2} \quad \text{etc.} \qquad\qquad [5.46]$$

Assume also that the heat of decomposition of compound J is not known. However, information is available on the heats of formation of the reactant, $(\Delta H_f)_J$, and of the products, $(\Delta H_f)_K$, $(\Delta H_f)_L$, and $(\Delta H_f)_M$. Calculate first the enthalpies of these compounds at the nominal reaction temperature, T_r.

$$(H_J)_{T_r} = \Delta(H_f)_J + \int_{T_0}^{T_r} C_J dT = \Delta(H_f)_J + \int_{T_0}^{T} \left(e_J + f_J T - \frac{g_J}{T^2} \right) dT$$

---

8   The heat of formation is the enthalpy of a compound at 298 K, usually in relation to the enthalpies of the constituent atoms or diatomic molecules (O_2, N_2, Br_2, etc.). The heat of formation for most compounds is a negative quantity, reflecting the fact that at room temperature compounds are, as a rule, more stable than the atoms or molecules of which they are made up. Note, however, that the stability of a compound depends not only on its enthalpy, but also on its temperature and entropy.

In the case of silicates, it is more convenient to express the heat of formation of a compound in relation to the heats of formation of the constituent oxides.

Information on the heats of formation of a large number of inorganic and organic compounds is available from a number of sources, e.g. from the already mentioned handbooks: by Perry (1950) and by Eitel (1952).

9   All processes to be discussed in this chapter are expected to take place at constant (atmospheric) pressure. The subscript p will be shown only if displaying it does not result in multiple subscripts.

$$= \Delta(H_f)_J + \left[e_J T + \frac{1}{2} f_J T^2 + \frac{g_J}{T} \right]_{T_o}^{T_r} \qquad [5.47]$$

$$(H_K)_{T_r} = \Delta(H_f)_K + \left[e_K T + \frac{1}{2} f_K T^2 + \frac{g_K}{T} \right]_{T_o}^{T_r} \quad \text{etc.} \qquad [5.48]$$

where $(H_J)_{T_r}$, $(H_K)_{T_r}$, etc. are the enthalpies of compounds J, K, etc., at the nominal reaction temperature, T_r (K), and T_o ($= 298$ K) is the standard reference temperature (for which the heats of formation are listed in the handbooks).

The heat of reaction is obtained by taking the difference between the sum of the enthalpies of the products (each multiplied by its stoichiometric coefficient) and the enthalpy of the reactant.

$$(\Delta H_r)_{T_r} = k(H_K)_{T_r} + l(H_L)_{T_r} + m(H_M)_{T_r} - (H_J)_{T_r} \qquad [5.49]$$

where $(\Delta H_r)_{T_r}$ (J kmol^{-1}) is the heat of reaction (decomposition of compound J) at the reaction temperature, T_r. When ΔH_r is positive, the reaction is endothermic (heat is absorbed in the process); when it is negative, the reaction is exothermic (heat is evolved).

The scheme of calculating the c_p versus T relation for the material consisting originally of compounds I and J is shown in Fig. 5.4. The calculated values of the specific heat are related at any temperature to 1 kg material *at the start of the heating process*.

The calculation procedure is quite straightforward below and above the temperature interval of the reaction. It is somewhat complex through the reaction zone, owing to the changing composition (and possibly mass) of the material, and to the absorption (or evolution) of latent heat by the reacting system. In that zone, the reaction progress variable, ξ, plays a prominent role.

If the system undergoes decomposition reaction(s) and the decomposition is accompanied by the release of gaseous products (usually H_2O or CO_2), the ξ versus T relation can be determined from a thermogravimetric test, and the $d\xi/dT$ versus T relation by graphic differentiation of the $\xi(T)$ curve. For compounds that experience physicochemical changes that do not show up in a thermogravimetric test, the ξ versus T and $d\xi/dT$ versus T relations have to be estimated, possibly on the basis of DTA records. It should be emphasized that if the c_p versus T relation is sought with a view to application in numerical heat flow studies, selecting the width of the temperature interval of reaction and the nature of the ξ versus T function is only of moderate significance. Inaccuracies resulting from somewhat arbitrary selections are usually smoothed out soon after the computed temperatures rise past the temperature of the reaction zone.

The temperature interval of the reaction is known to depend on the rate of heating. A DTA test conducted at a rate 5 °C min^{-1} may provide some guidance. If no guidance is available, the reaction zone may be selected to straddle the nominal reaction temperature, T_r, and extend at least 15 °C on either side.

Table 5.1 gives values of the ξ versus T and $d\xi/dT$ versus T functions for three

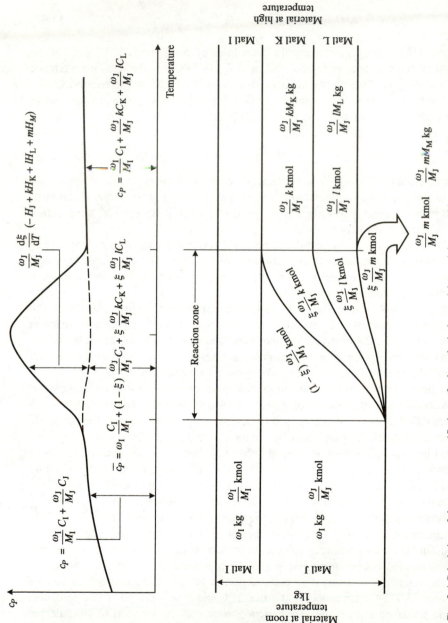

Fig. 5.4 Calculation of the apparent specific heat of a hypothetical material containing a stable compound (I) and an unstable compound (J)

Table 5.1 Alternatives for selecting the reaction progress variable and its derivative (see Figs 5.4 and 5.5); $(T_{r2} - T_{r1})$ = width of reaction zone

$\dfrac{T - T_{r1}}{T_{r2} - T_{r1}}$	Alternative 1		Alternative 2		Alternative 3	
	ξ	$(T_{r2} - T_{r1})\dfrac{d\xi}{dT}$	ξ	$(T_{r2} - T_{r1})\dfrac{d\xi}{dT}$	ξ	$(T_{r2} - T_{r1})\dfrac{d\xi}{dT}$
−0.1	0	0	0	0	0.002	0.038
0	0	0	0	0	0.010	0.122
0.1	0.1	1.0	0.02	0.4	0.031	0.328
0.2	0.2	1.0	0.08	0.8	0.081	0.701
0.3	0.3	1.0	0.18	1.2	0.176	1.204
0.4	0.4	1.0	0.32	1.6	0.321	1.666
0.5	0.5	1.0	0.50	2.0	0.500	1.856
0.6	0.6	1.0	0.68	1.6	0.679	1.666
0.7	0.7	1.0	0.82	1.2	0.824	1.204
0.8	0.8	1.0	0.92	0.8	0.919	0.701
0.9	0.9	1.0	0.98	0.4	0.969	0.328
1.0	1.0	1.0	1.00	0	0.990	0.122
1.1	1.0	0	1.0	0	0.998	0.038

selection alternatives. With all three, the reaction zone is regarded as symmetrical about the nominal reaction temperature, T_r (see Fig. 5.4). The width of the reaction zone $(T_{r2} - T_{r1})$ can be assigned more or less arbitrarily.

Alternative 3 in the table has been calculated on the assumptions that the $d\xi/dT$ versus T relation is of the shape of the normal frequency function, and that 98 per cent of the reaction is completed within the designated reaction zone, i.e. within the temperature interval $(T_{r2} - T_{r1})$. Although this alternative appears to be the most realistic, the table reveals that it is not much different from Alternative 2. In fact, even the much simpler Alternative 1, which assumes the constancy of $d\xi/dT$ over the entire reaction zone, may prove adequate for most practical applications.

Figure 5.5 shows the appearance of the apparent heat capacity versus temperature curve for the three alternatives. The figure also shows an exothermic reaction, handled using Alternative 1. It should be emphasized that the apparent heat capacity cannot be allowed to go negative. If the rate of heat evolution from the reaction is very high, the heat is either scattered to the surroundings or, if absorbed by the material, causes a very fast temperature rise.

The application of the procedure outlined above to the calculation of the specific heat of concrete will be shown in the next chapter.

Nomenclature

a, b	empirical constant, characteristic of the pore geometry, dimensionless
c	(apparent) specific heat, J kg^{-1} K^{-1}

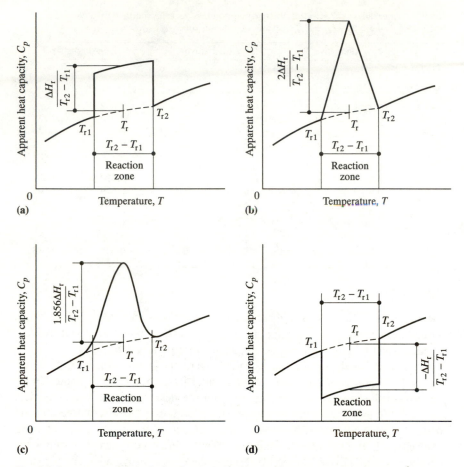

Fig. 5.5 Appearance of the calculated apparent heat capacity versus temperature curve for a hypothetical unstable material, as influenced by the selection of the $\xi(T)$ function: (a) endothermic reaction, $\xi(T)$ according to Alternative 1; (b) endothermic reaction, $\xi(T)$ according to Alternative 2; (c) endothermic reaction, $\xi(T)$ according to Alternative 3; (d) exothermic reaction, $\xi(T)$ according to Alternative 1

$\bar{c}$	sensible heat contribution to the apparent specific heat, J kg^{-1} K^{-1}
C	(apparent) heat capacity, J kmol^{-1} K^{-1}
$\bar{C}$	sensible heat contribution to the apparent heat capacity, J kmol^{-1} K^{-1}
e, f, g	empirical constants, various dimensions
E	Young's modulus (modulus of elasticity), Pa
G	modulus of rigidity, Pa
h	enthalpy, J kg^{-1}
H	molar enthalpy, J kmol^{-1}
ΔH_c	activation energy of creep, J kmol^{-1}
ΔH_f	heat of formation, J kmol^{-1}

ΔH_r	heat of 'reaction' (physicochemical change), $J\,kmol^{-1}$
HB	Brinell hardness, $kgf\,mm^{-2}$ (multiply by 9.8 to obtain values in MPa)
k	thermal conductivity, $W\,m^{-1}\,K^{-1}$
k, l, m	stoichiometric coefficients, dimensionless
K	bulk modulus, Pa
l	length, m
Δl	$= l - l_o$, m
m	exponent, dimensionless
M	molecular mass ('molecular weight'), $kg\,kmol^{-1}$
n	empirical constant, dimensionless
p	pressure, Pa
P	porosity, $m^3\,m^{-3}$
Q	heat received by unit kmole of the system, $J\,kmol^{-1}$
R	gas constant $= 8315\,J\,kmol^{-1}K^{-1}$
t	time, h (s)
T	temperature, K
U	molar internal energy, $J\,kmol^{-1}$
v	volume, m^3
V	specific volume (molar), $m^3\,kmol^{-1}$
W	mass, kg
x	mole fraction, $kmol\,kmol^{-1}$
Z	Zener–Hollomon parameter, $m\,m^{-1}\,h^{-1}$
α	coefficient of thermal expansion, $m\,m^{-1}\,K^{-1}$
$\bar{\alpha}$	coefficient of volume expansion, $m^3\,m^{-3}\,K^{-1}$
β	variable, defined by Eq. 5.6, dimensionless
γ	electric conductivity, $\Omega^{-1}\,m^{-1}$
δ	characteristic pore size, m
ϵ	strain (deformation), $m\,m^{-1}$; emissivity of pores, dimensionless
ϵ_{fo}	creep parameter, $m\,m^{-1}$
$\dot{\epsilon}_{rs}$	rate of secondary creep, $m\,m^{-1}\,h^{-1}$
θ	temperature-compensated time, h
μ	Poisson's ratio, dimensionless
ξ	reaction progress variable, dimensionless
π	any material property, various dimensions
ρ	density, $kg\,m^{-3}$
σ	stress, Pa; strength, Pa; Stefan–Boltzmann constant $= 5.67 \times 10^{-8}$ $W\,m^{-2}\,K^{-4}$
v	volume fraction, $m^3\,m^{-3}$
ω	mass fraction, $kg\,kg^{-1}$

Subscripts

a	of air
i	of the ith constituent
m	of the melting point
p	at constant pressure
r	of reaction (physicochemical change)

t	time-dependent (plastic)
T_r	at the (nominal) reaction temperature
u	ultimate
V	at constant volume
y	yield
o	at a reference temperature level; in reference condition; original

6 Concrete and its Constituents

Although it is referred to by a simple word, concrete is anything but a simple material. Structural engineers have always been keenly aware of the problems presented by the heterogeneity of the material, even though their interest is usually restricted to its mechanical properties under ordinary climatic conditions. Fire safety practitioners, who have a much wider range of interests in concrete, face an assortment of additional problems. Being a Type L/I (load-bearing/insulating) material, concrete gives rise to problems that concern its mechanical as well as thermal properties, and not only under ordinary climatic conditions but also under fire exposure conditions.

Of equal interest to both structural engineers and fire safety practitioners is the microstructure of the cement paste. The nature of the network of pores in the paste is a prominent factor in the strength-related properties of concrete. But it is also a decisive factor in the concrete's ability to hold moisture, and in its susceptibility to spall if exposed to fire. As pointed out in Chapter 4, the presence of moisture and the combined effect of high moisture content and low permeability have a substantial bearing on the performance of building elements in fire.

Those properties of concrete that relate to the performance of buildings in normal service have already been dealt with in many handbooks and thousands of papers and reports. The publications most frequently consulted in the preparation of this chapter were books by Bogue (1955), Lea (1970), Neville (1971), Neville and Brooks (1987), Ramachandran *et al.* (1981) and Taylor (1990), papers by Powers and Brownyard (1948) and Powers (1962), and the *ACI Manual of Concrete Practice* (American Concrete Institute 1989). As to properties specifically related to performance in fire, two comprehensive reports, one by Bennetts (1981) and the other by Schneider (1985), and the ACI *Guide* (ACI Committee 216 1989) proved valuable sources of information.

Owing to rapid developments since the 1960s in the science and technology of concrete, it would seem impossible to deal, in a book specializing in fire safety, with all the different kinds of concrete that are now available to the building designer. Only the 'all-purpose' concretes will be discussed in some detail. Nethertheless, the discussion will be presented in such a way as to provide foundations for evaluating the properties of some of the less conventional types.

6.1   The Portland Cement Paste

Concrete is usually looked upon as a two-component material, consisting of hydrated cement paste and aggregates or, alternatively, cement—sand mortar and coarse aggregate. The properties of concrete are determined by the properties of its components and the interfaces between them.

The most important ingredient of a concrete mixture is the portland cement: a pulverized, hydraulic material manufactured from the calcination product (clinker) of carefully controlled amounts of calcareous (lime-bearing) and argillaceous (clay-like) materials. The principal components of the portland cement are:

Tricalcium silicate	$3CaO . SiO_2$	C_3S
β-Dicalcium silicate	β-$2CaO . SiO_2$	β-C_2S
Tricalcium aluminate	$3CaO . Al_2O_3$	C_3A
Brownmillerite	$4CaO . Al_2O_3 . Fe_2O_3$	C_4AF

The last column lists the shorthand notations adopted for the oxide composition of the components of cement.[1]

The portland cement sets by the hydration of its components on addition of water. The product is called *cement paste* at any stage of the hydration process. The composition of the portland cement (primarily the amount of C_3S and C_3A) is occasionally adjusted to suit special requirements. Thus, portland cement mixes are offered for moderate heat evolution on hydration, for high early strength, for low heat evolution on hydration, and for high sulfate resistance. These special cement mixes are referred to as Type II, III, IV and V portland cements, respectively, whereas Type I is looked upon as the general-purpose cement.

Workability is an important factor in the design of the concrete mixtures. From the point of view of ease of placement, mixtures of good flowability, prepared with relatively high water—cement ratios, would seem to be preferable to stiff mixtures. Yet, the strength, durability and watertightness of concrete decrease with the amount of water added to the mixture. Fortunately, good workability can also be achieved at relatively low water—cement ratios with the use of certain additives, such as air-entraining agents and water-reducing admixtures.[2]

The strength of concrete and all the good properties that go with it (low permeability, durability, resistance to abrasion, resistance to sulfates, etc.) keep

---

1   The shorthand notations are: C for CaO, S for SiO_2, A for Al_2O_3, F for Fe_2O_3, and H for H_2O.

2   Organic foaming materials are used as air-entraining agents. They produce minute air bubbles during the mixing of concrete. Since the air cavities do not form a continuous network, the frost resistance of the concrete is not impaired. In fact, its frost resistance is improved, because less water is required to achieve the same workability, and therefore a less well developed porous network is left behind by the water not consumed in the hydration of cement.

The conventional water-reducing agents, also called plasticizers, ensure good workability of the mixture at water—cement ratios 5—15 per cent less than usually required. With the use of the so-called superplasticizers, up to 30 per cent reduction in the water requirement is possible.

improving with the reduction of water in the mixture, even though below a certain water–cement ratio (about 0.4 by mass) complete hydration of the cement is no longer possible. The use of high-strength concretes, prepared at very low water–cement ratios with the aid of water-reducing admixtures appears to be on the rise in some areas of the building industry.

The principal product of the hydration of portland cement is an impure, quasi-amorphous calcium silicate hydrate of somewhat indefinite composition, known as C–S–H gel (or tobermorite gel), which develops from the C_3S and C_2S components of the cement (Kalousek and Prebus 1958; Brunauer and Greenberg 1962; Copeland *et al.* 1962; Copeland and Kantro 1969; Taylor 1969). The C–S–H gel amounts to roughly 70 per cent of the weight of the fully hydrated paste. Another major component of the paste is calcium hydroxide ($Ca(OH)_2$; CH in shorthand), which is present in the form of fairly well developed crystals embedded in the C–S–H gel and amounts to about 20 per cent of the total weight.[3] The paste also contains a number of less easily identifiable colloidal hydrates formed from other compounds, and possibly some residues of the anhydrous cement. All these components of the cement paste are referred to simply as 'the gel'.

The stoichiometry of the hydration of portland cement is poorly understood. Even less is known about the dehydration of the cement paste at elevated temperatures. However, a fair insight into the behavior of the cement paste in fire can still be gained through conveniently constituted models covering the material, and the hydration and dehydration processes.

The composition of a Type I portland cement used in a series of experimental and theoretical studies (Harmathy 1970a) is shown in Table 6.1. This cement may be modeled to consist only of the two dominant components, C_3S and C_2S. The composition of such a 'model cement', calculated by retaining the actual C_3S–C_2S ratio, is also given in the table, in terms of mass fractions as well as mole fractions.

From the work of Brunauer and Greenberg (1962) it appears that the average composition of the C–S–H gel can be described by the formula $C_{1.62}SH_{1.5}$. Thus, the hydration scheme for the model cement is

$$0.568C_3S + 0.432C_2S + 2.448H \rightarrow mC_{1.62}SH_{1.5} + 0.948mCH$$
$$+ 0.568(1 - m)C_3S + 0.432(1 - m)C_2S + 2.448(1 - m)H \qquad [6.1]$$

where m (dimensionless) is the paste's maturity factor (or hydration progress variable). Although the hydration of the various components of portland cement are known to proceed at various rates, it is assumed in this model that a single value of m applies to all components, and that m depends only on the curing time, t (h), and the water–cement ratio, W_w/W_c (kg kg^{-1}), in the original cement (or

---

3   Concrete products to be subjected to high-pressure steam curing (autoclaving, performed in pressure vessels at a temperature of about 175 °C) are often prepared with finely ground silica (silica flour) added to the mixture. The silica reacts with the calcium hydroxide (CH) that forms on the hydration of cement. Consequently, the cement paste in autoclaved concrete products may not contain CH.

Table 6.1 Composition of a cement used in experiments, and of the 'model cement' (Harmathy 1970a)

Compound	Actual composition of cement (kg kg^{-1})	Composition of 'model cement' (kg kg^{-1})	(kmol kmol^{-1})
C$_3$S	0.470	0.635	0.568
β-C$_2$S	0.270	0.365	0.432
C$_3$A	0.116	0	0
C$_4$AF	0.090	0	0
Other	0.054	0	0
	1.000	1.000	1.000

Note Molecular masses: C$_3$S 228.33   C 56.08
C$_2$S 172.25   S 60.06

concrete) mixture. The family of curves in Fig. 6.1, based mainly on information published by Taplin (1959) for a particular cement paste, illustrates the typical appearance of the $m = m(t, W_w/W_c)$ function for hydration processes developing under favorable conditions (e.g. in a fog chamber).[4]

On complete hydration ($m = 1$), the model cement paste will hold[5]

$$\frac{2.448 \times 18.02}{0.568 \times 228.33 + 0.432 \times 172.25} = 0.216$$

kg of nonevaporable water per kg anhydrous cement.

Upon heating, the dehydration of the paste starts as soon as the desorption of evaporable water is completed (at about 105 °C), and continues uninterrupted up to temperatures in excess of 800 °C. The residue of the C−S−H gel consists probably of β- or α'-dicalcium silicate (β- or α'-C$_2$S) and wollastonite (β-CS).[6] The residue of the calcium hydroxide (CH) is calcium oxide (C).

In modeling the dehydration process, it is assumed that

(a) the only compounds present in the cement paste are those shown on the right-hand side of Eq. 6.1;
(b) there are only two dehydration reactions, one involving the C−S−H gel, and the other the CH;

---

4   In addition to the water−cement ratio and the age of the paste, m also depends on the fineness of the anhydrous cement particles, and the temperature and relative humidity of the environment. Note that if $W_w/W_c \leq 0.4$, complete hydration ($m = 1$) will never occur. The hydration process may also stop on account of self-desiccation. The rate of hydration slows down considerably if the equilibrium relative humidity in the pores drops below 0.95, and becomes zero at 0.8 (Powers and Brownyard 1948; Copeland and Bragg 1955).
5   The molecular masses of water, C$_3$S, and C$_2$S are 18.02, 228.33, and 172.25 respectively.
6   The various forms of polymorphous crystals are denoted by Greek letters.

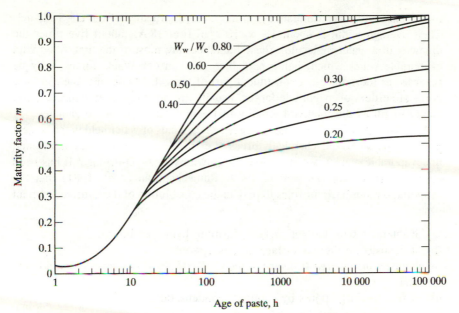

Fig. *6.1* Maturity factor for a particular cement paste (Taplin 1959; Harmathy 1967a)

(c) as the dehydration proceeds, it is invariably the final dehydration products that form;

(d) there is no chemical activity among the dehydration products; and

(e) the progress variables for the two dehydration reactions, ξ_{CSH} and ξ_{CH} (dimensionless), are unique functions of the temperature.

The dehydration reactions are modeled as

$$C_{1.62}SH_{1.5} \rightarrow [(1 - \xi_{CSH})C_{1.62}SH_{1.5} + 0.62\xi_{CSH}C_2S + 0.38\xi_{CSH}CS] + 1.5\xi_{CSH}H\uparrow \qquad [6.2]$$

and

$$CH \rightarrow [(1 - \xi_{CH})CH + \xi_{CH}C] + \xi_{CH}H\uparrow \qquad [6.3]$$

where the terms between the brackets represent the solid products of dehydration, and the last term is the gaseous product (water vapor) that leaves the system.

6.2 Density, Porosity, and Thermal Expansion of the Paste

The hydrated portland cement may be pictured as consisting of a *gel matrix* and a system of *capillary pores*. The latter is the residue of originally water-filled and, to a lesser extent, air-filled spaces that have not become invaded by the gel. However, the gel is itself a porous material. The *gel pores* are narrow, interconnected spaces (interstices) in the layered structure of the C−S−H gel. Their average width is about 18×10^{-10} m (i.e. 18 Å), about five times the diameter of a water molecule. These interstices are most of the time filled with evaporable water which, being held there by van der Waals forces, may be regarded as more or less a constituent part of the gel. In contrast, the capillary pores are orders of magnitude larger, usually are not fully interconnected, and empty or fill as the relative humidity of the ambient atmosphere changes.

As discussed in Chapter 4, the line between the kinds of water held in the cement paste by physical and chemical forces (i.e. the line between evaporable and nonevaporable water) is conventionally drawn by P- or D-drying.[7] It has been suggested (Feldman and Sereda 1969; Ramachandran *et al.* 1981) that the evaporable water has four roles to play in the constitution of the hydrated cement paste:

(a) it interacts with the gel surface, forming hydrogen bonds;
(b) it deposits on the gel surface by adsorption;
(c) it enters the layered structure of the C−S−H gel at low relative humidities; and
(d) it fills the large pores by capillary condensation.

According to Powers and Brownyard (1948) and Powers (1962), there are two kinds of evaporable water: *gel water* and *capillary water*. The latter is, for all intents, indistinguishable from free water. Following their line of reasoning, one may model the portland cement paste as a system consisting of the following four quasi-components:

● *Cement* (solid): pulverized, anhydrous clinker alone, or that part of the hydrated gel which has originated from the anhydrous powder.
● *Nonevaporable water* (solid): all kinds of water of constitution, including water of crystallization and water otherwise chemically combined.
● *Gel water* (liquid-like): evaporable water, held in the gel interstices by adsorption (in roughly two molecular layers) or by other forces of the van der Waals type.
● *Free water and capillary water* (liquid): water used in the original cement or concrete mixture, and evaporable water settled in the capillary pores mainly by capillary condensation.[8]

Modeling the cement paste as a quasi-four-component system offers simple and perspicuous ways of deriving expressions for such important properties of the paste as density, porosity and gel−space ratio.

Powers and Brownyard (1948) suggested that the following values could be assigned to the densities of the above four quasi-components:

---

7   P-drying is rarely used nowadays in scientific work.
8   Since the capillary pores are orders of magnitude larger than the gel pores, the amount of water held by adsorption on their surfaces is negligible compared with that held by capillary forces.

Cement (anhydrous or in hydration products): $\rho_c = 3150\ \mathrm{kg\ m^{-3}}$
Nonevaporable water: $\rho_n = 1220\ \mathrm{kg\ m^{-3}}$
Gel water: $\rho_{gw} = 1110\ \mathrm{kg\ m^{-3}}$
Free water and capillary water: $\rho_w = 1000\ \mathrm{kg\ m^{-3}}$

As only the dry and fully saturated conditions of the paste are of practical interest, the fifth component of the system, namely air, need not be considered separately. The air in the fresh paste, purposely entrained or unintentionally entrapped, will eventually enlarge the capillary pore system; it is convenient, therefore, to add the total volume of air voids to the volume of water in the original cement- or concrete-mixture. Thus, the 'adjusted' mass of water in the original mixture, W_{wa} (kg), is

$$W_{wa} = W_w (1 + A) \qquad [6.4]$$

where W_w = 'effective' mass of water in the original mixture (see Footnote 8 in Chapter 4), kg;

A = mass of water that would fill the voids created by entrained or entrapped air in the original mixture,[9] expressed as fraction of W_w, kg kg^{-1}.

The mass of nonevaporable water in the cement paste can be written as

$$W_n = W_c \left(\frac{\bar{W}_n}{W_c} \right) m \qquad [6.5]$$

where W_n = mass of nonevaporable water, determined by D-drying, kg;
$\bar{W}_n$ = mass of nonevaporable water on complete hydration, determined by D-drying, kg;
W_c = mass of cement in the original cement- or concrete- mixture, kg;

and m (dimensionless), as already discussed, is the paste maturity factor (or hydration progress variable).

$\bar{W}_n / W_c$ in Eq. 6.5 ranges from 0.17 to 0.25 (Powers 1962). For an average Type I portland cement, $\bar{W}_n / W_c \simeq 0.23$. Powers found that the mass of gel water, W_{gw} (kg), is proportional to the mass of nonevaporable water, W_n (kg), and, for an average Type I portland cement, the relation is

$$W_{gw} = 0.93\ W_n \qquad [6.6]$$

The density of cement paste can be defined in three different ways:

---

9   The recommended entrained air content is 3–6 per cent of the volume of concrete. The smaller the average size of the coarse aggregates the larger the applicable value. The entrapped air usually amounts to 0.2–3 per cent by volume. Again, larger values apply to smaller aggregates. To obtain A, divide these values by the volume fraction of water in the concrete mixture (usually between 0.17 and 0.24).

- *True density*, ρ_t (kg m^{-3}): the mass of the solid 'components' of the system (i.e. cement and nonevaporable water), divided by the volume of these components.

$$\rho_t = \frac{W_c + W_n}{\dfrac{W_c}{\rho_c} + \dfrac{W_n}{\rho_n}} = \frac{1 + \left(\dfrac{\overline{W}_n}{W_c}\right)m}{\dfrac{1}{\rho_c} + \dfrac{1}{\rho_n}\left(\dfrac{\overline{W}_n}{W_c}\right)m} \qquad [6.7]$$

- *Density of gel*, ρ_g (kg m^{-3}): the mass of the solid components of the system, divided by the volume of the gel.

$$\rho_g = \frac{W_c + W_n}{\dfrac{W_c}{\rho_c} + \dfrac{W_{gw}}{\rho_{gw}} + \dfrac{W_n}{\rho_n}} = \frac{1 + \left(\dfrac{\overline{W}_n}{W_c}\right)m}{\dfrac{1}{\rho_c} + \left(\dfrac{0.93}{\rho_{gw}} + \dfrac{1}{\rho_n}\right)\left(\dfrac{\overline{W}_n}{W_c}\right)m} \qquad [6.8]$$

- *Density (bulk) of paste*, ρ_{pt} (kg m^{-3}): the mass of the solid components, divided by the volume of the paste (sum of the volume of cement and adjusted volume of water in the original mixture[10]).

$$\rho_{pt} = \frac{W_c + W_n}{\dfrac{W_c}{\rho_c} + \dfrac{W_{wa}}{\rho_w}} = \frac{1 + \left(\dfrac{\overline{W}_n}{W_c}\right)m}{\dfrac{1}{\rho_c} + \dfrac{1}{\rho_w}\left(\dfrac{W_w}{W_c}\right)(1 + A)} \qquad [6.9]$$

There are also three different expressions for the porosity:

- *Porosity of gel*, P_g (m^3 m^{-3}):

$$P_g = \frac{\dfrac{0.93}{\rho_{gw}}\left(\dfrac{\overline{W}_n}{W_c}\right)m}{\dfrac{1}{\rho_c} + \left(\dfrac{0.93}{\rho_{gw}} + \dfrac{1}{\rho_n}\right)\left(\dfrac{\overline{W}_n}{W_c}\right)m} \qquad [6.10]$$

- *Capillary porosity*, P_{cp} (m^3 m^{-3}):

$$P_{cp} = \frac{\dfrac{1}{\rho_w}\left(\dfrac{W_w}{W_c}\right)(1 + A) - \left(\dfrac{0.93}{\rho_{gw}} + \dfrac{1}{\rho_n}\right)\left(\dfrac{\overline{W}_n}{W_c}\right)m}{\dfrac{1}{\rho_c} + \dfrac{1}{\rho_w}\left(\dfrac{W_w}{W_c}\right)(1 + A)} \qquad [6.11]$$

---

10 Volume changes, relative to the original volume of the mixture (plastic shrinkage, drying shrinkage, etc.), are not expected to amount to more than 0.5 per cent by volume and therefore can be disregarded in the calculation of densities and porosities.

- *Overall paste porosity, P_{pt} (m³ m⁻³):*

$$P_{pt} = \frac{\dfrac{1}{\rho_w}\left(\dfrac{W_w}{W_c}\right)(1 + A) - \dfrac{1}{\rho_n}\left(\dfrac{\bar{W}_n}{W_c}\right)m}{\dfrac{1}{\rho_c} + \dfrac{1}{\rho_w}\left(\dfrac{W_w}{W_c}\right)(1 + A)} \qquad [6.12]$$

Equations 6.10, 6.11 and 6.12 have been derived by means of Eq. 4.1.

Apparently, ρ_t, ρ_g and P_g are independent of the water–cement ratio in the original mixture. Assuming that the cement is an average Type I portland cement (for which, as mentioned, $\bar{W}_n/W_c = 0.23$), and that the hydration of cement has proceeded to completion ($m = 1$), the following values are obtained:

$$\rho_t = 2430 \text{ kg m}^{-3} \qquad \rho_g = 1760 \text{ kg m}^{-3} \qquad P_g = 0.28$$

Equation 6.11 indicates that the capillary porosity can be brought down to zero by reducing the water–cement ratio, W_w/W_c, in the original mixture. The condition of zero capillary porosity (i.e. the condition that the paste consists solely of gel) is

$$\frac{W_w}{W_c} = \frac{\rho_w}{1 + A}\left(\frac{0.93}{\rho_{gw}} + \frac{1}{\rho_n}\right)\left(\frac{\bar{W}_n}{W_c}\right)m \quad \text{for } P_{cp} = 0 \qquad [6.13]$$

Thus, with $\bar{W}_n/W_c = 0.23$ and $A = 0$, and with the values of ρ_w, ρ_{gw}, and ρ_n given earlier,

$$\frac{W_w}{W_c} = 0.38 \, m \qquad \text{for } P_{cp} = 0 \qquad [6.13a]$$

Since the capillary porosity cannot go negative, the complete hydration of cement ($m = 1$) is only possible if $W_w/W_c \geq 0.38$.

Reference will be made later to the so-called gel–space ratio, X (dimensionless). It is defined as the ratio of the gel volume to paste volume.[11]

$$X = \frac{\dfrac{1}{\rho_c} + \left(\dfrac{0.93}{\rho_{gw}} + \dfrac{1}{\rho_n}\right)\left(\dfrac{\bar{W}_n}{W_c}\right)m}{\dfrac{1}{\rho_c} + \dfrac{1}{\rho_w}\left(\dfrac{W_w}{W_c}\right)(1 + A)} \qquad [6.14]$$

The relation between the gel–space ratio and capillary porosity is

$$X = 1 - P_{cp} \qquad [6.15]$$

With the rise of temperature, the cement paste will lose first the evaporable water held by capillary forces in the larger pores, and then the water held by

---

11 This is the definition proposed by Powers in his 1959 paper. An earlier version of X (Powers 1949) did not include the term A, i.e. the correction of capillary porosity for air voids.

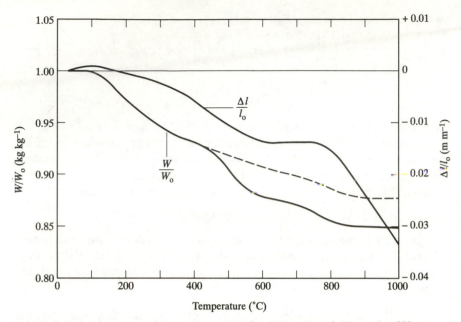

Fig. 6.2 Thermogravimetric and dilatometric curves for Cement Paste C (Harmathy 1970a)

van der Waals forces in the gel pores. By the time the temperature reaches the 105 °C level, there will be only nonevaporable water present. A further rise in temperature will cause a gradual decomposition of the C−S−H gel and, at about 500 °C, a rather rapid decomposition of the CH. Since the dehydration reactions are accompanied by loss of mass, thermogravimetric tests offer convenient means for determining the reaction progress variables as functions of temperature.

One of the two curves in Fig. 6.2 is the thermogravimetric curve of a cement paste prepared with a water−cement ratio of 0.5 from the cement described in Table 6.1. It will be referred to as Paste C, and its hypothetical equivalent prepared from the 'model cement' (see Table 6.1), as Model Paste C. As the thermogravimetric curve shows, the dehydration of the C−S−H gel starts at 100 °C, and goes on uninterrupted to about 850 °C, where the leveling-off of the curve indicates the completion of dehydration of the paste. A steep decline in the mass of the paste between 450 and 550 °C is caused by the dehydration of calcium hydroxide (CH).

According to the figure, the total loss of mass is $(1 - 0.849)/0.849 = 0.178$ kg per kg anhydrous residue. Assuming that the nonevaporable water has been the only constituent of the cement paste expelled by heating,[12] and knowing that at complete hydration the paste would hold 0.216 kg nonevaporable water per kg anhydrous cement, the maturity factor can be calculated as $0.178/0.216 =$

---

12 In reality, the paste also contains some $CaCO_3$, and a small fraction of the loss of mass is due to the decarbonization of this compound above 650 °C.

Table 6.2 Calculation of the composition of Model Paste C made from the 'model cement' with a water−cement ratio of 0.5 (see Eq. 6.1); $m = 0.824$ (Harmathy 1970a)

Compound	Number of moles in hydrated paste (kmol)		Mass in hydrated paste (kg)		Mass fraction (kg kg^{-1})
$C_{1.62}SH_{1.5}$	0.824×1.0	$= 0.824$	$0.824 \times 177.97 =$	146.65	0.610
CH	0.824×0.948	$= 0.781$	$0.781 \times 74.10 =$	57.87	0.240
C_3S	$(1-0.824) \times 0.568 = 0.100$		$0.100 \times 228.33 =$	22.83	0.095
C_2S	$(1-0.824) \times 0.432 = 0.076$		$0.076 \times 172.25 =$	13.09	0.055
				240.44	1.000

Note Molecular masses: $C_{1.62}SH_{1.5}$ 177.97 C_3S 228.33
$\qquad\qquad\qquad\qquad$ CH $\qquad$ 74.10 C_2S 172.25

0.824. The composition of Model Paste C for $m = 0.824$ is given in Table 6.2.

As to the dehydration process, three kinds of reaction (dehydration) progress variable can be defined:

- Overall reaction progress variable, ξ_{ov}, to be evaluated from the solid-line thermogravimetric curve in Fig. 6.2. The variable ξ_{ov} describes the course of dehydration as a function of temperature, without regard to the stoichiometric relations.
- Progress variable for the dehydration of the C−S−H gel, ξ_{CSH}. The broken-line extension to the thermogravimetric curve from 450 °C on is an estimate of the course of the dehydration for a hypothetical paste that does not contain CH.
- Progress variable for the dehydration of calcium hydroxide, ξ_{CH}. Information on ξ_{CH} can be obtained from Table 5.1, using, for example, Alternative 3. The nominal reaction temperature, T_r, is apparently about 500 °C, and the width of the reaction zone, $T_{r2} - T_{r1}$, may be chosen as 100 °C.

The variables ξ_{ov} and ξ_{CSH} can be calculated from the thermogravimetric curve as follows:

$$\xi_{ov} = \frac{1 - \dfrac{W}{W_o}}{1 - \dfrac{W_d}{W_o}} \qquad\qquad [6.16]$$

$$\xi_{CSH} = \frac{1 - \left(\dfrac{W}{W_o}\right)^+}{1 - \left(\dfrac{W_d}{W_o}\right)^+} \qquad\qquad [6.17]$$

where W = mass of the thermogravimetric specimen (at temperature T), kg;
    W_o = mass of the specimen at the reference temperature, (298 K, 25 °C), kg;
    W_d = mass of the specimen at the end of the dehydration process (i.e. above 850 °C), kg.

The + superscript is used as a reminder that the broken-line section of the thermogravimetric curve should be followed above 450 °C.

The reaction progress variables ξ_{ov} and ξ_{CSH}, and of the temperature derivative of ξ_{CSH}, i.e. $d\xi_{CSH}/dT$, are displayed as functions of the temperature in Fig. 6.3.

The dilatometric curve of Cement Paste C is also shown in Fig. 6.2. In the figure, $\Delta l = l - l_o$, where l (m) is the changed length of the dilatometric sample (on account of the temperature rise), and l_o (m) is the length of the sample at the reference (room) temperature. The dilatometric curve provides another set of information toward determining the density versus temperature and porosity versus temperature relations for the cement paste. Knowing the latter relation is important in understanding the effect of temperature on such structure-sensitive properties as thermal conductivity and strength.

In the course of dehydration, the amount of nonevaporable water present in the cement paste at a particular temperature can be described by means of the overall reaction progress variable, ξ_{ov}, as follows:[13]

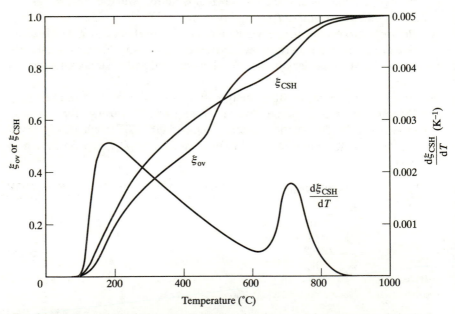

Fig. 6.3 Reaction (dehydration) progress variables, ξ_{ov}, ξ_{CSH}, and the temperature derivative $d\xi_{CSH}/dT$, for Cement Paste C (Harmathy 1970a)

13 In reality, as Piasta *et al.* (1984) suggested, the dehydration process may be somewhat more complex.

$$(W_n)_T = W_c \left(\frac{\bar{W}_n}{W_c} \right) m[1 - (\xi_{ov})_T] \qquad [6.18]$$

where Eq. 6.5 has been utilized, and the T subscript denotes the temperature level considered.

One may think at first that substituting this expression of W_n into Eq. 6.7 will yield an equation from which the variation of the true density of the paste with temperature can be calculated. Unfortunately, it is not at all certain that in Eq. 6.7 the densities ρ_c and ρ_n stay constant for the whole dehydration process. That the density of the solid residue of dehydration is not much different from ρ_c, i.e. from the density of the anhydrous cement from which the paste was prepared, is almost certain.[14] On the other hand, it is likely that, as the dehydration progresses, the density of the remaining nonevaporable water will increase. Yet, for lack of information, there seems to be no alternative to assuming that ρ_n remains constant at 1220 kg m^{-3}, and that the true density of the paste at elevated temperatures can be assessed from the following modified form of Eq. 6.7:

$$(\rho_t)_T = \frac{1 + \left(\dfrac{\bar{W}_n}{W_c} \right) m[(1 - (\xi_{ov})_T)]}{\dfrac{1}{\rho_c} + \dfrac{1}{\rho_n} \left(\dfrac{\bar{W}_n}{W_c} \right) m[(1 - (\xi_{ov})_T)]} \qquad [6.19]$$

where the meaning of the T subscripts is the same as before.

The bulk density versus temperature relation for the cement paste is obtained by means of Eqs 5.12 and 6.9, using values read from the thermogravimetric and dilatometric curves.

$$(\rho_{pt})_T = \left\{ \frac{1 + \left(\dfrac{\bar{W}_n}{W_c} \right) m}{\dfrac{1}{\rho_c} + \dfrac{1}{\rho_w} \left(\dfrac{W_w}{W_c} \right)(1 + A)} \right\}_o \times \frac{\left(\dfrac{W}{W_o} \right)_T}{\left[1 + \left(\dfrac{\Delta l}{l_o} \right)_T \right]^3} \qquad [6.20]$$

Hence, according to Eq. 4.1, the porosity of the paste can be calculated as

$$(P_{pt})_T = \frac{(\rho_t)_T - (\rho_{pt})_T}{(\rho_t)_T} \qquad [6.21]$$

The expression in Eq. 6.20 between the braces represents the density of the (D-dried) paste. The subscript o signifies reference (room) temperature, and the subscripts T denote values at temperature T.

---

14 The densities of the three major dehydration products, C_2S, CS and C (see Eqs 6.2 and 6.3), are 3280, 2920 and 3320 kg m^{-3}, respectively, as compared with 3150 kg m^{-3} for the anhydrous cement.

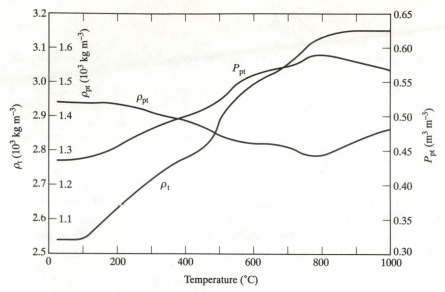

Fig. 6.4 True density (ρ_t), bulk density (ρ_{pt}), and porosity (P_{pt}) of Cement Paste C (calculated): note the different scales for the three properties (Harmathy 1970a)

Figure 6.4 shows the true density, bulk density and porosity of Paste C, calculated using Eqs 6.19, 6.20, and 6.21, and information obtained from Figs 6.2 and 6.3.

The coefficient of thermal expansion of the cement paste in the vicinity of room temperature cannot be unambiguously defined. The actual expansion is the combined effect of true expansion caused by an increase in the amplitude of atomic vibrations, and apparent expansion associated with the movement of moisture from the capillaries into the gel pores (Dettling 1964; Zoldners 1971). The true expansion is essentially constant, and is roughly the same under oven-dry and fully water-saturated conditions, ranging from 9.3×10^{-6} to 12.6×10^{-6} $m\,m^{-1}\,K^{-1}$ (Verbeck and Helmut 1969). The total expansion is highest somewhere between 45 and 70 per cent saturation (dependent on the age of the paste), and may vary from less than 9.3×10^{-6} to more than 21.6×10^{-6} $m\,m^{-1}\,K^{-1}$.

6.3   Specific Surface and Permeability of the Paste:

Owing to the fineness of the pore structure of the gel, the specific surface of the cement paste, $\bar{S}$ ($m^2\,kg^{-1}$), is very large. Since it is related to the hydrated material in the gel, it increases as the paste matures. According to Powers and Brownyard (1948), $\bar{S}$ is proportional to the mass of nonevaporable water in a unit mass of the solid content of paste:

$$\bar{S} \simeq 10^6 \ \frac{\left(\dfrac{\bar{W}_n}{W_c}\right) m}{1 + \left(\dfrac{\bar{W}_n}{W_c}\right) m} \qquad\qquad [6.22]$$

where the constant 10^6 is of $m^2\,kg^{-1}$ dimension, and is applicable to an average Type I portland cement.

The paste in autoclaved concrete products is coarse and largely microcrystalline. Its specific surface is only about 5 per cent of that of the paste in nonautoclaved concretes.

As discussed in Chapter 4, the overall paste porosity and the specific surface together characterize the amount of moisture that a cement paste can hold in equilibrium with an atmosphere of given relative humidity. The shape of the sorption isotherm for a typical cement paste has been shown in Fig. 4.2.

For some time following the placement of concrete, the capillary pores (though still saturated with water) form an interconnected network. As the cement paste matures, the hydration products invade the water-filled spaces and disrupt the continuity of that network. If the water–cement ratio is not too high, the *segmentation* of the capillary pores may begin within a few days; if, on the other hand, it is large, greater than about 0.7, segmentation will never occur (Powers *et al.* 1959).

After segmentation, the capillary pores will become interconnected through the much finer gel pores. Studies conducted by Powers and coworkers have indicated that the coefficient of permeability drops quite rapidly, sometimes as much as one order of magnitude a day, as the hydration products consume the pore space originally occupied by water.

Although, as established earlier, the porosity of the gel is 28 per cent, its permeability is very low, about $7 \times 10^{-23}\,m^2$ (Powers 1958). According to Powers (1964), the permeability of mature cement paste to water varies as

$$\frac{P_{pt}^2}{1 - P_{pt}} \exp \frac{1 - P_{pt}}{P_{pt}}$$

On the other hand, Nyame and Illston (1980) claimed that the permeability is related to the the maximum continuous pore size, as determined by the mercury injection technique, rather than to the porosity of the paste.

Typical values of the paste's permeability are of the order of $10^{-19}\,m^2$.

6.4   Thermal Conductivity of the Paste

Because of the poor crystalline order in the C–S–H gel, the thermal conductivity of the cement paste is very low and comparable with that of porous glassy materials. It is, for all intents, independent of the chemical makeup and age of

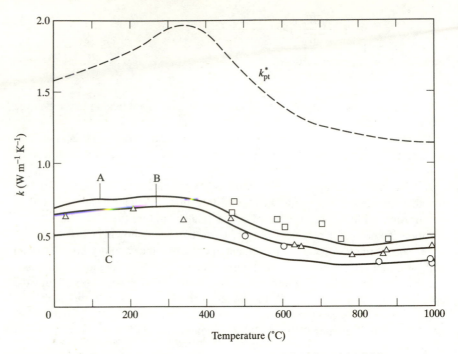

Fig. 6.5 Thermal conductivities of Cement Pastes A, B, and C (points and solid-line curves), and thermal conductivity of a hypothetical poreless cement paste (broken-line curve): □, Paste A; △, Paste B; ○, Paste C. (Harmathy 1970a)

the paste and, at moderately elevated temperatures, increases slightly with the rise of temperature.

Experimental values of the thermal conductivity of three cement pastes[15] are plotted in Fig. 6.5 (Harmathy 1970a, 1983b). The pastes were prepared at three water−cement ratios (0.25, 0.33 and 0.5) from the cement described in Table 6.1. The last-mentioned paste has been referred to as Paste C. The first and second will be called Pastes A and B, and their hypothetical equivalents, prepared from the 'model cement' (also shown in Table 6.1), Model Pastes A and B.

A dry portland cement paste consists of two continuous phases: the solid components of the paste and the air in the capillary and gel pores. As discussed in Chapter 5, the Hamilton−Crosser formula (Eqs 5.29 and 5.30) offers great flexibility in formulating the thermal conductivity of two-phase systems with a variety of phase configurations. It was noted earlier that 1.5 is the value of n (in Eq. 5.30) likely to be applicable to a system with two continuous phases. With $n = 1.5$, the following expression is obtained from Eqs 5.29 and 5.30 for the thermal conductivity of the cement paste, k_{pt} (W m^{-1} K^{-1}):

---

15 The values were obtained using a variable-state method (Harmathy 1964). They are in substantial agreement with data reported by Carman and Nelson (1921), and Harada *et al.* (1972).

$$k_{pt} = k_{pt}^* \frac{0.5k_{pt}^* + k_a - 0.5P_{pt}(k_{pt}^* - k_a)}{0.5k_{pt}^* + k_a + P_{pt}(k_{pt}^* - k_a)}$$ [6.23]

where k_{pt}^* = thermal conductivity of a hypothetical poreless cement paste, $W\,m^{-1}\,K^{-1}$;

k_a = thermal conductivity of air,[16] $W\,m^{-1}\,K^{-1}$;

and P_{pt} is the overall paste porosity (Eq. 6.21). All the variables in Eq. 6.23 are functions of the temperature.

The experimental values plotted in Fig. 6.5 and values of P_{pt} derived from thermogravimetric and dilatometric curves (such as those in Fig. 6.2) have been used to determine, by trial and error, the supposedly unique $k_{pt}^* = k_{pt}^*(T)$ function. That function is shown as a broken line in the same figure. The decline of the curve above 400 °C seems to indicate that the structure of the paste turns more and more crystalline at temperatures above 400 °C. The three solid-line curves have been obtained by means of Eq. 6.23, using the broken-line curve for k_{pt}^*.

The $k_{pt}^* = k_{pt}^*(T)$ function and Eq. 6.23 may be used for developing information on the thermal conductivity of cement pastes prepared with other water—cement ratios. Equation 6.23 may also be used for the calculation of the conductivity of fully saturated pastes, if the conductivity of air, k_a ($\simeq$ 0.026 $W\,m^{-1}\,K^{-1}$ at room temperature), is replaced by the conductivity of water, k_w ($\simeq$ 0.616 $W\,m^{-1}\,K^{-1}$ at room temperature). One will find that the presence of water has a substantial effect on the conductivity of the paste. For example, for a paste with 38 per cent porosity the thermal conductivity is about 1.1 $W\,m^{-1}\,K^{-1}$ in fully saturated condition, while in oven-dry condition it is only 0.59 $W\,m^{-1}\,K^{-1}$.

When the paste is not fully saturated, the conductive heat transport is augmented by energy transport associated with the migration of moisture. From the work of Cammerer (1942) and Loudon and Stacy (1966) it appears that the presence of 10 per cent (by volume) evaporable water in the paste may more than double the thermal conductivity of the paste.

6.5   Specific Heat of the Paste

The calculation of the specific heat of a material consisting of stable and unstable compounds was dealt with in detail in Chapter 5. The following input information is needed for the calculations:

- composition of material,
- nature and temperature of the reactions that the unstable components are expected to undergo,

---

16 Because of the fineness of the pore structure, the radiative heat transfer through the pores need not be considered.

- heat capacities as functions of temperature for all components of the system (stable and unstable compounds, and products of reactions),
- latent heats of the reactions or, alternatively, heats of formation of the compounds participating in the reactions,
- progress variables for the reactions.

Model Paste C will be used to illustrate the calculation procedure. Among its components (see Table 6.2) two, the C−S−H gel ($C_{1.62}SH_{1.5}$) and CH, become unstable at elevated temperatures, and two, C_3S and β-C_2S (the unhydrated residues of cement), remain stable throughout the entire temperature interval of interest.[17] The decomposition products of the two unstable components of the paste have been identified (see Eqs 6.2 and 6.3), and the decomposition process has been explored through a thermogravimetric test (Fig. 6.2). The material balance for the process is illustrated in Fig. 6.6.

The reaction progress variable for the decomposition of the C−S−H gel, ξ_{CSH}, and its derivative, $d\xi_{CSH}/dT$, have been plotted as functions of temperature in Fig. 6.3. Information on ξ_{CH} and $d\xi_{CH}/dT$ can be taken from Table 5.1, assuming that the nominal reaction temperature, T_r, is 500 °C (see Fig. 6.2), and the width of the reaction zone, $T_{r2} - T_{r1}$, is, say, 100 °C. Data yet to be obtained are the heat capacity versus temperature relations for all components of the system and, since the heats of the dehydration reactions are not known, the heats of formation of the compounds participating in the reactions.

As discussed in Chapter 5, the heat capacity (at constant pressure), C_p ($J\,kmol^{-1}\,K^{-1}$), is usually described by an equation of the form

$$C_p = e + fT - \frac{g}{T^2} \tag{6.24}$$

where e, f and g are empirical constants of various dimensions. It was also mentioned earlier that two handbooks, Perry's (1950) and Eitel's (1952), and a series of publications by Kelley (1934, 1936, 1949, 1960) are probably the most comprehensive sources of data on heat capacities and heats of formation.

Table 6.3 lists all the information needed for the calculation of the enthalpy versus temperature relation for all the compounds that appear in the material balance for the model cement paste in the course of its heating from 25 to 1000 °C (see Fig. 6.6). The heats of formation, ΔH_f ($J\,kmol^{-1}$), are relative to those of the constituent oxides. ΔH_i ($J\,kmol^{-1}$) stands for the heats associated with polymorphic transformations. Most of the data have been assembled from the sources mentioned above, and from a report by Wagman et al. (1982).

Some explanation is needed for the values concerning the C−S−H gel. For lack of information, the heat capacity versus temperature relation for the gel was calculated using the method mentioned in Chapter 5 (Harmathy 1970a). According

---

17 From the point of view of the present treatment, they can be looked upon as more or less stable compounds. In reality, at about 700 °C the β-C_2S is expected to change into α'-C_2S.

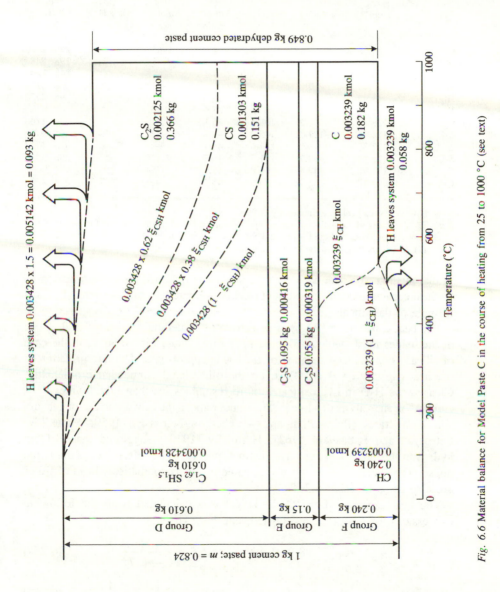

Fig. 6.6 Material balance for Model Paste C in the course of heating from 25 to 1000 °C (see text)

Table 6.3 Thermodynamic properties of the compounds that appear in the material balance for Model Paste C in the course of heating from 298 to 1273 K (25 to 1000 °C)

Compound	M (kg kmol^{-1})	ΔH_f (MJ kmol^{-1})	e (10^3 J kmol^{-1} K^{-1})	f (J kmol^{-1} K^{-2})	g (10^9 J K kmol^{-1})	ΔH_i (MJ kmol^{-1})	Range of validity (K)
C	56.08	0	48.8	4.5	0.65		298–2000
α-S	60.06	0	46.9	34.3	1.13	1.21	298–848
β-S	60.06		60.3	8.1	—		848–2000
H(l)	18.02	0	75.5	—	—		298–373
H(g)	18.02	43.9	30.5	10.3	—		298–2750
H(hs) [a]	18.02		27.2	27.6	0.38		?
C$_3$S	228.33	−114.2	208.6	36.1	4.25		298–1800
β-C$_2$S	172.25	−127.2	145.9	40.8	2.62	1.84	298–970
α'-C$_2$S	172.25		134.6	46.1	—		970–1710
β-CS	116.14	−88.7	111.5	15.1	2.73		298–1450
C$_{1.62}$SH$_{1.5}$	177.97	−142.4	166.8	83.0	2.75		?
CH	74.10	−65.3	79.8	45.2	—		298–700

[a] For H$_2$O in hydrosilicates, estimated by author.

Notes (1) The heats of formation of the compounds are referred to the heats of formation of the constituent oxides (H$_2$O as liquid water and SiO$_2$ as α-quartz) at 298 K.

(2) ΔH_i is the heat of polymorphic inversion.

to that method, the heat capacity of a silicate is obtained by the summation of the heat capacities of the constituent oxides, weighted according to the chemical formula of the silicate.

The heat capacities to be used in Harmathy's method are, in general, those of the oxides most stable at room temperature. Thus, for example, in the case of silica, using the heat capacity of α-SiO$_2$ is appropriate. The interpretation of the heat capacity of H (H$_2$O) is, on the other hand, somewhat problematic. Clearly, the H$_2$O in hydrosilicates more resembles ice than liquid water.

Based on an analysis of a limited amount of data on the heat capacities of calcium silicate hydrates (Babuschkin and Mtschedlow-Petrossian 1959; Mchedlov-Petrosyan and Babushkin 1962), Harmathy (1970a) suggested that H$_2$O in hydrosilicates be looked upon as a distinct 'phase', to be referred to as the H(hs) phase. The values of e, f, and g applicable to this (hypothetical) form of water are also listed in Table 6.3.[18]

The heat capacity of the C−S−H gel was calculated from the following expressions for CaO (C), α-SiO$_2$ (S), and H$_2$O [H(hs)]:

$$C \qquad C_C = 48.8 \times 10^3 + 4.5T - \frac{0.65 \times 10^9}{T^2} \qquad [6.25a]$$

$$S \qquad C_S = 46.9 \times 10^3 + 34.33T - \frac{1.13 \times 10^9}{T^2} \qquad [6.25b]$$

18 The heat capacity of H$_2$O in hydrosilicates is apparently about 23 per cent lower than the heat capacity of ice.

$$\text{H(hs)} \qquad C_H = 27.2 \times 10^3 + 27.6T - \frac{0.38 \times 10^9}{T^2} \qquad [6.25c]$$

Thus, since the chemical formula of the C−S−H gel is $C_{1.62}SH_{1.5}$,

$$C_{CSH} = 1.62C_C + C_S + 1.5C_H$$

$$= 166.8 \times 10^3 + 83.0T - \frac{2.75 \times 10^9}{T^2} \qquad [6.26]$$

The heat of formation for the C−S−H gel, $(\Delta H_f)_{CSH}$ (J kmol^{-1}), was evaluated from the heats of hydration of C_3S and β-C_2S. Lerch and Bogue (1934) found[19] that at room temperature $(\Delta H_r)_{C_3S} = -114.6$ MJ kmol^{-1} and $(\Delta H_r)_{C_2S} = -44.7$ MJ kmol^{-1} (both exothermic). Hence, the heat of hydration for the model cement $(\Delta H_r)_{pt}$ is (see Table 6.1)

$$(\Delta H_r)_{pt} = -0.568 \times 116.4 - 0.432 \times 44.7 = -84.4 \text{ MJ kmol}^{-1}$$
$$[6.27]$$

Recalling that at the standard reference temperature (298 K) the heat of reaction is equal to the difference between the heats of formation of products and the reactants, one can write (see Eq. 6.1 and Table 6.3)

$$-84.4 = (\Delta H_f)_{CSH} + 0.948 \times (-65.3) - 0.568 \times (-114.2) -$$
$$0.432 \times (-127.2) - 2.448 \times 0 \qquad [6.28]$$

from which

$$(\Delta H_f)_{CSH} = -142.4 \text{ MJ kmol}^{-1}$$

This is the value shown in Table 6.3.

The heats of dehydration of the C−S−H gel and the CH at temperatures other than 298 K are to be calculated in the way shown in Eqs 5.47 to 5.49. Some effort may be saved by plotting the enthalpy, H (J kmol^{-1}), as a function of temperature for all compounds participating in the reactions, using equations similar to Eqs 5.47 or 5.48 and the information in Table 6.3. Figure 6.7 is such a plot. The heat of reaction (dehydration) for a compound (the C−S−H gel or CH) is then obtained by reading the difference between the enthalpies of the products (each multiplied by its stoichiometric coefficient) and the reactant (see Eq. 5.49).

In calculating the apparent specific heat of the cement paste, it is convenient to divide the components of the system into three groups, as shown in Fig. 6.6.

- *Group D*: includes the C−S−H gel ($C_{1.62}SH_{1.5}$) and its decomposition products (see Eq. 6.2);

---

19 According to Brunauer *et al.* (1954), the actual heats of hydration are about 20 MJ kmol^{-1} less than the stated values; 20 MJ kmol^{-1} represents the heat of adsorption of water on the gel. This finding is of no little significance to the fire safety practitioner. It implies that a fire-exposed structure will absorb roughly one-quarter of the so-called heat of dehydration of the paste even before its temperature starts rising above 105 °C.

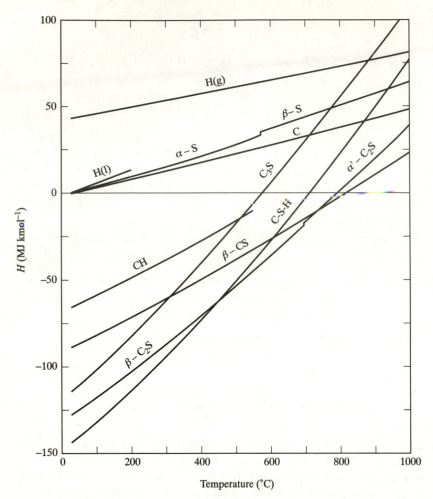

Fig. 6.7 Enthalpy of the various constituents of Model Paste C and its dehydration products (Harmathy 1970a)

- *Group E*: includes the two stable compounds, C_3S and β-C_2S;
- *Group F*: includes the CH and its decomposition products (see Eq. 6.3).

According to the scheme shown in Fig. 5.4, the following equations are applicable:

- *Group D*:
  From 298 to 373 K (25 to 100 °C)

$$c_D = \frac{0.610}{177.97} C_{CSH} \qquad\qquad [6.29]$$

From 373 to 1130 K (100 to 857 °C)

$$c_D = \frac{0.610}{177.97} \left\{ [(1-\xi_{CSH})C_{CSH}+0.62\xi_{CSH}C_{C_2S}+0.38\xi_{CSH}C_{CS}] \right.$$

$$\left. + \frac{d\xi_{CSH}}{dT}(-H_{CSH}+0.62H_{C_2S}+0.38H_{CS}+1.5H_{H(g)}) \right\} \qquad [6.30]$$

From 1130 to 1273 K (857 to 1000 °C)

$$c_D = \frac{0.610}{177.97}(0.62C_{C_2S}+0.38C_{CS}) \qquad [6.31]$$

- *Group E*:
  From 298 to 1273 K (25 to 1000 °C)

$$c_E = \frac{0.095}{228.33}C_{C_3S} + \frac{0.055}{172.25}C_{C_2S} \qquad [6.32]$$

- *Group F*:
  From 298 to 723 K (25 to 450 °C)

$$c_F = \frac{0.240}{74.10}C_{CH} \qquad [6.33]$$

From 723 to 823 K (450 to 550 °C)

$$c_F = \frac{0.240}{74.10}\left\{ [(1-\xi_{CH})C_{CH}+\xi_{CH}C_C]+\frac{d\xi_{CH}}{dT}[-H_{CH}+H_C+H_{H(g)}] \right\} \qquad [6.34]$$

From 823 to 1273 K (550 to 1000 °C)

$$c_F = \frac{0.240}{74.19}C_C \qquad [6.35]$$

Naturally, at any temperature

$$c_{pt} = c_D + c_E + c_F \qquad [6.36]$$

In Eqs 6.29 to 6.36,

c_{pt}, c_D, c_E and c_F are the apparent (or, in temperature intervals, of physicochemical stability, actual) specific heats for the cement paste and its Group D, E and F constituents, respectively, $J\,kg^{-1}\,K^{-1}$;

C = heat capacity of the compound indicated by the subscript, $J\,kmol^{-1}\,K^{-1}$;

H = enthalpy of the compound indicated by the subscript (to be read from Fig. 6.7), $J\,kmol^{-1}$;

ξ = reaction (dehydration) progress variable for the decomposition of C–S–H or CH (as indicated by the subscript), dimensionless;

T = temperature, K.

All c values signify specific heats at constant pressure, and all C values heat capacities at constant pressure. (To avoid double subscripting, the p subscript indicating the constancy of pressure has been omitted.)

Curve C in Fig. 6.8 shows the $c_p = c_p(T)$ function calculated for Model Paste C, using Eqs 6.29 to 6.36, and information presented in Tables 5.1 and 6.3, and in Figs 6.3 and 6.7. Curves A and B represent Model Pastes A and B made, as mentioned before, from the same 'model cement' with 0.25 and 0.33 water−cement ratios, respectively. Figure 6.9 displays the experimental curves for the actual Pastes A, B, and C. There seems to be a good qualitative agreement between the theoretical and experimental results.

The large peaks of the experimental curves between 100 and 150 °C indicate that the test samples contained a substantial amount of moisture, which they probably picked up during their pulverization for DSC tests. The curves also show that in the DSC tests the dehydration of $Ca(OH)_2$ took place in the 400−450 °C temperature interval, as compared to the 450−550 °C interval in the thermogravimetric and dilatometric tests (Fig. 6.2). This discrepancy is probably attributable to differences in the rates of heating and in the sample sizes. The greater sharpness of the peaks in Fig. 6.9 is due to the narrower reaction zone. Discrepancies of this kind are not expected to lead to major errors in numerical heat flow calculations.

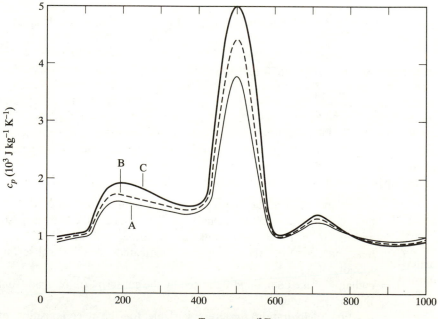

Fig. 6.8 Specific heat of Model Pastes A, B, and C (Harmathy 1970a)

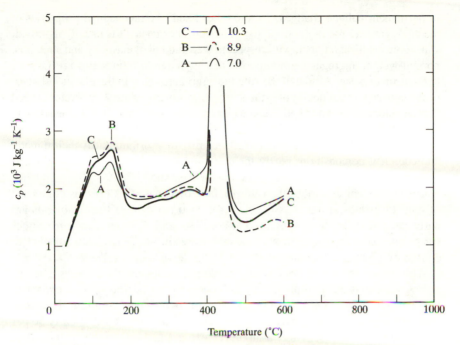

Fig. 6.9 Specific heat of Cement Pastes A, B, and C (Harmathy 1983b)

6.6   Aggregates

Diluting the cement paste with aggregates, while unquestionably justified from
an economic point of view, has some merits and demerits as far as the engineering
aspects of the practice are considered. The fine aggregates improve the workability
of the concrete mixture, and the fine and coarse aggregates together are beneficial
for the shrinkage, the creep, and usually also the elastic behavior of the material.
On the other hand, the interfaces between the cement paste and the aggregates
are known to be weak regions, conducive to the development and propagation
of microcracks, which may lead to premature failure.

The fine aggregate are commonly referred to as sand. A good sand consists
of a variety of particle sizes between 0.06 and 4.8 mm. In normal-weight concretes,
gravel or crushed stone varying in size from 5 to 20 mm is most frequently used
as coarse aggregate. In lightweight concretes, appropriately graded fractions of
the same lightweight aggregate are commonly used for both fine and coarse
material. However, in order to improve the workability of the concrete mixture,
normal-weight fines are sometimes preferred.

The soundness of the material is an important aspect of selecting normal-weight
aggregates. The ASTM C 88 test method (American Society for Testing and
Materials 1983) describes a simple procedure for determining the aggregate's
ability to withstand alternate freezing, thawing, wetting and drying without large

or permanent volume changes. On the other hand, the mechanical properties of the aggregate are not high on the list of selection criteria. It is usually assumed, with some justification, that, with respect to their moduli of elasticity and strengths, normal-weight aggregates are superior and lightweight aggregates are inferior to the cement paste. Although the role the aggregates play in the elastic behavior of the concrete is not negligible, the strength of normal-weight concretes indeed depends more on the bond between the paste and the aggregate than on the strength of the aggregate.

The names given to various stones reflect partly their origin, partly their mineral and chemical composition, partly their texture and structure. A name alone is not a reliable guide to suitability for use in concrete.

It is significant that more than 95 per cent of the Earth's crust is composed of only 12 groups of minerals (Table 6.4). Two per cent of these minerals are carbonates (calcite, aragonite, and dolomite), and 1.5 per cent are oxides (magnetite and titanomagnetite); all the others are silicates. Practically all silicates consist of 13 oxides, which are shown in the lower part of Table 6.8.[20]

With respect to their geological origin, natural stones are classified as igneous, sedimentary and metamorphic. Igneous rocks are by far the most abundant in the Earth's crust. They belong chiefly to two types: granitic (granites, granodiorites, quartz diorites) and basaltic (basalts and andesites). Nearly 50 per cent of all the crustal rocks are basalt (the coarse-grained variant of which is called gabbro), which consists mainly of plagioclase feldspars, pyroxenes, olivines, and possibly of amphiboles, epidotes and micas.

Table 6.4 gives the ranges for the true density of the 11 principal groups of minerals. The density of common rocks varies within a relatively narrow range: between 2500 and 3150 kg m^{-3}. A typical value is 2650 kg m^{-3}.

As has been pointed out repeatedly, porosity is one of the most important characteristics of building materials. Hundreds of data surveyed by Wolff (in Touloukian et al. 1981) reveal that the usual range of porosities is 0−5 per cent for igneous rocks, 0−10 per cent for sedimentary rocks, and 0.5−5 per cent for metamorphic rocks. There are some exceptions, however. Among the igneous rocks, rhyolites have a porosity ranging from 7 to 15 per cent, tuffs from 15 to 41 per cent, and pumice from 30 to 87 per cent. Among the sedimentary rocks, the porosity of sandstones usually ranges from 10 to 40 per cent, and that of shales from 0 to 40 per cent. And among the metamorphic rocks, the porosity of some schists may be as high as 47 per cent. Consequently, the permeability of various rocks varies within a very wide range: from 5×10^{-20} to 3×10^{-8} m^2.

In a comprehensive study, Griffith (1937) surveyed the properties of approximately 100 typical American rocks. It is clear from his study that normal-weight aggregates can indeed be expected to have mechanical properties superior to those of the cement paste. Although he found rocks with compressive strengths

---

20 The reader may find it beneficial to acquire a cursory knowledge of the terminology in mineralogy. The textbooks by Hurlbut (1959) and Mason and Berry (1968) are recommended.

Table 6.4 The principal minerals in the Earth's crust (last column based on information from Ronov and Yaroshevskiy 1967)

Minerals	Characteristic member of group	True density (10^3 kg m^{-3})	Abundance (%)
Plagioclase feldspars	Anorthite $CaAl_2Si_2O_8$	2.62–2.76	39
Potash feldspars	Orthoclase $KAlSi_3O_8$	2.56	12
Quartz	α-Quartz SiO_2	2.0–2.65	12
Pyroxenes	Enstatite $MgSiO_3$	3.1–3.9	11
Amphiboles	Anthophyllite $Mg_7Si_8O_{22}(OH)_2$	2.9–3.6	5
Micas	Muscovite $KAl_2(AlSi_3O_{10})(OH)_2$	2.5–3.4	5
Clay minerals and chlorite	Kaolinite $Al_4Si_4O_{10}(OH)_8$	2.0–3.3	4.6
Olivines	Forsterite Mg_2SiO_4	3.2–4.2	3.0
Calcite and aragonite	Calcite $CaCO_3$	2.71–2.95	1.5
Magnetite and titanomagnetite	Magnetite Fe_3O_4	4.5–5.2	1.5
Dolomite	Dolomite $CaMg(CO_3)_2$	2.85	0.5
Other minerals			4.9
			100.0

less than 70 MPa in all three groups, seven of the ten weakest rocks belonged in the group of sedimentary rocks. In another survey, Singh (in Touloukian *et al.* 1981) found that the strength of rocks may range from 4.6 to 238 MPa (with sandstones at both extremes of the scale), and that about one-third of the rocks have strengths less than 70 MPa.

The modulus of elasticity of rocks can also be expected to be higher than that of the cement paste, although 125 of the 270 rocks surveyed by Haas (in Touloukian *et al.* 1981) exhibited moduli less than 30×10^3 MPa. The extreme values were 0.21×10^3 MPa for a claystone and 118×10^3 MPa for a marble. Poisson's ratio is usually between 0.1 and 0.3, although it may be as low as 0.01 and as high as 0.61.

In regard to the thermal compatibility between the aggregates and the cement paste, the thermal expansion of the various rocks is an important piece of information. As Table 6.5 shows, at room temperature the coefficient of thermal

Table 6.5 Coefficients of thermal expansion for a few minerals
(Kozu *et al.* 1929; Griffith 1937; Dettling 1964; Venecanin 1990)

Mineral	Coefficient of thermal expansion $(10^{-6}\ m\ m^{-1}\ K^{-1})$
Quartz ∥[a]	7.1
Quartz ⊥	13.2
Calcite ∥	23.8
Calcite ⊥	−7.4
Igneous rocks	
Rhyolite — granite series	1.8−11.9
Andesite — diorite series	4.1−10.3
Basalt — gabbro series	3.6−9.0
Sedimentary rocks	
Breccias and conglomerates	5.4−12.1
Sandstones	4.3−13.9
Shales	9.5−11.0
Limestones	−5.2−12.2
Dolomites	6.7−8.6
Metamorphic rocks	
Gneisses and schists	2.3−7.9
Quartzites and slates	6.3−13.1
Marble	−2.2−16.0

[a] ∥ Parallel with optic axis; ⊥ perpendicular to optic axis.

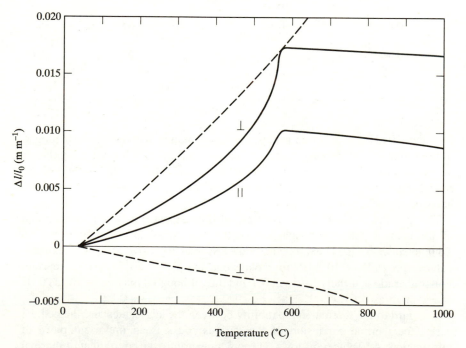

Fig. 6.10 Dilatometric curves for quartz (solid lines) and calcite (broken lines): ∥, parallel with optic axis; ⊥, perpendicular to optic axis (Sosman 1927; Kozu *et al.* 1929)

expansion for most rocks is usually lower than that of the cement paste, and ranges from 1.8×10^{-6} to $13.9 \times 10^{-6}\,\mathrm{m\,m^{-1}\,K^{-1}}$. However, for some carbonate rocks, namely limestone and marble, it may be negative in one direction, and large and positive in another direction.

The dilatometric curves of quartz and calcite are shown in Fig. 6.10 (Sosman 1927; Kozu *et al.* 1929), and those of ten Canadian rocks in Fig. 6.11 (Geller *et al.* 1962). Some characteristics of the ten Canadian rocks are listed in Table 6.6. For other rocks, the coefficient of thermal expansion at elevated temperatures may be estimated (within the ranges of physicochemical stability) on the strength of the observation that the ratio of specific heat to coefficient of volume thermal expansion is nearly constant at all temperatures (Chapter 5).

From the point of view of performance in fire, granites are apparently the least

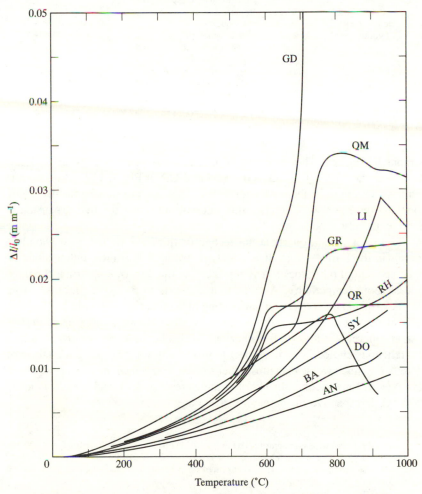

Fig. 6.11 Dilatometric curves for ten rocks described in Table 6.6 (Geller *et al.* 1962)

Table 6.6 Some characteristics of the ten rocks whose dilatometric curves are shown in Fig. 6.11 (Geller *et al*. 1962)

Symbol	Name	Geological origin	Composition	Density (kg m^{-3})	Grain size (mm)
AN	Anorthosite	Igneous	Almost all plagioclase feldspars	2770	0.05–20
BA	Basalt	Igneous	Mainly epidotes, pyroxenes and plagioclase feldspars	3040	0.005–0.08
DO	Dolomite	Sedimentary	Almost all dolomite	2490	0.1–7
GD	Granodiorite	Igneous	Plagioclase feldspars, quartz, amphiboles and micas	2750	0.05–4
GR	Granite	Igneous	Mainly potash- and plagioclase-feldspars, quartz	2620	0.05–5
LI	Limestone	Sedimentary	Mainly calcite	2700	0.002–2
QM	Quartz monzonite	Igneous	Mainly potash- and plagioclase-feldspars and quartz	2645	0.5–7
QR	Quartz rock		All quartz	2650	0.2–10
RH	Rhyolite	Igneous	Mainly potash- and plagioclase-feldspars, and quartz	2640	0.05–3
SY	Syenite	Igneous	Mainly potash- and plagioclase-feldspars, and amphiboles	2715	0.1–10

desirable aggregates. They undergo an inordinate expansion between 650 and 800 °C (see the curves for rocks GD, QM and GR in Fig. 6.11).[21] In contrast, anorthosites seem to be the ideal aggregates. They are physicochemically stable throughout the 25–950 °C temperature interval, and their thermal expansion at 950 °C is merely 1 per cent.

As Table 6.4 shows, quartz is the second or third among the most abundant minerals in the Earth's crust. It is widely distributed in nature, either alone or as a component of other rocks. Among its crystalline forms, α-quartz is the most stable at room temperature. At 575 °C it converts to β-quartz. Because of its sudden expansion on conversion, amounting to 0.86 per cent by volume (Sosman 1927), quartz has been believed to be an infelicitous choice for aggregate in fire-resistant concretes. Although for other reasons (to be discussed) quartz is indeed not a felicitous choice, it is, in respect to propensity for disrupting the fabric of concrete, not much worse than most other stones. The cement paste matrix, which holds together the concrete, shrinks on heating (see Fig. 6.2), and is badly cracked at 575 °C, irrespective of the nature of the aggregate.

21 The mineral referred to as granodiorite (GD in Fig. 6.11) displayed 11.5 % expansion at 800 °C. It contained 50.5 % plagioclase feldspars, 20.6 % quartz, 12.0 % hornblende (amphibole group) and 14.0 % biotite (mica group). The expansion of other granites studied by Geller *et al*. ranged from 3 to 4 % at that temperature.

Table 6.7 Some information on the decomposition of four carbonate minerals[a]

Mineral	Heat of decomposition (MJ kg^{-1})	Temperature interval of decomposition (°C)
Calcite, $CaCO_3$	1.76	700–950[b]
Magnesite, $MgCO_3$	1.15	500–710
Dolomite, $CaMg(CO_3)_2$	1.64	$\begin{cases} 700–800^c \\ 800–950 \end{cases}$
Siderite, $FeCO_3$	0.50 / −0.72	$\begin{cases} 450–600^d \\ 600–750 \end{cases}$

[a] Information on temperature intervals based on data by Todor (1976).
[b] The presence of other minerals may appreciably lower the temperature interval shown.
[c] The decomposition proceeds in two steps: the first involves the $MgCO_3$, the second the $CaCO_3$. Extra amounts of $MgCO_3$ or $CaCO_3$, not bonded to the dolomitic lattice, decompose in their regular temperature intervals.
[d] The reaction proceeds in two steps: first the siderite decomposes into ferrous oxide and CO_2, then the ferrous oxide oxidizes in an exothermic reaction.

As Fig. 6.7 and Table 6.3 indicate, the heat absorbed in the polymorphic inversion of quartz is not significant enough to be noticeable in the performance of concrete building elements in fire. Some other aggregates are also susceptible to physicochemical changes that are potentially beneficial for fire performance. The allegedly superior performance of concretes made with carbonate aggregates, such as calcite ($CaCO_3$), magnesite ($MgCO_3$), and dolomite ($CaCO_3 . MgCO_3$), is commonly attributed to absorption of heat in the decomposition of these minerals. Although, as Table 6.7 shows, the heats of decomposition are substantial, the temperature intervals of the decomposition reactions are too high to be of real benefit for the concrete's performance.

The (apparent) specific heats of quartz and calcite are plotted as functions of temperature in Fig. 6.12. Table 6.8 lists the constants e, f and g (see Eq. 6.24) for a number of important minerals and the oxide components of silicates.

The heat capacity of silicates not shown in the table can be estimated by Harmathy's method (discussed in Chapter 5 and earlier in this chapter) from the heat capacities of the constituent oxides. For example, the heat capacity of jadeite ($NaAlSi_2O_6 = \frac{1}{2} Na_2O . Al_2O_3 . 4SiO_2$, a member of the pyroxene group) can be obtained by adding up the values of e, f, and g for Na_2O, Al_2O_3, and SiO_2, weighted as indicated by the chemical formula. The result is

$$C_p = 184.1 \times 10^3 + 86.3T - \frac{4.03 \times 10^9}{T^2} \; J\,kmol^{-1}\,K^{-1}$$

Thus, at room temperature $C_p = 164.4 \times 10^3 \, J\,kmol^{-1}\,K^{-1}$, i.e. (since the molecular mass of jadeite is $202.09 \, kg\,kmol^{-1}$), $c_p = 813 \, J\,kg^{-1}\,K^{-1}$. The value reported by Kelley (1949) is $792 \, J\,kg^{-1}\,K^{-1}$.

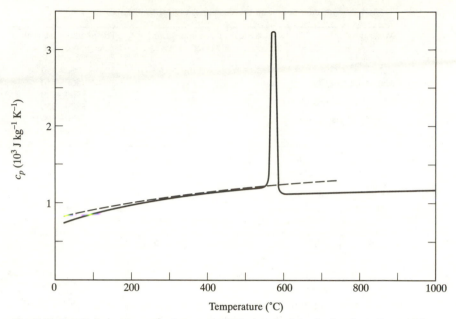

Fig. 6.12 Specific heat (apparent) of quartz (solid line) and calcite (broken line) (Perry 1950; Harmathy 1983b)

It has been claimed (Touloukian *et al.* 1981) that for the great majority of rocks the product of density and specific heat, ρc_p (the 'volume specific heat'), lies within 20 per cent of 2.3 MJ m^{-3} K^{-1}. For jadeite, which has a density of 3300 kg m^{-3}, the actual value of ρc_p is 2.61 MJ m^{-3} K^{-1}.

In assessing the specific heat of rocks, it should be remembered that the heat capacity of physicochemically stable compounds is not sensitive to the material's microstructure or to minor changes in its composition.

The thermal conductivity is, in contrast, an extremely structure-sensitive property. Luckily, the porosity of normal-weight aggregates is, as a rule, less than 5 per cent and, therefore, the conductivity of most rocks depends mainly on their mineral composition, crystalline character, and grain structure. As was pointed out in Chapter 5, the conductivity decreases with increasing temperature for materials with well developed crystalline order, and increases slightly for materials with disordered or vitreous structure.

The variation of the thermal conductivity with temperature is shown for quartz and calcite in Fig. 6.13, and for 13 polycrystalline rocks and two glasses in Fig. 6.14 (Birch and Clark 1940). Some characteristics of the 13 rocks and two glasses are listed in Table 6.9. The designer may use this information as a guide in assessing the conductivity of normal-weight concrete aggregates. It should be borne in mind, however, that the crystalline—amorphous nature and texture of a mineral are almost as important factors in its conductivity as its composition. If the

Table 6.8 Heat capacities of a few rock-forming minerals (see also Table 6.3), and the oxide components of silicates and carbonates (Kelley 1949; Perry 1950; Eitel 1952)

Compound	Name	M (kg kmol^{-1})	e (10^3 J kmol^{-1} K^{-1})	f (J kmol^{-1} K^{-2})	g (10^9 J K kmol^{-1})	Range of validity (K)
$CaAl_2Si_2O_8$	Anorthite	278.14	269.5	57.3	7.07	298–1700
$KAlSi_3O_8$	Orthoclase	278.25	267.1	54.0	7.13	298–1400
$MgSiO_3$	Enstatite	100.38	86.1	50.2	1.70	298–800
Mg_2SiO_4	Forsterite	140.70	117.9	—	—	298
Fe_3O_4	Magnetite	231.55	167.0	78.9	4.19	298–1100
$CaCO_3$	Calcite	100.09	82.3	49.7	1.29	298–1033
$CaMg(CO_3)_2$	Dolomite	184.42	167.8	—	—	298–372
$MgCO_3$	Magnesite	83.43	70.7	—	—	298
Al_2O_3		101.94	114.8	12.8	3.54	298–1800
CaO		56.08	48.8	4.5	0.65	298–1800
CO_2		44.01	43.3	11.5	0.82	273–1200
FeO		71.85	51.8	6.8	0.16	298–1200
Fe_2O_3		159.70	97.3	72.1	1.29	298–1100
$H_2O(l)$		18.02	75.5			298–373
$H_2O(g)$		18.02	30.5	10.3		293–2750
$H_2O(hs)$[a]		18.02	27.2	27.6	0.38	?
K_2O[b]		94.19	97.6	33.6	—	?
MgO		40.32	42.6	7.3	0.62	298–2100
MnO		70.93	46.5	8.1	0.37	298–1100
Na_2O		61.99	65.7	22.6	—	298–1100
P_2O_5		141.96	35.1	225.9	—	298–631
SiO_2		60.06	46.9	34.3	1.13	298–848
TiO_2		79.90	75.2	1.2	1.82	298–1800

[a] For H_2O in hydrosilicates; estimated by author.

[b] Estimated by author.

conductivity of the aggregate is an essential aspect of concrete design, the designer is well advised to check his estimate of the conductivity of aggregates by some crude experimental technique.[22]

Table 6.10 gives some information on the most common lightweight aggregates. Natural aggregates, such as diatomite, pumice and scoria, are used mainly in the areas where they are won. Most of the manufactured aggregates are produced from clays, shales and slates, by 'bloating' them for 15–20 min at an appropriate temperature, usually between 1000 and 1250 °C (Conley *et al.* 1948). Perlite and vermiculite aggregates are also produced by bloating, the former from a kind of volcanic deposit, the latter from a micaceous mineral. Expanded slag is a foamed material manufactured from molten blast-furnace slag by the application of water.

22 A method suitable for measuring the thermal diffusivity of small specimens was described by Harmathy (1971).

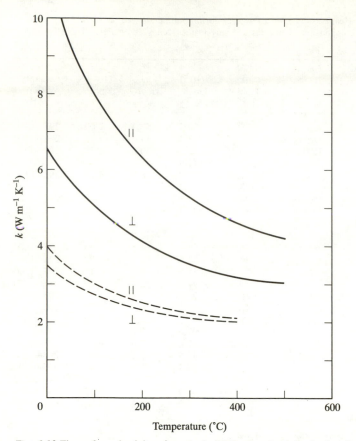

Fig. 6.13 Thermal conductivity of quartz (solid lines) and calcite (broken lines): ∥, parallel with optic axis; ⊥, perpendicular to optic axis (Birch and Clark 1940)

Cinders are residues of the high-temperature combustion of coal or coke, and are usually not processed.

For lack of strength, concretes made with perlite and vermiculite are regarded principally as insulating materials. All other aggregates are capable of yielding concretes of at least 7 MPa strength, and therefore are suitable for use in building blocks. However, expanded clay, shale, slate and slag aggregates are, in general, strong enough to be used in structural grade concretes (Shideler 1957).

The thermal conductivity of lightweight aggregates depends primarily on their porosities. Since both the solid matrix and the pores are essentially continuous, the Hamilton−Crosser formula with $n = 1.5$ is applicable. Thus, combining Eqs 5.29 and 5.30, the thermal conductivity of the aggregate, k_A (W m^{-1} K^{-1}), is

$$k_A = k_{sm} \frac{0.5k_{sm} + k_a - 0.5P_A(k_{sm} - k_a)}{0.5k_{sm} + k_a + P_A(k_{sm} - k_a)}$$ [6.37]

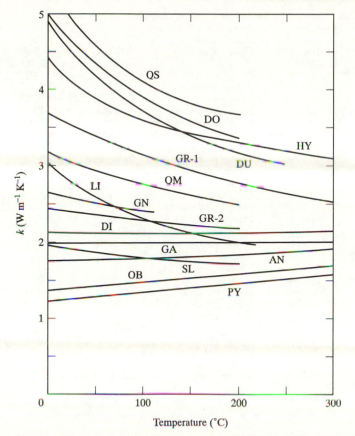

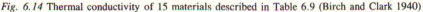

Fig. 6.14 Thermal conductivity of 15 materials described in Table 6.9 (Birch and Clark 1940)

where k_{sm} = thermal conductivity of the solid matrix, $W\,m^{-1}\,K^{-1}$;

$\quad\ \ k_a$ = thermal conductivity of air, $W\,m^{-1}\,K^{-1}$;

$\quad\ \ P_A$ = porosity of the aggregate, $m^3\,m^{-3}$

and k_A, k_{sm} and k_a are functions of the temperature.

Since the characteristic pore size may be in the millimeter range, correcting the thermal conductivity of air for radiative heat transfer is usually necessary at higher temperatures.[23].

With lightweight aggregates, the solid matrix is, as a rule, of a glassy or semiglassy structure. The designer may choose the conductivity of one of the glasses or low-conductivity crystalline rocks shown in Fig. 6.14 as representative of the solid matrix.

As to the specific heat of lightweight aggregates, for those produced from

---

23 See the second term on the right-hand side of Eq. 5.32.

Table 6.9 Some characteristics of the 15 materials whose thermal conductivities are plotted in Fig. 6.14 (Birch and Clark 1940)

Symbol	Name	Geological origin	Composition	Density $(kg\ m^{-3})$	Mean grain size (mm)
AN	Anorthosite	Igneous	Almost all plagioclase feldspars	2700	0.5
DI	Diabase	Igneous	Mainly plagioclase feldspars and pyroxenes	2960	0.5
DO	Dolomite	Sedimentary	Carbonate group	2830	0.01
DU	Dunite	Igneous	Almost all olivines	3250–3270	1.0
GA	Gabbro	Igneous	Mainly plagioclase feldspars, pyroxenes and olivines	2860–2880	3.0
GN	Gneiss	Metamorphic	Layered mineral, mainly feldspar and quartz	2640	0.2
GR-1	Granite 1	Igneous	Mainly potash feldspars and quartz	2610	1.5–2.0
GR-2	Granite-2	Igneous	Mainly potash- and plagioclase-feldspars and quartz	2640	0.5
HY	Hypersthenite	Igneous	Pyroxene group	3290	2.0
LI	Limestone	Sedimentary	Carbonate group	2610	0.001–0.01
OB	Obsidian	Igneous	Glassy potash feldspar and quartz	2440	
PY	Pyrex	Artificial		2230	
QM	Quartz monzonite	Igneous	Mainly potash- and plagioclase-feldspars and quartz	2640	1.0
QS	Quartzitic sandstone	Sedimentary	Mainly quartz	2640–2650	0.3
SL	Slate	Metamorphic	Layered, clay minerals	2760	

Table 6.10 Some characteristics of common lightweight aggregates

Aggregate	Geological origin of raw material	True density $(kg\ m^{-3})$	Porosity $(m^3\ m^{-3})$	Unit weight $(kg\ m^{-3})$	Density of concrete $(kg\ m^{-3})$
Diatomite	Sedimentary	2000–2300	0.25–0.40	250–500	
Pumice	Igneous	2600–2700	0.30–0.60	500–900	700–1400
Scoria	Igneous			700–900	900–1400
Cinders			0.20–0.60	800–1100	1100–1400
Expanded clay, shale, slate	Sedimentary, Metamorphic	2600–2800	0.20–0.50	650–1000	1400–1800
Expanded slag		2000–2500	0.25–0.45	400–1100	1000–1700
Expanded perlite	Igneous	2600–2800	0.70–0.90	50–200	400–1000
Expanded vermiculite	Sedimentary	2200–2700	0.70–0.90	50–150	300–800

igneous rocks the designer may assume that the solid matrix consists mainly of plagioclase feldspars (e.g. anorthite), and use Eq. 6.24 with values of e, f and g listed in Table 6.8. For those made from sedimentary minerals, using the specific heat of brick (see next chapter) may prove sufficiently accurate.

6.7   Mechanical Properties of Concrete

Since the mechanical properties of the concrete depend on those of the cement paste and the aggregates, one may assume that determining them is a simple matter of applying the appropriate mixture rules. Unfortunately, that assumption is rarely correct, because the properties of concrete also depend on the bond between the paste and the aggregates. It has been shown (Ferran 1956; Lyubimova and Pinus 1962) that regions of increased porosity exist in the vicinity of paste—aggregate interfaces. Internal stresses lead to the formation of microcracks, primarily at these interfaces, but also in the paste itself. These microcracks are distributed throughout the material even before the application of load (Hsu *et al.* 1963; Shah and Slate 1968), but remain stable up to a load amounting to at least 30 per cent of the ultimate.

As the modulus of elasticity of concrete is usually measured at low or moderate stresses, it is one of those properties of concrete that depend least on the presence of microcracks.

According to Hirsch (1962), the modulus of elasticity of the cement paste, E_{pt} (Pa), is in the range of 7×10^3 to 28×10^3 MPa, and decreases with the increase of water—cement ratio. Helmuth and Turk (1966) showed that E_{pt} depends on the paste's capillary porosity, P_{cp} (see Eq. 6.11), as[24]

$$\frac{E_{pt}}{E_g} = (1 - P_{cp})^3 \tag{6.38}$$

where E_g (Pa) is the modulus of elasticity of the gel. Their experimental data have indicated that E_g is approximately equal to 30×10^3 MPa.

The moduli of elasticity of normal-weight concretes are known to be generally higher, and those of lightweight concretes lower, than the modulus of elasticity of the paste. The Hamilton—Crosser formula (see Eqs 5.5 and 5.6) seems to be an obvious choice for expressing the modulus of concrete in terms of the moduli of its constituents.[25]

Concrete is a three-component material, consisting of one continuous component (the cement paste) and two discontinuous components (the fine and coarse aggregates). Since the Hamilton—Crosser formula is applicable only to two-component systems, the calculation of the modulus of elasticity of concrete must

24 It may be of interest to note that E_{pt} and P_{cp} are equally well correlatable by means of Eq. 5.17, using the values $E_g = 33 \times 10^3$ MPa and $a = 4.2$.

25 More sophisticated expressions have been suggested by Hirsch (1962) and Hansen (1968).

be performed in two steps. The modulus of the mortar (paste and fine aggregate) is to be determined first (with the paste as continuous and the fine aggregate as discontinuous component), and then the modulus of concrete (with the mortar as continuous and the coarse aggregate as discontinuous component). As mentioned in Chapter 5, Eq. 5.6 with $n = 3$ is applicable to any system consisting of a continuous and a discontinuous component, irrespective of the geometry of the latter.

To simplify the discussion, it will be assumed here that the fine and coarse aggregates consist of the same mineral. The modulus of elasticity of a concrete made with a single aggregate can be expressed as

$$E_C = E_{pt} \frac{2E_{pt} + E_A - 2v_A(E_{pt} - E_A)}{2E_{pt} + E_A + v_A(E_{pt} - E_A)} \qquad [6.39]$$

where E_C = modulus of elasticity of concrete, Pa;
$\quad\quad E_A$ = modulus of elasticity of aggregates, Pa;
$\quad\quad v_A$ = volume fraction of aggregates, $m^3 m^{-3}$;

and E_{pt} is the modulus of elasticity of the cement paste. Assuming that $E_A = 1.5E_{pt}$ for normal-weight aggregates, and $E_A = 0.5E_{pt}$ for lightweight aggregates, and that $v_A = 0.68$, one will find that $E_C = 1.32E_{pt}$ for normal-weight concrete, and $E_C = 0.64E_{pt}$ for lightweight concrete. In general, values of E_C calculated from Eq. 6.39 compare favorably with the experimental values reported by Hirsch (1962).

Unlike the modulus of elasticity, the strength of concrete is strongly affected by the microcracks that develop at the paste—aggregate interfaces and in the paste itself. Consequently, the strength of concrete depends primarily on the strength of the paste, and is largely independent of the strength of aggregates.

A number of empirical formulas have been suggested for correlating the strength of concrete with some characteristics of the paste. The best known among them are:[26]

- *Abrams's formula* (Abrams 1918; Popovics 1990):

$$\sigma_C = \frac{K_1}{B^{w_w/w_c}} \qquad [6.40]$$

- *Feret's formula* (Feret 1897):

$$\sigma_C = K_2 \left[\frac{\dfrac{W_c}{\rho_c}}{\dfrac{W_c}{\rho_c} + \dfrac{W_w}{\rho_w}(1 + A)} \right]^2 \qquad [6.41]$$

---

26 It may be mentioned that Eq. 5.19 would also be suitable for such a correlation. Apparently, concrete scientists prefer their own empiricism.

- *Powers's formula* (Powers 1949, 1958, 1962):

$$\sigma_C = \beta X^r \qquad\qquad [6.42]$$

In Eqs 6.40, 6.41 and 6.42,

σ_C	=	compressive strength of concrete, Pa;
K_1, K_2	=	empirical factors that depend on some characteristics of the cement, the age of the paste, the admixtures, and the curing and testing conditions, Pa;
B	=	empirical factor, dimensionless;
β	=	empirical factor, characteristic of the type of cement, Pa;
r	=	empirical constant, also characteristic of the type of cement, dimensionless.

All the other symbols have been defined in Section 6.2 of this chapter.

Since $X = 1$ for a concrete with zero capillary porosity (see Eq. 6.15), β in Eq. 6.42 can be taken to represent the strength of the cement gel itself.

According to Powers (1962), for cement−sand mortars r is in the range of 2.44−3.08, and β ranges from 90 to 128 MPa, depending on certain (indeterminate) characteristics of the cement. In an earlier study, Powers (1949) found that, at identical values of X, the mortar is somewhat weaker than the neat cement paste. The loss of strength that occurs on the addition of fine aggregate is presumably due to the formation of microcracks at the paste−aggregate interfaces. Yet, since the strength of (conventionally produced) concretes does not normally exceed 40 MPa, one is led to believe that the presence of coarse aggregate is mainly responsible for the marked difference in the strengths of the cement paste and concrete. Hsu *et al.* (1963) suggested that the failure of concrete is due to the joining up of the 'bond cracks' (at the coarse aggregate and mortar interfaces) with the 'mortar cracks', as the load reaches 70−90 per cent of the ultimate load.

Since microcracking does not seem to have a significant effect on the elastic behavior of concrete, one may wonder if there is any objective ground for correlating the modulus of elasticity of concrete with its strength. Yet such correlations do exist.[27] The most recent among them is one proposed by Francis *et al.* (1991):

$$E_C = 46.8 \, \rho_C^{1.5} \sigma_C^{0.5} \qquad\qquad [6.43]$$

where ρ_C (kg m^{-3}) is the density of concrete and 46.8 is a dimensional constant. This formula, which is only slightly different from that recommended in the ACI Building Code (ACI Committee 318 1989), seems to be valid for both normal-weight and lightweight concretes with densities within the 1440−2480 kg m^{-3} range.

---

27 In fact, if one compares Eqs 6.38 and 6.42, keeping in mind that in the latter equation $r \simeq 3$ and, according to Eq. 6.15, $X = 1 - P_{cp}$, one may conclude that σ_C is directly proportional to E_C. Of course, such a conclusion would be erroneous, and would only prove that there is a great deal of latitude in the interpretation of the test results.

Concrete is roughly ten times as strong in compression as in tension. Failure in tension is also due to the presence of microcracks, but apparently the cracks join up and enlarge more readily under tension. The ratio of tensile strength to compressive strength decreases with an increase in the compressive strength. The expression recommended by the ACI (ACI Committee 318 1989) is

$$(\sigma_C)_t = 620 \; \sigma_C^{0.5} \qquad\qquad [6.44]$$

where $(\sigma_C)_t$ (Pa) is the strength of concrete in tension, and 620 is a dimensional constant.

With a rise of temperature, the behavior of concrete becomes more and more influenced by the growing number of microcracks, created by the shrinkage of the cement paste and the expansion of the aggregates. Although Figs 6.2 and 6.11 and Tables 6.5 and 6.6 may provide some guidance for assessing the extent of crack formation, it is rarely possible to predict the behavior of any concrete at elevated temperatures without laboratory studies.

Both the modulus of elasticity and the strength of concrete decrease as the temperature rises. However, as Fig. 6.15 indicates, the modulus of elasticity declines more rapidly than the strength. This finding comes as no surprise. The propagation of microcracks has been recognized as the principal cause of compressive failure even at room temperature, so that the additional cracking caused by thermal stresses at elevated temperatures will have a less dramatic effect on the strength of concrete than on its elasticity.

The variation of the modulus of elasticity with temperature is plotted in Fig. 6.15(a) for various normal-weight and lightweight concretes (Bennetts 1981). In the figure, E_C and $(E_C)_0$ stand, respectively, for the modulus at a temperature T and at the reference (room) temperature. Apparently, lightweight concretes stand up to high temperatures somewhat better than their normal-weight counterparts, probably because some lightweight aggregates are able to adjust to the shrinkage of the cement paste.

Following the pioneering work of Malhotra (1956), a great deal of effort has been expended on investigating the effect of elevated temperatures on the strength and stress−strain characteristics of concrete. Zoldners (1960), Abrams (1971), Harada et al. (1972), Weigler and Fischer (1972), Anderberg and Thelandersson (1973, 1976), Schneider et al. (1980), Dias et al. (1990), and many others made important contributions to the knowledge in this field. The work of a number of authors was reviewed in comprehensive reports by Bennetts (1981) and Schneider (1985), and in an ACI Guide (ACI Committee 216 1989).

Three methods have been generally accepted for determining the effect of elevated temperatures on the compressive strength of concrete specimens:

• The specimen is heated without load to the target temperature, allowed to cool to room temperature, and then loaded to failure. The strengths so determined will be referred to as UR (specimen heated Unstressed, tested at Room temperature) strengths. Clearly, this kind of information finds

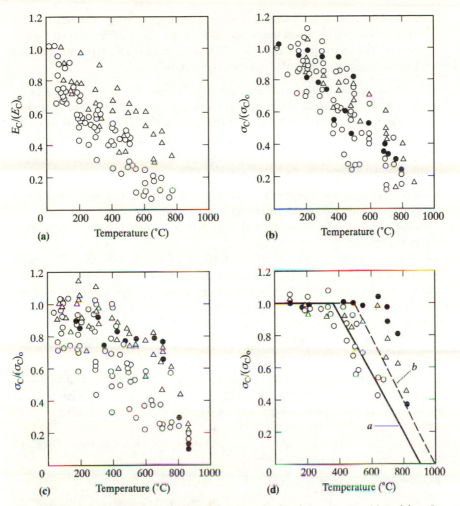

Fig. 6.15 Modulus of elasticity and compressive strength of various concretes: (a) modulus of elasticity; (b) UR strength; (c) UE strength; (d) SE strength. In (a): o represents normal-weight concretes; Δ, lightweight concretes. In (b), (c) and (d): o represents silicate aggregate concretes; ●, carbonate aggregate concretes; Δ, lightweight concretes. In (d) (recommended by The Institution of Structural Engineers): curve *a* is for normal-weight concrete; curve *b*, lightweight concrete (Bennetts 1981; Bobrowski 1978)

application mainly by those who are concerned with the behavior of the material after fire rather than during fire.

- The specimen is heated without load to the target temperature, allowed to attain reasonably uniform temperature throughout, and then loaded to failure. The result of such a test will be termed UE (specimen heated Unstressed, tested at Elevated temperature) strength.

- The specimen is subjected to a constant load (amounting to 25−55 per cent of the load that causes failure at room temperature) during both heating and 'heat-soaking' at the target temperature. Then the load is increased until failure occurs. The strength so measured will be referred to as SE (specimen heated Stressed, tested at Elevated temperature) strength.

The variations with temperature of the UR, UE and SE strengths of a large number of concretes made with silicate, carbonate, and lightweight aggregates are shown in Figs 6.15(b), 6.15(c), and 6.15(d) (Bennetts 1981). In the figures σ_C and $(\sigma_C)_o$ stand, respectively, for the compressive strength at a temperature T and at the reference (room) temperature.

The following conclusions have been drawn:

- The fractional loss of strength for a concrete exposed to elevated temperatures is largely independent of its strength at room temperature; consequently it is also independent of the type of cement and the water−cement ratio used in the mixture.
- The aggregate−paste ratio has a significant effect. The decline of strength is less for lean (low cement content) mixes than for rich mixes.
- Among the various kinds of strength, the UR strength is the lowest and the SE strength is the highest. The SE strength does not seem to be affected by the magnitude of load applied during heating.[28]
- The fractional loss of strength is, in general, lowest for concretes made with lightweight aggregates, and highest for those made with silicate aggregates.

In view of the better performance of lightweight concretes at elevated temperatures, the Institution of Structural Engineers (Bobrowski 1978) has allowed higher elevated-temperature design stresses (in relation to those applicable at room temperature) for lightweight concretes than for normal-weight concretes. The full-line curves in Fig. 6.15(d) reflect the Institution's recommendation. They are described by the following equations:

- For normal-weight concrete

$$\frac{\sigma_C}{(\sigma_C)_o} = \begin{cases} 1 & \text{if } 20 \leq T \leq 350\,°C \\ 1.622 - 0.00178T & \text{if } 350 \leq T \leq 911\,°C \end{cases} \qquad [6.45]$$

- For lightweight concrete

$$\frac{\sigma_C}{(\sigma_C)_o} = \begin{cases} 1 & \text{if } 20 \leq T \leq 500\,°C \\ 2 - 0.002T & \text{if } 500 \leq T \leq 1000\,°C \end{cases} \qquad [6.46]$$

Figures 6.16(a), 6.16(b) and 6.16(c) show typical stress−strain curves in compression at elevated temperatures (Harmathy and Berndt 1966; Schneider 1977) for cement paste (with a water−cement ratio of 0.33), normal-weight concrete (with crushed quartz aggregates), and lightweight concrete (with expanded

---

28 However, high-strength concretes prepared with water-reducing admixtures may fail in an explosive manner when heated under load (Castillo and Durrani 1990).

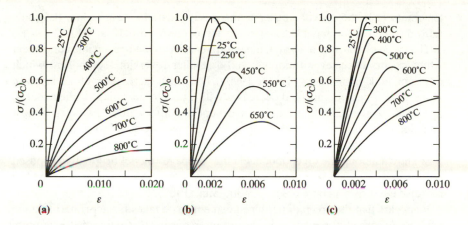

Fig. 6.16 Stress–strain curves in compression (normalized): (a) neat portland cement paste (water–cement ratio 0.33); (b) normal-weight concrete (with crushed quartz aggregate); (c) lightweight concrete (with expanded shale aggregate) (Harmathy and Berndt 1966; Schneider 1977)

shale aggregates). The tests were conducted on specimens heated to the target temperature without load, and loaded while hot. Those in Figs 6.16(a) and 6.16(c) represent averaged and 'smoothed' values. To facilitate comparison, 'normalized' stresses (i.e. stresses relative to the compressive strengths of the materials at room temperature[29]) are shown. In all three figures, σ (Pa) stands for stress, $(\sigma_C)_0$ (Pa) for compressive strength at reference (room) temperature, and ϵ (m m^{-1}) for strain (deformation).

The effect of elevated temperatures on the strength of bond between concrete and steel is known from the work of Hertz (1980), Diederichs and Schneider (1981), and others. Schneider (1985) summarized the available information as follows:

- In the case of deformed bars and rusted plain bars, the bond strength varies with the temperature roughly in the same way as the compressive strength of concrete. In the case of new plain bars, however, the bond strength declines faster than the compressive strength.
- The type of aggregate is an important factor. The lower the thermal expansion of the concrete, the higher the bond strength at elevated temperatures.
- The water–cement ratio for the concrete mixture and the diameter of the reinforcing bars have little effect on the bond strength.

Maréchal's (1972a) experimental studies indicate that Poisson's ratio for normal-weight concretes is expected to decrease almost linearly with the rise of

---

29 The compressive strength of the three materials were as follows: cement paste, 71 MPa; normal-weight concrete, 26.5 MPa; lightweight concrete, 18 MPa.

temperature. In the case of a concrete made with quartzite[30] aggregate, Poisson's ratio declined from about 0.27 at room temperature to 0.10 at 400 °C.

The creep of concrete at room temperature consists of an irreversible component which occurs over several months, and a smaller reversible component which develops in days or weeks (Taylor 1990). The creep increases with an increase in the water–cement ratio and decreases with age before loading. If the material is heated to 110 °C before loading and kept dry afterwards, the creep almost completely disappears. It will reappear on re-exposure to a moist atmosphere.

There is no general agreement on the mechanism of creep deformation. The best known among the creep models is the 'seepage' model originally proposed by Lynam (1934) and later modified by Powers (1968). A number of other models are discussed in a book edited by Wittmann (1982).

It appears that the creep of hardened cement paste takes place primarily in the strongly adsorbed water layers in the gel interstices. The finding that maximum creep occurs somewhere between 50 and 70 °C (Nasser and Neville 1967; Nasser 1971; Geymayer 1972) is attributed to the increased mobility of this 'load-bearing' water.

The elevated-temperature creep of normal-weight concretes was investigated by Cruz (1968), Maréchal (1969, 1972b) and Gross (1975), and that of lightweight concretes by Schneider et al. (1980). Two families of curves, showing the elastic and plastic deformation of a normal-weight concrete, are displayed in Fig. 6.17 (Gross 1975). The specimens were heated without load to the target temperature, then loaded to either 20 or 40 per cent of the compressive strength of concrete.[31]

Anderberg and Thelandersson (1976) and Schneider (1976) claimed that the strain exhibited by concrete heated under conditions characteristic of fire exposure consists of four components:

$$\epsilon = \epsilon_{th} + \epsilon_{\sigma} + \epsilon_{cr} + \epsilon_{tr} \qquad [6.47]$$

where ϵ = total strain (measurable by strain gages), $m\,m^{-1}$;

ϵ_{th} = thermal strain, i.e. thermal expansion or shrinkage (without load), as read from the dilatometric curve (see, for example, Fig. 6.18), $m\,m^{-1}$;

ϵ_{σ} = quasi-instantaneous stress-related strain, i.e. strain read from stress–strain curves recorded at (stabilized) constant temperatures (see, for example, Fig. 6.16), $m\,m^{-1}$;

ϵ_{cr} = (time-dependent) creep strain, i.e. strain read from creep–strain versus time curves obtained at various constant stresses and (stabilized) constant temperatures (see, for example, Fig. 6.17), $m\,m^{-1}$;

ϵ_{tr} = transient strain which accounts for the physicochemical changes taking place on heating at constant stress, $m\,m^{-1}$.

---

30 Metamorphic rock with high quartz content.
31 The compressive strength of the concrete was 42 MPa at room temperature.

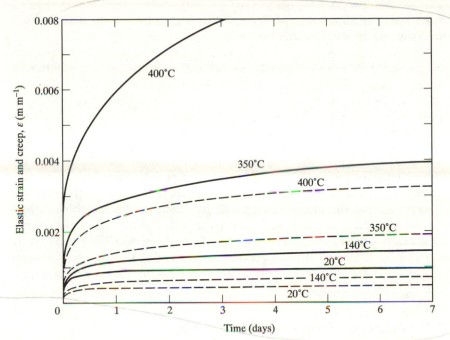

Fig. 6.17 Elevated-temperature elastic and creep deformation of a particular normal-weight concrete loaded to 40 per cent (solid lines) and 20 per cent (broken lines) of its cold strength (Gross 1975)

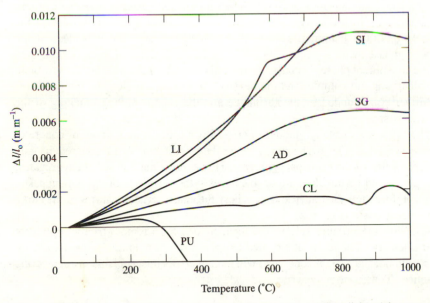

Fig. 6.18 Dilatometric curves for three normal-weight concretes and three lightweight concretes. Aggregates: LI, limestone; SI, siliceous rock; AD, andesite; SG, expanded slag; CL, expanded clay; PU, pumice (Harada *et al.* 1972; Harmathy and Allen 1973)

According to Anderberg and Thelandersson, ϵ_{cr} and ϵ_{tr} can be described by the following empirical equations:

$$\epsilon_{cr} = b_1 \frac{\sigma}{(\sigma_C)_T} t^{1/2} \exp[j(T - 293)] \qquad [6.48]$$

$$\epsilon_{tr} = b_2 \frac{\sigma}{(\sigma_C)_o} \epsilon_{th} \qquad [6.49]$$

where σ is stress (Pa), $(\sigma_C)_T$ and $(\sigma_C)_o$ (Pa) are strengths at temperature T (K) and at room temperature, respectively, t (s) is time, and b_1 (s$^{-1/2}$), b_2 (dimensionless), and j (K^{-1}) are empirical constants characteristic of the material.

Anderberg and Thelandersson suggested that, under conditions arising during a fire exposure, ϵ_{tr} is a very important contribution to the total strain, being considerably greater than ϵ_σ and ϵ_{cr} together. From the work of other researchers (e.g. Bresler and Iding 1983) it appears, however, that satisfactory analytical simulations of the behavior of fire-exposed concrete structures can be achieved without explicitly taking account of the transient strain.

✳ 6.8   Thermal Properties of Concrete

Assuming that no reaction takes place between the paste and the aggregates, the thermogravimetric curve for concrete can be constituted from those of its components, taking into account the components' mass fractions at room temperature. Unfortunately, the simple mixture rule is not applicable to the calculation of the dilatometric curve. As Eq. 5.10 indicates, the expanded volume of a polyphase material depends not only on the coefficients of thermal expansion of the constituent phases, but also on the bulk moduli of the phases. The problem of calculating the volume (and the density) of concrete at elevated temperatures is further compounded by the shrinkage and the accompanying cracking of the cement paste.

The aggregates usually amount to 60−75 per cent of the volume of concrete. Consequently, as Fig. 6.18 shows,[32] the dilatometric curve of a concrete usually resembles that of the principal aggregate. However, some lightweight aggregates, e.g. perlite and vermiculite, are unable to resist the shrinkage of the cement paste, and therefore their dilatometric curves bear the characteristic features of the curve for cement paste (Fig. 6.2).

Although the mixture rule in the form of Eq. 5.13 is not strictly applicable to the density of concrete at elevated temperatures, it is unlikely that its use (at least with normal-weight concretes) will lead to errors unacceptable in engineering practice. Thus, one may write

---

32 Andesite is an igneous rock composed chiefly of plagioclase feldspars. Quartz is usually absent.

$$\rho_C \simeq v_{pt}\rho_{pt} + v_{fA}\rho_{fA} + v_{cA}\rho_{cA} \qquad [6.50]$$

where ρ_C, ρ_{pt}, ρ_{fA}, and ρ_{cA} (kg m^{-3}) stand for the densities of concrete, cement paste, fine aggregate, and coarse aggregate, respectively, at a temperature T; v_{pt}, v_{fA}, and v_{cA} (m m^{-1}) are the volume fractions (at room temperature) of paste, fine aggregate, and coarse aggregate, respectively.

The specific heat of concrete can be expressed with the aid of the mixture rule,

$$c_C = \omega_{pt}c_{pt} + \omega_{fA}c_{fA} + \omega_{cA}c_{cA} \qquad [6.51]$$

where c_C, c_{pt}, c_{fA}, and c_{cA} (J kg^{-1} K^{-1}) stand for the specific heats at constant pressure[33] of concrete, cement paste, fine aggregate, and coarse aggregate, respectively, at a temperature T; ω_{pt}, ω_{fA}, and ω_{cA} (kg kg^{-1}) are the mass fractions (at room temperature) of paste, fine aggregate, and coarse aggregate, respectively. The way of calculating c_{pt}, c_{fA}, and c_{cA} has already been discussed (see sections 6.5 and 6.6).

When calculating the temperature of concrete structures, the 'volume specific heat', i.e. the product of ρ_C and c_C, is the required input information. Once ρ_C and c_C are known as functions of temperture, determining the $(\rho c_p)_C$ versus T function is a trivial task.

Figure 6.19 shows calculated curves of the volume specific heat (in oven-dry condition) for four hypothetical concretes (Concretes 1 to 4). The principal

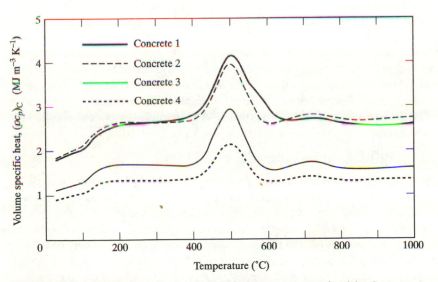

Fig. 6.19 Volume specific heats for four hypothetical concretes: normal-weight, Concretes 1 and 2; lightweight, Concretes 3 and 4 (Harmathy 1970a)

---

33 As usual, the p subscripts indicating constant pressure have been omitted in order to avoid the crowding of subscripts.

Table 6.11 Principal characteristics of the aggregates in the four 'limiting' concretes (Harmathy 1970a)

	Aggregate in			
	Concrete 1	Concrete 2	Concrete 3	Concrete 4
			Aggregate C	Aggregate D
General description	Crystalline quartz with coarse grain structure	Crystalline anorthosite with coarse grain structure	Crystalline material with fine grain structure	Amorphous material
Volume fraction in concrete, $m^3\,m^{-3}$	0.73	0.73	0.73	0.82
Bulk density, $kg\,m^{-3}$	2650	2700	1450	1150
Porosity, $m^3\,m^{-3}$	0	0	0.505	0.608
Specific heat at room temperature, $J\,kg^{-1}\,K^{-1}$	740	785	730	730
Thermal conductivity, $W\,m^{-1}\,K^{-1}$	6.22 at room temperature, decreases as temperature rises; see Fig. 6.13 for quartz $\perp$	1.75 at room temperature, increases as temperature rises; see Fig. 6.14 for anorthosite	0.56 at room temperature, increases as temperature rises; see Fig 6.14 for diabase	0.23 at room temperature, increases as temperature rises; see Fig. 6.14 for obsidian

Cement paste for all four concretes: Model Paste C, water–cement ratio = 0.5, maturity factor = 0.824 (see also Table 6.2).

characteristics of the aggregates used in the four concretes are presented in Table 6.11. These concretes, as will soon be pointed out, were conceived to represent 'limiting cases' among normal-weight and lightweight concretes.

The ever-present moisture can be treated as an unstable component of concrete. It undergoes the following 'reaction' in the vicinity of 100 °C:

$$H_2O(1) \rightarrow (1 - \xi_w)H_2O(1) + \xi_wH_2O(g) \uparrow \qquad [6.52]$$

where ξ_w (dimensionless) is the progress variable for the vaporization of moisture, and l and g signify liquid and gaseous (vapor) phases, respectively. The arrow indicates that the water vapor leaves the system.

According to the earlier adopted convention, ξ_w and $d\xi_w/dT$ are regarded as functions of temperature. The nominal 'reaction' temperature, T_r, may be chosen as 105 °C (378 K), and the width of the reaction zone, $T_{r2} - T_{r1}$, as 30 °C. Information on the ξ_w versus T and $d\xi_w/dT$ versus T relations can be obtained from Table 5.1.

---

34 It is tacitly assumed here that the moisture in sorbed (quasi-liquid) phase is an immobile component of the solid. When dislodged by vaporization, it becomes mobile and leaves the solid along an isothermal plane. Consequently, there is no heat transport associated with the movement of water vapor.

The expressions for the volume specific heat of concrete, corrected for the presence of moisture, are as follows:[34]

- From 298 to 363 K (25 to 90 °C),

$$(\rho c)_C = \rho_C[c_C + mc_w] \tag{6.53}$$

- From 363 to 393 K (90 to 120 °C),

$$(\rho c)_C = \rho_C \left\{ c_C + m \left[(1 - \xi_w)c_w + \frac{d\xi_w}{dT}\Delta h_w \right] \right\} \tag{6.54}$$

where m = moisture content of concrete (related to oven-dry condition), kg kg^{-1};

c_w = specific heat of water = $4190\,\text{J kg}^{-1}\text{K}^{-1}$;

Δh_w = heat of vaporization of water (at 100 °C) = $2.27\,\text{MJ kg}^{-1}$.

The thermal conductivity of concrete can be calculated by means of the Hamilton–Crosser formula (Eqs 5.29 and 5.30). For reasons discussed in connection with the modulus of elasticity, the calculations have to be performed again in two steps. The applicable expression is analogous to Eq. 6.39. For simplicity, it is again assumed that both aggregates consist of the same mineral.

$$k_C = k_{pt} \frac{2k_{pt} + k_A - 2v_A(k_{pt} - k_A)}{2k_{pt} + k_A + v_A(k_{pt} - k_A)} \tag{6.55}$$

where k_C, k_{pt}, and k_A ($\text{W m}^{-1}\text{K}^{-1}$) are the thermal conductivities of concrete, cement paste, and aggregates, respectively, at a temperature T; v_A (m m^{-1}) is the volume fraction of the aggregates (at room temperature).

The method of assessing the thermal conductivity of the paste was discussed in connection with Eq. 6.23. The conductivity versus temperature relation for normal-weight aggregates is to be determined experimentally or estimated, making use of Fig. 6.14 and Table 6.9. That for lightweight aggregates can be estimated by means of Eq. 6.37.

The thermal conductivity (in oven-dry condition) versus temperature relations for the four hypothetical concretes (Concretes 1 to 4) mentioned earlier are displayed in Fig. 6.20. The curves were developed by calculations (Harmathy 1970a).

As Table 6.11 shows, the aggregates in Concrete 1 consisted of crystalline quartz, and those in Concrete 2 consisted of anorthosite. Since, according to Figs 6.13 and 6.14, quartz has the highest and anorthosite the lowest thermal conductivity among common rocks, Concretes 1 and 2 may be looked upon as representing concretes with maximum and minimum conductivities in the normal-weight group.

The aggregates in Concretes 3 and 4 are hypothetical porous materials. They are referred to as Aggregate C and Aggregate D in Table 6.11. The thermal conductivity of the solid matrix, k_{sm} (see Eq. 6.37), was assumed to be similar to the conductivity of diabase for Aggregate C, and similar to the conductivity of obsidian for Aggregate D (see Fig. 6.14). The conductivities of the porous

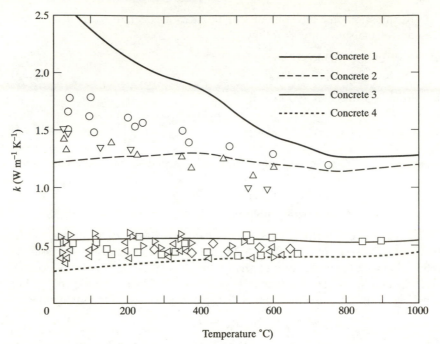

Fig. 6.20 Thermal conductivity of four hypothetical concretes (Concretes 1, 2, 3, and 4) and a number of real concretes. Aggregates: ○, various gravels; △, siliceous rocks; ▽, carbonate rocks; □, expanded shale; ◇, expanded clay; ▷, expanded slag; ◁, pumice (Harmathy 1970a; Harmathy and Allen 1973),

aggregates were calculated using Eq. 6.37, in which the conductivity of air, k_a, was corrected for radiative heat transfer (see Eq. 5.32).

Concretes 3 and 4 were conceived to represent concretes with maximum and minimum conductivities in the lightweight group.

To allow the reader to assess the tenability of the above speculations and the calculation techniques employed, the results of a multitude of tests are also shown as discrete points in Fig. 6.20. It appears that the thermal conductivity of lightweight concretes can indeed be expected to lie roughly within the area bounded by the curves representing Concretes 3 and 4. In the normal-weight group, the conductivities of most concretes seem to fall in the lower half, or even below the area bounded by the conductivities of Concretes 1 and 2.

There is no well founded method for taking the effect of moisture on the thermal conductivity of concrete into account. A crude way of dealing with the problem would be to use a substantially increased (e.g. doubled) 'effective' value for the thermal conductivity of the cement paste, k_{pt}, in Eq. 6.55 for the 25−105 °C temperature interval.

6.9   'Types' of Concrete

It has been conventional in fire safety engineering practice to divide the normal-weight concrete aggregates into two groups: siliceous aggregates and calcareous aggregates. The dictionary defines *siliceous* as containing silica, and *calcareous* as 'containing lime or limestone'. Since the majority of rocks found in nature, namely silicates that do not contain free quartz, and carbonates of magnesium, iron, manganese, etc. are not covered by this grouping, it would be more appropriate to refer to the two major groups as *silicate* and *carbonate* aggregates.

In reports of fire tests performed on normal-weight concrete building elements, the material is usually identified as either siliceous aggregate concrete or calcareous (carbonate) aggregate concrete. The writer of a book, whose function is to survey the available information, has no other choice but to go along with the prevailing practice, even though the practice may not be justified.

From numerous test reports, the fire performance of building elements made from carbonate aggregate concretes appears to be considerably better than that of elements made from siliceous aggregate concretes. The reason for the invariably superior performance of carbonate aggregate concretes is not clear. Reviewing the information presented in Table 6.5, and Figs 6.10 and 6.11 on thermal expansion, Table 6.8 and Fig. 6.12 on heat capacity (specific heat), and Fig. 6.14 on thermal conductivity, one finds hardly anything that would point to the inherent superiority of carbonate aggregate concretes in fire. In fact, many silicate aggregate concretes, especially those made with basaltic rocks, are expected to yield better performance in high-temperature applications.

It has been customary to attribute the superior performance of carbonate aggregate concretes to the fact that the heat absorbed in the decomposition of the aggregate slows down the penetration of heat into the building element. However, as Table 6.7 indicates, the temperatures of decomposition are fairly high; hence only elements exposed to long fires can possibly benefit from the decomposition process. It is not unlikely, therefore, that concretes made with aggregates containing some limestone or dolomite, and found to offer good performance in fire resistance tests, are automatically labeled as carbonate aggregate concretes.

The Supplement to the National Building Code of Canada (Associate Committee on the National Building Code 1990) has grouped the various concretes, on the basis of the constituent aggregates, into eight 'types' (Table 6.12). As long as certain dimensional requirements are fulfilled, building elements fabricated with these types of concretes are allowed to be used in fire-resistant buildings without experimental proof of fire resistance. This kind of grouping of concretes has allegedly evolved from experience. Relying on generic information is, naturally, of great convenience in fire safety design. However, the designer should be aware of possible pitfalls associated with the use of information based on undefined material properties and unclear terminology.

Nomenclature

A mass of water that would fill the voids created by entrapped or entrained air, expressed as fraction of mass of water in the original mixture, $kg\,kg^{-1}$

b empirical constants, various dimensions

Table 6.12 Types of concrete according to the National Building Code of Canada (Associate Committee on the National Building Code 1990)

Type	Description
S	Concrete in which the coarse aggregate is granite, quartzite, siliceous gravel or other dense material containing at least 30 per cent quartz, chert or flint
N	Concrete in which the coarse aggregate is cinders, broken brick, blast furnace slag, limestone, calcareous gravel, trap rock, sandstone or other dense material containing not more than 30 per cent of quartz, chert or flint
L	Concrete in which all the aggregate is expanded slag, expanded clay, expanded shale or pumice
L_1	Concrete in which all the aggregate is expanded shale
L_2	Concrete in which all the aggregate is expanded slag, expanded clay or pumice
L40S	Concrete in which the fine portion of the aggregate is sand and low-density aggregate, and the sand does not exceed 40 per cent of the total volume of all aggregates
$L_1$20S	Type L_1 concrete in which the fine portion of the aggregate is sand and low-density aggregate, and the sand does not exceed 20 per cent of the total volume of all aggregates
$L_2$20S	Type L_2 concrete in which the fine portion of the aggregate is sand and low-density aggregate, and the sand does not exceed 20 per cent of the total volume of all aggregates

B	empirical factor, dimensionless
c	specific heat, $J\,kg^{-1}\,K^{-1}$
C	heat capacity, $J\,kmol^{-1}\,K^{-1}$
e, f, g	empirical constants, various dimensions
E	modulus of elasticity (Young's modulus), Pa
H	enthalpy, $J\,kmol^{-1}$
Δh_w	latent heat of vaporization of water, $J\,kg^{-1}$
ΔH_f	heat of formation, $J\,kmol^{-1}$
ΔH_i	heat of polymorphic inversion, $J\,kmol^{-1}$
ΔH_r	latent heat of reaction (hydration, dehydration), $J\,kmol^{-1}$
j	empirical constant, K^{-1}
k	thermal conductivity, $W\,m^{-1}\,K^{-1}$
k^*	thermal conductivity of a (hypothetical) poreless material, $W\,m^{-1}\,K^{-1}$
K	empirical factor, Pa
l	length, m
Δl	$= l - l_o$, m
m	maturity factor (hydration progress variable), dimensionless; moisture content, related to oven-dry condition, $kg\,kg^{-1}$
M	molecular mass, $kg\,kmol^{-1}$
P	porosity, $m^3\,m^{-3}$
r	empirical constant, dimensionless
S	specific surface, $m^2\,kg^{-1}$
t	time, s (or h)

T	temperature, K (or °C)
W	mass, kg
$\overline{W}$	mass on complete hydration, kg
X	gel−space ratio, $m^3\,m^{-3}$
β	empirical factor, dimensionless
ϵ	strain (deformation); total strain, $m\,m^{-1}$
ξ	reaction (dehydration, vaporization) progress variable, dimensionless
ρ	density, $kg\,m^{-3}$
σ	stress; with subscript, strength, Pa
υ	volume fraction, $m^3\,m^{-3}$
ω	mass fraction, $kg\,kg^{-1}$

Subscripts

a	of air
A	of aggregate
c	of cement
cA	of coarse aggregate
cp	of capillary space
cr	creep
C	of CaO; of concrete
CH	of $Ca(OH)_2$
CSH	of C−S−H gel
d	at the end of dehydration process
D, E, F	of components of cement paste in the D, E, F groups
fA	of fine aggregate
g	of gel
gw	of gel water
H	of H_2O
n	of nonevaporable water
o	at reference temperature (298 K; 25 °C)
ov	overall
p	at constant pressure
pt	of cement paste
r	of reaction
sm	of solid matrix
S	of SiO_2
t	true (density); tensile (stress)
th	thermal
tr	transient
T	at temperature T
w	of/for water (or moisture)
wa	of water (adjusted)
σ	(quasi-instantaneous) strain-related

Superscript

+	for thermogravimetric curve in partly broken line

7 Steel

The areas of application of steel and concrete in the building industry overlap. Decisions on which of the two receives the bigger share in one or another project are usually made on economic considerations, influenced mainly by the local cost and availability of material and skilled labor. In many respects, however, the two materials complement each other; a modern building is hardly conceivable without some use of both concrete and steel. Having decided in favor of a concrete-framed structure, the building designer still has to specify steel among the construction materials, if only as reinforcement for concrete.

7.1 The Kinds of Steel

Carbon steel is an alloy of two principal components: iron (Fe) and carbon (C) or, more accurately, *ferrite* (α-iron) and *cementite* (iron carbide, Fe_3C). Ferrite is a soft, ductile, malleable material; cementite is hard and brittle.

Carbon steels are usually thought of as belonging in one of three groups:

(1) *low-carbon* steels, containing up to about 0.25 per cent (by mass) carbon,
(2) *medium-carbon* steels, with carbon contents roughly between 0.30 and 0.50 per cent, and
(3) *high-carbon* steels, with 0.55 to 0.95 per cent carbon.

Most steels used in the building industry belong in the low- and medium-carbon groups.

The properties of steels in both the low- and medium-carbon groups vary over a wide range, depending on their composition, the heat treatment they received, and the cold work imparted to them. In order to understand their characteristics, it is advisable to follow up their temperature histories, starting with solidification from melt.[1]

---

1 The terms *killed, semikilled* refer to the practice of reducing the oxygen content (absorbed during the metallurgical process) of the molten metal, so that no reaction can occur between carbon and oxygen during solidification. In general, silicon and aluminum are used as deoxidizing agents. As the names imply, the removal of oxygen is not complete in the case of semikilled metals. The main purpose of deoxidation is to ensure a greater degree of uniformity of the chemical composition of steel. *Rimmed* and *capped* steels are not deoxidized; they can be recognized from the lower amount of silicon present, typically less than 0.05 per cent. Structural steels are usually semikilled.

The freezing of liquid iron containing 0.1 to 0.5 per cent C begins somewhere near 1520 °C, and is completed as the temperature drops to about 1460 °C. The solid present in the 900−1460 °C interval is called *austenite* (γ-iron). It is slightly harder and less ductile than ferrite. Much of the hot work on steel (rolling, forging, shaping) is done while the material is in its austenitic phase.

As the temperature further drops to below the 780−900 °C level (depending on the carbon content), ferrite begins to precipitate at the boundaries of the austenite grains. Ferrite grains form, and the carbon concentration in the remaining austenite increases, tending toward its *eutectoid* concentration, 0.8 per cent. About 727 °C the separation of the ferrite is completed and, with some further drop of temperature, the eutectoid austenite decomposes into 88 per cent (by mass) ferrite and 12 per cent cementite. These two phases, however, coexist within the same grain boundaries in the form of alternate layers of ferrite and cementite. This eutectoid structure, which has hardness and ductility values between those of the ferrite and cementite, is called *pearlite*. Thus, a slowly cooled low- or medium-carbon steel consists of a continuous matrix of ferrite grains and islands of pearlite grains.

High-carbon steels are used in the manufacture of steel tendons for prestressed concrete. The carbon content of these tendons is usually in the vicinity of the eutectoid concentration, 0.8 per cent. A eutectoid melt solidifies between 1480 and 1390 °C, and remains in the austenitic phase until its temperature drops to 727 °C. At that temperature it decomposes into pearlite.

If the cooling of steel from 727 °C to room temperature is very fast, i.e. the material is *quenched*, the eutectoid austenite cannot decompose in an orderly fashion. A new phase is formed, called *martensite*, which is essentially ferrite supersaturated with carbon atoms. This solid solution is unstable at any temperature, and tends to decompose into ferrite and cementite. However, at room temperature the rate of its decomposition is virtually zero. Martensite is a hard material with very little ductility. To improve its ductility (at the expense of its hardness), it has to be *tempered* at a temperature somewhere between 150 and 700 °C, depending on the desired hardness and ductility.

With low-carbon steels, very little eutectoid austenite is left behind after the precipitation of ferrite. Consequently, these steels cannot be hardened by quenching. Steels of good hardenability usually contain 0.35−0.55 per cent carbon, a fair amount (up to 1.75 per cent) of manganese, and possibly small amounts of other alloying elements.

The mechanical properties of steel can be improved by cold work: rolling, drawing, or machining. Cold work also changes the shape of the stress−strain curves. As Fig. 7.1 shows, with an increasing amount of cold work the yielding plateau (characteristic of hot-worked unhardened steels and a few nonferrous alloys) gradually disappears, and the yield and ultimate strengths (especially the yield strength) increase. These changes are associated partly with strain hardening, and partly with strain aging (to be discussed).

Low-carbon reinforcing wires and high-carbon or low-alloy prestressing wires,

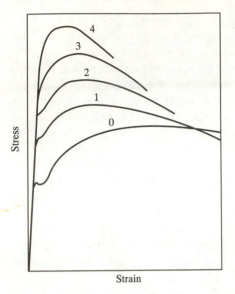

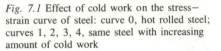

Fig. 7.1 Effect of cold work on the stress–strain curve of steel: curve 0, hot rolled steel; curves 1, 2, 3, 4, same steel with increasing amount of cold work

strands, and bars are usually manufactured by *cold drawing*, i.e. drawing steel rods or bars at room temperature through dies, often in two or more steps.[2] In each step, the diameter of the rod or bar is reduced by up to 12 per cent. Drawing produces an elongated grain structure, which markedly improves the strength of the material but lowers its ductility. In order to restore its ductility between successive steps of cold work, *process annealing* is sometimes employed.

The properties of carbon steels are listed for *as-rolled, normalized*, or *annealed* conditions. Normalization involves 'austenitizing' the material by raising its temperature into the 830–925 °C range. The purpose of normalization is to improve its machinability, to refine and homogenize its grain structure, and to eliminate some residual stresses. Annealing consists of holding the material at a suitable temperature between 790 and 900 °C, and cooling it at an appropriate rate. It is done primarily to soften the material and to promote dimensional stability.

Stress-relief heat treatment is used to eliminate the stresses locked into some steel forms during the manufacturing sequences. Cold-drawn prestressing wires and strands are usually stress-relieved at a temperature level somewhere between 600 and 675 °C. The length of treatment depends on the temperature.

Steel also contains a small amount of alloying elements other than carbon, such as manganese (Mn), silicon (Si), aluminum (Al), copper (Cu), titanium (Ti), sulfur (S), phosphorus (P), and possibly some others. They are, as a rule, not intentionally added and do not influence the properties of steel as significantly as does carbon. A steel is regarded as nonalloyed carbon steel if the secondary alloying elements do not exceed the following percentage limits:

2   Steels with carbon contents higher than 0.55 per cent are annealed before cold drawing.

Mn	1.65	Cu	0.4
Si	0.5	S	0.06
Al or Ti	0.1	P	0.09

In North America, any steel with less than 1.65 Mn, 0.6 Si, and 0.6 Cu is regarded as carbon steel (Boyer and Gall 1985).

Steels with any noncarbon alloying element up to a limit of 5 per cent are referred to as *low-alloy* steels. Manganese, silicon, copper, nickel, chromium, molybdenum, vanadium, niobium, titanium, and zirconium are commonly used as alloying elements. Concrete reinforcing bars and prestressing wires and strands are often made from low-alloy steels.

Designation is the term used for classifying the various types of steel on the basis of chemical composition. The most widely used designation system is that developed by the American Iron and Steel Institute (AISI) and the Society of Automotive Engineers (SAE) (Boyer and Gall 1985). In the AISI—SAE system, a type of steel is usually identified by a four-digit number. The first two digits denote the major alloying element(s). In the case of ordinary carbon steels, the designation number begins with 10. The last two digits (XX) indicate the approximate carbon content in hundredths per cent. (For example, the designation number 1020 stands for a low-carbon steel with carbon from 0.18 to 0.23 per cent, and manganese from 0.3 to 0.6 per cent.) In the British system (British Standards Institution 1983), the designation number for carbon steels shows the amounts of both the manganese and carbon. (For example, BS 080M36 stands for a medium-carbon steel with manganese from 0.60 to 1.00 per cent, and carbon from 0.32 to 0.40 per cent.) The 'material numbers' developed in Germany (Deutsches Institut für Normung 1961) consist of seven digits and are capable of giving a fairly complete description of any steel product.[3]

Standard specifications are documents that describe the requirements certain products must meet. They may be umbrella-type specifications covering a variety of steel products, or they may be concerned with specific products. Table 7.1 lists a number of national and international standards related to the building industry.

Good weldability is an important requirement in many applications of steel. It is usually taken for granted that steels with carbon contents up to 0.22 per cent present no weldability problems. A document developed by the American Welding Society (American Welding Society 1979) gives some guidelines for the weldability of various concrete reinforcing bars in terms of the 'carbon equivalent' (*CE*) defined as

---

3   For steel, the first digit is 1. The second and third digits indicate the general type of steel. For example, 00 stands for commercial and basic-quality carbon steels. The fourth and fifth digits refer to a kind of steel specified either in DIN 17100 (Deutsches Institut für Normung 1980; Böge 1983) or by the Committee for Iron and Steel. The sixth and seventh digits denote the steelmaking process (e.g. open-hearth steel, semikilled) and the treatment of the steel product (e.g. normalized), respectively.

Table 7.1 Selected list of standards for steel products used in the construction industry

Steel product	Standard					
	Canada	France	Germany	UK	USA	International
Structural steel	CSA G40.20 CSA G40.21	NF A 35-501 NF A 03-115	DIN 17100 DIN 1025	BS 4/1 BS 4360	ASTM A 6 ASTM A 36	ISO 630 ISO 657 EU 25 EU 113
Reinforcing steel	CSA G30.3 CSA G30.12 CSA G30.14 CSA G30.16	NF A 35-015 NF A 35-016 NF A 35-017 NF A 35-019	DIN 488/1 DIN 488/2 DIN 488/6	BS 4449 BS 4461 BS 4482	ASTM A 615[a] ASTM A 616[a] ASTM A 706 ASTM A 496	ISO 1035/1 EU 80
Reinforcing steel fabric	CSA G30.5 CSA G30.15	NF A 35-022	DIN 488/4	BS 4483	ASTM A 185 ASTM A 497	
Prestressing steel	CSA G279	NF A 35-054	DIN 4227/6	BS 4486 BS 4757 BS 5896	ASTM A 421 ASTM A 722 ASTM A 416	EU 138

[a] Billet-steel bars (A 615) are the all-purpose reinforcing bars in the USA.
Rail-steel bars (A 616) are usually specified if no bending is required.

$$CE = \%\,C + \frac{\%\ Mn}{6} + \frac{\%\ Cu}{40} + \frac{\%\ Ni}{20} + \frac{\%\ Cr}{10} - \frac{\%\ Mo}{50} - \frac{\%\ V}{10} \qquad [7.1]$$

If the composition of the steel is not known, $CE = 0.75$ should be assumed. According to the document, little or no preheat is required for the welding of reinforcing bars if $CE \leq 0.55$.

7.2   Mechanical and Thermal Properties

Being a Group L (load-bearing) material, steel is of interest to the fire safety practitioner mainly with regard to its mechanical properties. These properties depend primarily on the chemical makeup of steel, especially on its carbon content and, to a lesser extent, on the manufacturing process and heat treatment.

Figure 7.2 illustrates how the ultimate strength, σ_u (Pa), yield strength, σ_y (Pa), elongation at rupture (for test specimens of 50 mm gage length), λ_r (m m^{-1}), and Brinell hardness, HB (kgf mm^{-2}), vary with carbon content for 10XX steels. The curves represent a 'smoothed' assembly of data published in the *Metals Handbook* (Boyer and Gall 1985) for as-rolled, normalized, and annealed steels. The utility of the information presented in the figure is difficult to assess since, in addition to carbon content, a variety of factors are known to affect the mechanical properties of steel.

Apparently, with respect to the principal mechanical properties, there are no major differences between as-rolled and normalized steel products. On the other hand, annealing markedly reduces the strength and increases the ductility of carbon steels. It has a similar effect on the properties of low-alloy steels.

In the case of quenched-and-tempered carbon and low-alloy steels, the strength and ductility of the material can be adjusted within a wide range by the selection of the tempering temperature between 205 and 650 °C. Typical values for the strength and ductility of quenched-and-tempered steel are listed in the *Metals Handbook* (Boyer and Gall 1985).

The effect of elevated temperatures on the mechanical properties of steel is usually determined from three types of tests:

(a)   short-term tensile tests (or hardness tests) following the elevated-temperature exposure,
(b)   short-term elevated-temperature tests, and
(c)   long-term elevated-temperature tests.

Clearly, the results of short-term tests performed after elevated-temperature exposure (i.e. residual property tests) are of interest to fire safety practitioners mainly for assessing the reusability of the steel components of buildings following fires. Studies conducted by Abrams and Erlin (1967), Day *et al.* (1961), Smith *et al.* (1981), and Holmes *et al.* (1982) on the lasting effects of elevated temperatures on the mechanical properties of carbon and low-alloy steels can be briefly summarized as follows.

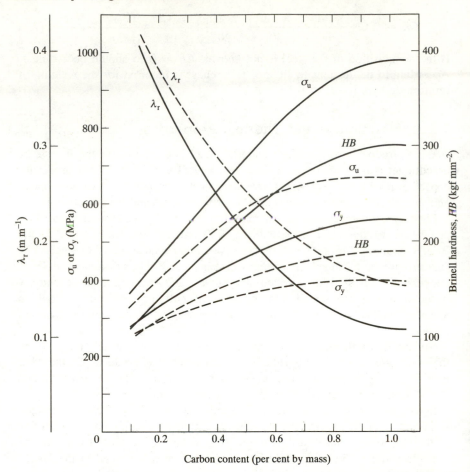

Fig. 7.2 Effect of carbon content on the mechanical properties of steel (———— as rolled or normalized; — — — — annealed): σ_u, ultimate strength; σ_y, yield strength; λ_r, elongation at rupture; *HB*, Brinell hardness

Exposure of hot-rolled, normalized, and annealed steels to temperatures up to 600 °C for a limited length of time has little effect on the residual strength and microstructure of the material.[4] If the temperature rises above 600 °C but remains below 727 °C, a microstructural change known as *spheroidization* will take place, in which the cementite lamellae of pearlite coalesce into globular particles. Spheroidization will permanently soften the material. Above 727 °C, the ferrite–pearlite structure transforms into a coarse-grained ferrite–austenite structure and, at somewhat higher temperatures, into coarse-grained austenite.

---

4   However, if the material is strained in the 200–350 °C interval, strain aging (to be discussed) may cause a permanent increase in the strength and hardness of the material.

On cooling, a material with coarse-grained ferrite—pearlite structure will appear, with a strength lower than that of the original steel. Since the coarsening of grains is independent of the original grain size, the loss of strength due to exposure to temperatures above 727 °C is greater for fine-grained steels than for coarse-grained steels.

Superimposed on these changes are those resulting from a tendency of the material to assume a state corresponding to minimum free energy. A steel composed of distorted, elongated grains brought about by cold work is in an unstable state. As the temperature rises above 450 °C, the grains tend to reassume their equiaxed shapes, and thereby the material loses the excess strength imparted to it by cold work. On cooling, the strength of the material will become considerably lower, usually comparable with that of hot-rolled steels (Abrams and Erlin 1967).

Substantial permanent loss of strength following fire exposure can also be expected with quenched-and-tempered steels. As mentioned earlier, martensite is an unstable phase of iron; on heating it tends to decompose into ferrite and cementite. The total or partial decomposition of martensite at elevated temperatures will result in a material with lower strength and higher ductility.

Information on the residual yield strength of steel may be needed for designing the repair of fire-damaged building elements (Concrete Society Working Party 1978). The simplified curves shown in Fig. 7.3 for hot-rolled and cold-worked steel (Tovey 1986) are believed to be sufficiently accurate for design purposes. In the figure, σ_y stands for the yield strength after the material has been heated to a temperature T, and $(\sigma_y)_0$ for the yield strength before heating. The modulus of elasticity (E, Pa) and Poisson's ratio (μ, dimensionless) are structure-insensitive properties. For low- and medium-carbon steels and low-alloy steels, Young's modulus is about 210×10^3 MPa at room temperature. It deceases with the rise of temperature, as shown in Fig. 7.4. (In the figure, E_0 (Pa) is the modulus of elasticity at reference (room) temperature.) Poisson's ratio is about 0.29 at room temperature, 0.30 at 500 °C, and 0.34 at 750 °C (Boyer and Gall 1985).

Displayed in Figs 7.5 and 7.6 are the stress—strain (σ, Pa, versus ϵ, m m^{-1})

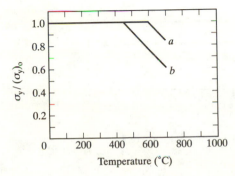

Fig. 7.3 Residual yield strength of steels after being heated to the given temperature: curve a, typical hot-rolled steel; curve b, typical cold-worked steel (Tovey 1986)

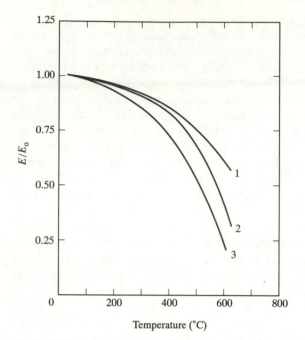

Fig. 7.4 Effect of temperature on the modulus of elasticity of steel. Curves: 1, structural steel (Harmathy 1967d); 2, prestressing steel (Richter and Sager 1980); 3, reinforcing steel (Anderberg 1978)

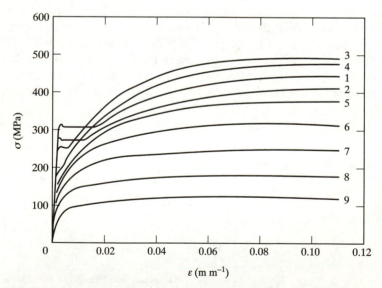

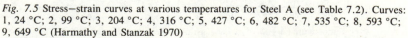

Fig. 7.5 Stress–strain curves at various temperatures for Steel A (see Table 7.2). Curves: 1, 24 °C; 2, 99 °C; 3, 204 °C; 4, 316 °C; 5, 427 °C; 6, 482 °C; 7, 535 °C; 8, 593 °C; 9, 649 °C (Harmathy and Stanzak 1970)

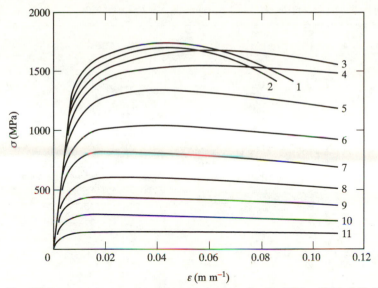

σ (MPa)

ε (m m⁻¹)

Fig. 7.6 Stress—strain curves at various temperatures for Steel B (see Table 7.2). Curves:
1, 21 °C; 2, 93 °C; 3, 204 °C; 4, 257 °C; 5, 310 °C; 6, 377 °C; 7, 432 °C; 8, 488 °C;
9, 538 °C; 10, 593 °C; 11, 649 °C (Harmathy and Stanzak 1970)

curves at room temperature and at elevated temperatures for two kinds of steel:
an ASTM A 36 structural steel, to be referred to as Steel A, and an ASTM A
421 wire for prestressed concrete, to be referred to as Steel B. Some information
concerning the two materials is given in Table 7.2.

Two features of the family of curves for the structural steel are worthy of notice:

(1) The yielding plateau becomes less and less noticeable with the rise of
    temperature, and disappears[5] about 300 °C.
(2) After a slight decline at moderately elevated temperatures, the ultimate
    strength of the material increases substantially in the 180—370 °C interval,[6]
    whereas its yield strength declines steadily as the temperature rises.

The increase of the strength of as-rolled steels in this temperature interval, known
as blue brittleness, is a result of *strain aging*. It is caused by the immobilization
of dislocations in the ferrite phase by the precipitation of carbon and nitrogen
atoms. Since the stage for this phenomenon is set by the plastic deformation
occurring during the test itself, strain aging has little effect on the yield strength,
which is measured at very small (0.2 per cent) plastic deformation. Susceptibility
for strain aging is lowest among normalized aluminum-killed steels, which have
a relatively low concentration of free nitrogen (Smith G V 1972).

---

5   This phenomenon is associated mainly with the partial coalescence of dislocations on account
    of their greater mobilities.
6   The peak shifts to higher temperatures with increasing strain rates.

Table 7.2 Information concerning Steels A and B (Harmathy and Stanzak 1970)

	Steel A ASTM A 36	Steel B ASTM A 421
Chemical analysis:		
C	0.19	0.794
Mn	0.71	0.780
Si	0.09	0.187
P	0.007	0.012
S	0.03	0.031
Manufacture:	Made by the basic open-hearth process, Si semikilled, hot-rolled into 12.7 mm plate	Made by the basic open-hearth process, Si–Al killed, cold-drawn in four steps into 7.0 mm wire, stress-relieved

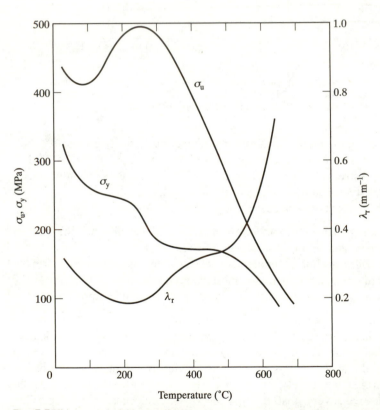

Fig. 7.7 Ultimate and yield strengths (σ_u and σ_y) and elongation at rupture (λ_r) for Steel A (see Table 7.2), as functions of temperature (Harmathy and Stanzak 1970)

Since cold-worked steels have undergone strain aging during the manufacturing process, they exhibit little or no increase in their ultimate strengths in tensile tests performed in the 180–370 °C temperature interval (Fig. 7.6).

The effect of temperature on the ultimate and yield strengths (σ_u and σ_y) and

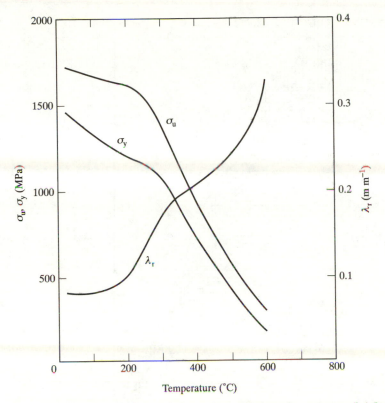

Fig. 7.8 Ultimate and yield strengths (σ_u and σ_y) and elongation at rupture (λ_r) for Steel B (see Table 7.2), as functions of temperature (Harmathy and Stanzak 1970)

elongation at rupture (λ_r) is displayed in Figs 7.7 and 7.8 for the two steels described in Table 7.2. As mentioned earlier, the grain structure of cold-worked steels tends to assume on heating an equiaxed arrangement, and thereby the material loses its excess strength associated with the elongated grains. Consequently, the loss of ultimate and yield strengths with the rise of temperature is more pronounced for cold-worked steels than for hot-rolled steels. This fact can be clearly discerned by comparing the information presented in Figs 7.9 and 7.10. In these figures, the decline of yield strength and ultimate strength with the rise of temperature is shown for hot-rolled (Fig. 7.9) and cold-worked (Fig. 7.10) steels, based on a multitude of data available from the literature (e.g. Simmons and Cross 1955; Brockenbrough and Johnston 1968; Bobrowski 1978; Harmathy and Stanzak 1970; Bennetts 1981; Holmes *et al.* 1982; Anderberg 1983). In the figures, σ_y and σ_u stand for yield and ultimate strengths at a temperature T, and $(\sigma_y)_o$ and $(\sigma_u)_o$ for these strengths at reference (room) temperature. Because of the substantial scatter of the data,[7] individual

---

7   The scatter is associated mainly with differences in the composition of the material and with the manufacturing procedure.

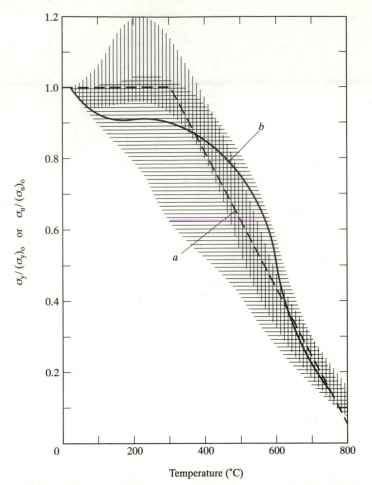

Fig. 7.9 Ultimate and yield strengths of hot-rolled steel. Vertically hatched area, ultimate strength; horizontally hatched area, yield strength; curve *a*, curve recommended by The Institution of Structural Engineers (Bobrowski 1978) for both yield strength and ultimate strength; curve *b*, Brockenbrough—Johnston (1968) curve for yield strength

experimental points are not shown. The horizontally hatched areas encompass yield strength values, and the vertically hatched areas ultimate strength values.

Shown separately as curve *b* in Fig. 7.9 is the Brockenbrough—Johnston curve, which has been adopted in several fire safety design guides (to be referenced in Chapter 14) to represent the yield strength of reinforcing steels. Curves *a* in Figs 7.9 and 7.10 are those recommended by the Institution of Structural Engineers (Bobrowski 1978) for both yield strength and ultimate strength. They can be described by the following equations:

- For low-carbon reinforcement bars, high-yield reinforcement bars, and high-strength alloy steel bars:

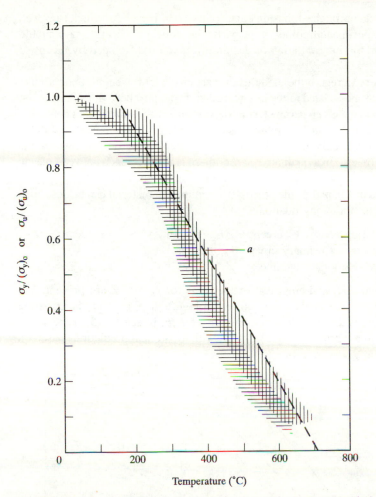

Fig. 7.10 Ultimate and yield strengths of cold-worked steels. Vertically hatched area, ultimate strength; horizontally hatched area, yield strength; curve *a*, curve recommended by The Institution of Structural Engineers (Bobrowski 1978) for both yield strength and ultimate strength

$$\frac{\sigma_y}{(\sigma_y)_o} = \frac{\sigma_u}{(\sigma_u)_o} = \begin{cases} 1 & \text{if } 20 < T \le 300\,^{\circ}\text{C} \\ 1.57 - 0.0019T & \text{if } 300 < T < 826\,^{\circ}\text{C} \end{cases} \quad [7.2]$$

- For prestressing wires and strands

$$\frac{\sigma_y}{(\sigma_y)_o} = \frac{\sigma_u}{(\sigma_u)_o} = \begin{cases} 1 & \text{if } 20 < T \le 150\,^{\circ}\text{C} \\ 1.27 - 0.0018T & \text{if } 150 < T < 706\,^{\circ}\text{C} \end{cases} \quad [7.3]$$

There seems to be a rather disturbing degree of scatter in the $\sigma_y/(\sigma_y)_o$ versus temperature relations for hot-rolled low-carbon steels in the 0.5–0.9 range of $\sigma_y/(\sigma_y)_o$, which is of primary interest to the fire safety practitioner. The wide use of the Brockenbrough–Johnston curve (curve b in Fig. 7.9), which runs well above most experimental points, is of particular concern. Clearly, care should be exercised in the interpretation of design information developed by means of that curve.

It is customary to regard the modulus of elasticity, yield strength, and ultimate strength as quantities related to the instantaneous rheological behavior of materials, and to look upon the creep parameters as descriptors of time-dependent behavior. However, since the plastic deformation is never truly instantaneous, the modulus of elasticity is the only material property not dependent on the rate of the deformation process. In a qualitative sense, there is no difference between plastic deformation developing in short-term tests and in creep tests.

As discussed in Chapter 5, the creep deformation of a material can be described by means of the following quantities:

Activation energy of creep ΔH_c (J kmol^{-1})
Parameter of primary creep ϵ_{to} (m m^{-1})
Zener–Hollomon parameter Z $(\text{m m}^{-1} \text{h}^{-1})$

The values of ΔH_c and empirical expressions for ϵ_{to} and Z are presented in Table 7.3 for the two steels described in Table 7.2. In Fig. 7.11, creep strain (ϵ_t, m m^{-1}) versus temperature-compensated time (θ, h; see Eq. 5.21) relations developed from experiments are compared with theoretical predictions based on the information given in Table 7.3.

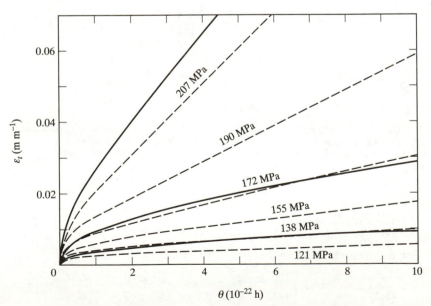

$\theta \, (10^{-22} \, \text{h})$

Fig. 7.11 Creep strain versus temperature-compensated time (ϵ_t versus θ) for Steel A (see Table 7.2). Curves: ——— experimental; — — — theoretical, based on information in Table 7.3 (Harmathy 1967d)

Table 7.3 Creep characteristics of Steels A and B (Harmathy and Stanzak 1970)

Steel	$\Delta H_c/R$ (K)	$\epsilon_{to}(\sigma)$ (m m^{-1})	$Z(\sigma)$ (m m^{-1} h^{-1})
A (ASTM A 36)	38 890	$\epsilon_{to} = 3.258 \times 10^{-17}\sigma^{1.75}$	$Z = 2.365 \times 10^{-20}\sigma^{4.7}$ if $\sigma \leq 103.4 \times 10^6$ Pa $Z = 1.23 \times 10^{16}\exp(4.35 \times 10^{-8}\sigma)$ if $103.4 \times 10^6 < \sigma \leq 310 \times 10^6$ Pa
B (ASTM A 421)	30 560	$\epsilon_{to} = 8.845 \times 10^{-9}\sigma^{0.67}$	$Z = 1.952 \times 10^{-10}\sigma^3$ if $\sigma \leq 172.4 \times 10^6$ Pa $Z = 8.21 \times 10^{13}\exp(1.45 \times 10^{-8}\sigma)$ if $172.4 \times 10^6 < \sigma \leq 690 \times 10^6$ Pa

Note σ in Pa.

ΔH_c is a structure-insensitive quantity. It is roughly equal to the activation energy of self-diffusion. For iron (ferrite), the latter is 241 to 256 MJ kmol^{-1} (Smithells and Brandes 1976). If no experimental information is available, ΔH_c may be chosen as 250 MJ kmol^{-1} for use in Eqs 5.21 and 5.24. It is advisable, however, to regard ΔH_c as an empirical constant to be evaluated from experiments.

The quantities ϵ_{to} and Z are extremely structure-sensitive. The scatter of their values is awesome even for supposedly identical steels. Microstructural changes that may develop at elevated temperatures (e.g. spheroidization) make it difficult to predict the creep characteristics of any material from its behaviour at room temperature. Manganese, silicon, molybdenum, vanadium, and niobium have been found to increase resistance to deformation at elevated temperatures.

Of the two creep parameters, the Zener–Hollomon parameter is the more important. The final, critical stage of creep can be assessed using Eq. 5.25 which, as discussed in Chapter 5, is applicable under variable temperature and slowly variable stress conditions. In that equation, Z is the only creep parameter.

The Zener–Hollomon parameter is a function of the applied stress, σ. The form of the function is usually

$$Z = A\sigma^B \qquad [7.4a]$$

or

$$Z = Ae^{B\sigma} \qquad [7.4b]$$

where A and B are empirical constants of various dimensions. The Z versus σ function for a steel can be determined by measuring the rate of secondary creep, $\dot{\epsilon}_{ts}$ (m m^{-1}h^{-1}), at various stresses and temperatures, and then plotting $\dot{\epsilon}_{ts} \exp (\Delta H_c/RT)$ (where R is the gas constant = 8315 J kmol^{-1} K^{-1}) against σ on log–log or semilog paper. Data sufficient for the evaluation of the Z versus σ relations for many carbon- and low-alloy steels are available in a number of publications (e.g. McKinney and Clark 1938; Simmons and Cross 1955; Mantell 1958).

The Z versus σ relation can also be estimated from the results of short-term tensile tests performed at elevated temperatures. As Figs 7.5 and 7.6 show, above

300 °C the stress—strain curves level off and run nearly parallel with the strain (ϵ) axis. Along these parallel sections, the speed of the testing machine's crosshead is approximately equal to rate of secondary creep at the test temperature and at the stress read along the vertical (σ) axis.

The fire safety practitioner need not be overly concerned about the accuracy (or rather, lack of accuracy) of information on the creep parameters. During a fire exposure, the rate of temperature rise in the key steel components of a building element is the principal factor controlling the element's behavior. According to Eq. 5.24, the creep rate increases exponentially with the temperature. Once the temperature of the key components has exceeded a certain limit (which depends on the acting stress), failure by excessive deformation will occur in a short time (see Fig. 14.13). Relatively large differences in the value of the Zener—Hollomon parameter will usually affect the time of the element's failure to the extent of no more than 5 to 15 minutes.

The density of steel (ρ) is usually taken as 7850 kg m^{-3}. The density of ferrite (pure α-iron) is somewhat higher (7871 kg m^{-3}), and that of cementite (Fe$_3$C) is somewhat lower (about 7400 kg m^{-1}).

The coefficient of thermal expansion (α) for an average carbon steel is 11.4×10^{-6} m^{-1}K^{-1} at room temperature. The thermal expansion of various carbon and low-alloy steels depends on the carbon content and heat treatment, but the differences rarely amount to more than 5 per cent.

Figure 7.12 shows the dilatometric curve for a steel containing 0.31 per cent carbon and 0.65 per cent manganese (Honda 1917). (In the figure, $\Delta l = l - l_o$, and l and l_o (m) are the changed and original lengths of the steel sample, respectively.) Apparently, the transformation of the ferrite—pearlite structure into austenite is accompanied by a substantial contraction.

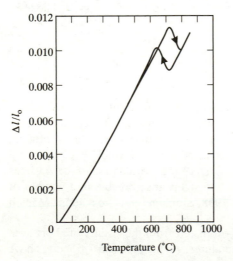

Fig. 7.12 Dilatometric curve for a medium-carbon steel (Honda 1917)

According to Kelley (1949), the specific heats (c_p, J kg^{-1} K^{-1}) for the various components of steel are as follows.

- Ferrite:

$$c_p = 252.5 + 0.532T + \frac{3.22 \times 10^6}{T^2} \qquad 298 \leq T \leq 1185 \text{ K} \quad [7.5]$$

- Austenite:

$$c_p = 363.5 + 0.225T \qquad 1000 \leq T \leq 1768 \text{ K} \qquad [7.6]$$

- Cementite:

$$c_p = 586.4 + 0.052T \qquad 298 \leq T \leq 1173 \text{ K} \qquad [7.7]$$

In Eqs 7.5, 7.6, and 7.7, T (K) is temperature.

The heat of the ferrite → austenite (α-iron → γ-iron) transformation is about 0.046 MJ kg^{-1}.

Assuming that steel is mixture of ferrite and cementite, the specific heat of steel may be assessed using the mixture rule on a mass fraction basis.[8] However, since steel also contains a number of noncarbon alloying elements, meticulousness is not really warranted, and the specific heat of steel may be chosen simply as 3−10 per cent higher than that of ferrite.

The thermal conductivity of steel, k (W m^{-1} K^{-1}), depends mainly on the amount of alloying elements and on the heat treatment. The thermal conductivity of pure iron and the ranges of variation for the thermal conductivities of the various steels are given in Fig. 7.13.

Further information on the properties of steel, assembled specifically for the benefit of fire safety practitioners, have been presented in review reports by Bennetts (1981) and Anderberg (1983).

Nomenclature

A, B	empirical constants, various dimensions
c_p	specific heat at constant pressure, J kg^{-1} K^{-1}
CE	carbon equivalent, dimensionless
E	modulus of elasticity (Young's modulus), Pa
HB	Brinell hardness, kgf mm^{-2}
ΔH_c	activation energy of creep, J kmol^{-1}
k	thermal conductivity, W m^{-1} K^{-1}
l	length of test sample, m

---

8   See, for example, Eq. 6.51. The carbon content of cementite, Fe$_3$C, is 6.69 per cent. Consequently, the mass fraction of cementite in a steel of, say, 0.3 per cent carbon content is 0.3/6.69 = 0.045, and the mass fraction of ferrite is 1 − 0.045 = 0.955. For such a steel, the specific heat is 1.6 per cent higher than that of ferrite.

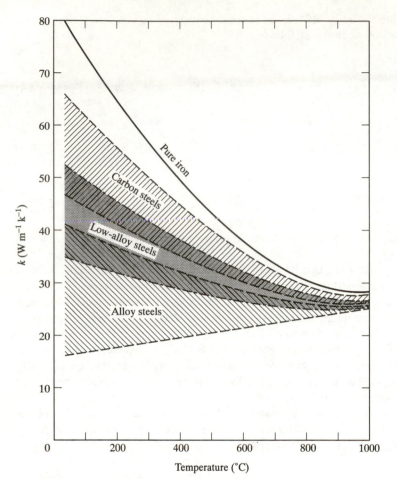

Fig. 7.13 Thermal conductivity of various steels at elevated temperatures

Δl	$= l - l_o$, m
R	gas constant $= 8315 \, \text{J kmol}^{-1} \, \text{K}^{-1}$
T	temperature, K, °C
Z	Zener–Hollomon parameter, m m^{-1} h^{-1}
α	coefficient of thermal expansion, m m^{-1} K^{-1}
ϵ	strain, m m^{-1}
ϵ_t	time-dependent (creep) strain, m m^{-1}
ϵ_{to}	parameter of primary creep, m m^{-1}
$\dot{\epsilon}_{ts}$	rate of secondary creep, m m^{-1} h^{-1}
θ	temperature-compensated time, h
λ_r	elongation at rupture, m m^{-1}
μ	Poisson's ratio, dimensionless
ρ	density, kg m^{-3}
σ	stress, strength, Pa

Subscripts

o	at reference (room) temperature; before heating
u	ultimate (strength)
y	yield (strength)

8   Wood, Brick, Gypsum, Plastics

The number of building materials has increased dramatically during the past few decades. Only a few, those most extensively used in recent years, can be dealt with in this book in some detail. The properties of concrete and steel have been discussed in Chapters 6 and 7. Four other groups of materials — wood, brick, gypsum, and plastics — will be dealt with in this chapter. Wood and plastics will be looked at again in the next chapter, with a view to their roles as fuels that affect the nature of building fires.

8.1   Wood

Although still favored in residential construction, wood structures have lost ground in all other fields to the two contemporary materials: concrete and steel.

Wood is a building material in the L/I/F (load-bearing/insulating/fuel) group. However, its use as a load-bearing material must not be taken for granted. Although more than 180 species are grown in the United States, only about 25 species groups have been assigned working stresses. The two groups most extensively used as structural lumber are the Douglas firs and the southern pines.

The oven-dry density (ρ) of the commercially important woods ranges from 300 kg m^{-3} (white cedar) to 700 kg m^{-3} (hickory, black locust). The density of the Douglas firs varies between 430 and 480 kg m^{-3}, and that of the southern pines between 510 and 580 kg m^{-3}. The true density of the solid material that forms the walls of wood cells (ρ_t) is about 1500 kg m^{-3}, regardless of species. Consequently, the porosity (P) of the various species lies between 0.5 and 0.8 m^3 m^{-3}.

The specific surface of wood, owing to its complex pore structure, is very high. Therefore, wood can hold more moisture than any other building material, except cement paste. The moisture content of seasoned wood in equilibrium with an atmosphere of 65 per cent relative humidity at room temperature is 0.12 kg kg^{-1}, or 0.04–0.08 m^3 m^{-3}, depending on species. In fact, 12 per cent moisture content is looked upon as characteristic of the 'standard' air-dry condition of wood.

On account of its cellular makeup, wood is a markedly anisotropic material.

Most of its properties depend on whether they are measured along or across the grain.

The modulus of elasticity, E (Pa), of air-dry, clear wood along the grain varies from 5.5×10^3 to 15.0×10^3 MPa, and its ultimate (crushing) strength, σ_u (Pa), from 13 to 70 MPa. According to the US Forest Products Laboratory (Forest Products Laboratory 1974), these properties are related and roughly proportional to the oven-dry density of wood, ρ (kg m^{-3}), regardless of species:

$$E = 23.3 \times 10^6 \rho \qquad\qquad [8.1]$$

$$\sigma_u = 0.084 \times 10^6 \rho \qquad\qquad [8.2]$$

The ratio of modulus of elasticity across the grain to that parallel to the grain is commonly 1:20, and the ratio of compressive strengths in these two directions may be as low as 1:7. The effect of temperature on the modulus of elasticity and compressive strength is shown in Fig. 8.1 (Schaffer 1984). [In the figure, E_o (Pa) and $(\sigma_u)_o$ (Pa) are modulus of elasticity and ultimate strength, respectively, at reference (room) temperature.]

The allowable working stresses in compression (along the grain) depend on the species and commercial grade of wood, and usually vary between 7 and 15 MPa.

The coefficient of linear thermal expansion (α) ranges from 3.2×10^{-6} to 4.6×10^{-6} m m^{-1} K^{-1} along the grain, and from 21.6×10^{-6} to 39.4×10^{-6} m m^{-1} K^{-1} across the grain (Wangaard 1958). The dilatometric curve (parallel with the grain) of an unidentified species of pine of 400 kg m^{-3} oven-dry density is shown in Fig. 8.2(a). The material contracts rapidly above 300°C where, as revealed by its thermogravimetric curve [Fig. 8.2(b)], its thermal decomposition greatly accelerates. [In the figures, $\Delta l = l - l_o$, where l and l_o (m) are the changed and original lengths, respectively, of the dilatometric specimen, and W and W_o (kg) are the changed and original masses, respectively, of the thermogravimetric specimen.]

The thermal conductivity (k) of wood across the grain usually lies between 0.1

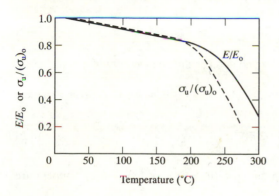

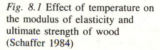

Fig. 8.1 Effect of temperature on the modulus of elasticity and ultimate strength of wood (Schaffer 1984)

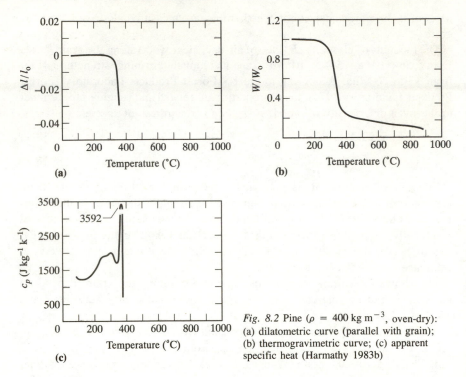

Fig. 8.2 Pine ($\rho = 400$ kg m^{-3}, oven-dry): (a) dilatometric curve (parallel with grain); (b) thermogravimetric curve; (c) apparent specific heat (Harmathy 1983b)

and 0.15 W m^{-1} K^{-1}. It is claimed to be a function of density and moisture content (Forest Products Laboratory 1974).

$$k = 0.0238 + 0.144 \times 10^{-3}\rho(1.39 + 2.80m) \qquad [8.3]$$

where m (kg kg^{-1}) is moisture content, related to oven-dry mass.

An expression for the thermal conductivity of oven-dry wood can also be derived by combining Eqs 5.31 (relating conductivity with porosity) and 4.1 (relating porosity with density; $\rho_t = 1500$ kg m^{-3}), and determining the constants in the resulting equation from an analysis of experimental data. The end result is

$$k = 0.2 \; \frac{\rho}{540 + 0.64\rho} \qquad [8.4]$$

where 0.2 (W m^{-1} K^{-1}) may be looked upon as the conductivity of a hypothetical 'nonporous' wood.

The specific heat, c_p (J kg^{-1} K^{-1}), of oven-dry wood is (Forest Products Laboratory 1974):

$$c_p = 1126 + 4.5(T - 273) \qquad [8.5]$$

where T (K) is temperature.

Figure 8.2(c) shows the apparent specific heat as a function of temperature for the pine already mentioned. The accuracy of the curve, which was developed by DSC, is somewhat questionable. However, it provides useful information on the nature of the decomposition reactions that take place between 150 and 370 °C.

8.2   Brick

Building brick belongs in the L/I (load-bearing/insulating) group of materials. It is rather unique among building materials in that it does not undergo substantial physicochemical changes on heating.

Brick is manufactured from clay minerals [kaolinite, $Al_2(Si_2O_5)(OH)_4$; halloysite, $Al_2(Si_2O_5)(OH)_4 \cdot 2H_2O$; pyrophyllite, $Al_2(Si_2O_5)_2(OH)_2$; montmorillonite, $\left(Al_{1.67} \begin{array}{c} Na_{0.33} \\ Mg_{0.33} \end{array}\right)(Si_2O_5)_2(OH)_2$; etc.] won from surface deposits or mined from sedimentary shale formations. The finely ground clay is mixed with water, the amount of which depends on the method of molding and may vary between 6 and 30 per cent (by mass). After its formation, the ware is first dried to a moisture content of about 3 per cent, then placed in a kiln and fired at approximately 1100 °C.

The density (ρ) of brick ranges from 1660 to 2270 kg m^{-3}, depending on the raw materials used and on the molding and firing technique. The true density of the material is somewhere between 2600 and 2800 kg m^{-3}. Applying Eq. 4.1, values from 19 to 36 per cent are obtained for the porosity of brick.

The modulus of elasticity (E) of brick is usually between 10×10^3 and 20×10^3 MPa. Its compressive strength varies in a very wide range, from 9 to 110 MPa. 50 MPa may be regarded as the average (McBurney and Lovewell 1933). This value is an order of magnitude higher than the stresses usually allowed in the design of grouted brickwork.

At room temperature, the coefficient of thermal expansion (α) of brick is about 5.5×10^{-6} m m^{-1} K^{-1}. Fig. 8.3(a) shows the dilatometric curve of a medium-quality brick of 1935 kg m^{-3} density (Harmathy 1983b). The calculated porosity of that brick is about 0.27 (see Eq. 4.1). Displayed in Figs 8.3(b) and 8.3(c) are the thermal conductivity[1] and specific heat of the same brick. The information presented in Fig. 8.3(b) may be extended to other bricks, using the relation between thermal conductivity and porosity (Eqs 5.31 and 5.32).

The specific heat of the material can also be assessed on theoretical ground, assuming that the brick was manufactured from kaolinite.[2] On firing, kaolinite will turn into a mixture of anhydrous compounds (Kingery et al. 1976), probably (based on the chemical formula of the kaolinite) into 63.9 per cent (by mass) mullite ($3Al_2O_3 \cdot 2SiO_2$) and 36.1 per cent (vitreous) silica (SiO_2). Using

---

1   The thermal conductivity values reported in this chapter were measured by means of a variable state method (Harmathy 1964).

2   As mentioned earlier, the specific heat is not sensitive to minor changes in the composition of the material.

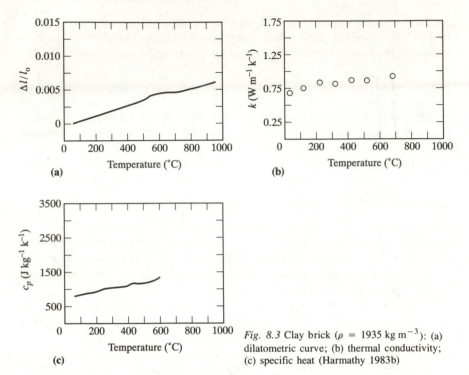

Fig. 8.3 Clay brick ($\rho = 1935 \text{ kg m}^{-3}$): (a) dilatometric curve; (b) thermal conductivity; (c) specific heat (Harmathy 1983b)

handbook information on the heat capacities of these two compounds, and Harmathy's method as discussed in Chapter 5 and in connection with Eq. 6.26, the following expression can be developed for the specific heat (c_p, $\text{J kg}^{-1} \text{K}^{-1}$) of the mixture (supposedly brick):

$$c_p = 710.87 + 0.512T - \frac{8.676 \times 10^6}{T^2} \qquad [8.6]$$

At room temperature (298 K), this equation yields $c_p = 765.8 \text{ J kg}^{-1} \text{K}^{-1}$. This value compares favorably with experimental information.

8.3   Gypsum

Gypsum products, from the viewpoint of fire safety, belong in the I (insulating) group. The raw material used in the manufacture of gypsum products is either calcium sulfate hemihydrate ($CaSO_4 \cdot \frac{1}{2}H_2O$) or calcium sulfate anhydrite ($CaSO_4$).[3] They are produced from gypsum rock ($CaSO_4 \cdot 2H_2O$), the former by heating the (ground and pulverized) rock at temperatures not exceeding 180 °C, and the latter at temperatures above 180 °C. The calcination residues, when finely

---

3   They may contain a small amount of minerals other than $CaSO_4 \cdot \frac{1}{2}H_2O$ or $CaSO_4$, such as silica, alumina, limestone, iron oxide.

powdered and (possibly) treated with some additives, are called plaster of Paris ($CaSO_4 \cdot \frac{1}{2}H_2O$) and Keene's cement ($CaSO_4$), respectively. Mixed with water, these materials revert to calcium sulfate dihydrate ($CaSO_4 \cdot 2H_2O$). The formation of an interlocking mass of gypsum crystals is responsible for the hardening of the mixture.

Gypsum products are extensively employed in building construction. Plaster of Paris, with the addition of aggregates (such as sand, perlite, vermiculite, wood fiber), is used in wall plaster as base coat, and Keene's cement (neat or mixed with lime putty) as finishing coat.

Although the compressive strength (σ_u) of the dihydrate may be as high as 13 MPa, the strength of gypsum is rarely relied on in the building design. Nowadays gypsum is used predominantly in the form of boards, including wallboard, lath, formboard and sheathing. The core of the boards is fabricated with plaster of Paris, to which weight- and set-controlling additives are added. It is faced on both sides and along the edges with paper.

The core is highly porous,[4] but its specific surface is, on account of the highly crystalline microstructure of the material, relatively low. Consequently, the moisture that the material can hold (adsorbed to the pores or held in the pore cavities) is not too significant under normal atmospheric conditions.

The coefficient of thermal expansion (α) of gypsum products may vary between 11.0×10^{-6} and $17.0 \times 10^{-6} \, \text{m m}^{-1} \text{K}^{-1}$ at room temperature, depending mainly on the nature and amount of aggregate used. The thermal conductivity of gypsum products is difficult to assess, owing to large variations in their porosities and in the nature of aggregates; $0.25 \, \text{W m}^{-1} \text{K}^{-1}$ may be regarded as a typical value for plaster boards of about $700 \, \text{kg m}^{-3}$ density.

The specific heat (c_p, $\text{J kg}^{-1} \text{K}^{-1}$) of gypsum ($CaSO_4 \cdot 2H_2O$) is (Perry 1950)

$$c_p = 1137 \qquad\qquad 282 \le T \le 393 \, \text{K} \qquad [8.7]$$

and that of calcium sulfate anhydrite ($CaSO_4$) is

$$c_p = 569 + 0.68T - \frac{4.8 \times 10^6}{T^2} \qquad 273 \le T \le 1373 \, \text{K} \qquad [8.8]$$

The heat of complete dehydration (between 125 and 200 °C) is 0.61×10^6 J per kg gypsum. Thanks to the substantial absorption of energy in the dehydration process, a gypsum layer applied to the surface of a building element is capable of markedly delaying the penetration of heat into the underlying load-bearing components. Gypsum is, therefore, an ideal fire protection material.

The variation of the apparent specific heat of neat gypsum with temperature was mapped out by Harmathy (1961), based on information published by Kelley et al. (1941), West and Sutton (1954) and Ljunggren (1960).

The dilatometric and thermogravimetric curves for a so-called 'fire-resistant'

---

4   The true density of gypsum ($CaSO_4 \cdot 2H_2O$) is $2320 \, \text{kg m}^{-3}$. The density of gypsum boards is usually between 650 and 800 kg m^{-3}.

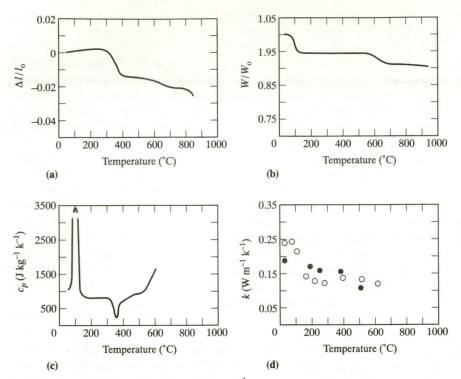

Fig. 8.4 Core of a gypsum board ($\rho = 678$ kg m^{-3}): (a) dilatometric curve; (b) thermogravimetric curve; (c) apparent specific heat; (d) thermal conductivity. Open circles, values obtained in the heating cycle; solid circles, values obtained in the cooling cycle (Harmathy 1983b)

gypsum board of 678 kg m^{-3} density are shown in Figs 8.4(a) and 8.4(b), respectively (Harmathy 1983b). Figure 8.4(c) shows the apparent specific heat versus temperature relation for the material, and Fig. 8.4(d) gives some information on its thermal conductivity.

Owing to physicochemical and microstructural changes brought about by prolonged heating, the elevated-temperature thermal conductivities obtained by stepwise increasing the specimen's temperature usually differ from those obtained during the cooling cycle. The points plotted in Fig. 8.4(d) as open and solid circles represent information obtained during the heating and cooling cycles, respectively.[5]

8.4   Plastics

With respect to their application in the construction industry, plastics belong in the I/F (insulating/fuel) group. Although many of them have higher strengths per

---

5   In the case of physicochemically unstable materials, there is no justification for connecting the experimental points with a solid line.

(a)   (b)   (c)

Fig. 8.5 Schematic illustration of the three kinds of polymer molecules: (a) linear; (b) branched; (c) crosslinked

unit mass than some metals,[6] pure plastics are rarely considered for use as highly stressed components of building elements. Their high creep rates at ordinary temperatures and at stresses well below their yield strengths impose serious limitations on their structural applications. Lack of dimensional stability, sensitivity to temperature changes, and propensity for moisture absorption present some further obstacles to their acceptance as fully fledged engineering materials.

Plastics are materials consisting of very large molecules, *polymers*, formed by polymerization from simple molecules, *monomers* or simply *mers*.[7] The monomers are linked by strong covalent bonds into long chains. Such *linear* polymers are the most common building blocks of plastics [Fig. 8.5(a)]. Somewhat more complex are the *branched* polymers [Fig. 8.5(b)] which may result from side reactions taking place during polymerization. Both kinds of molecular chains are held by weak van der Waals forces in a solid mass, which is often likened to spaghetti in a bowl. If the molecular chains in that mass are fully disordered, as in poly(methyl methacrylate), the polymeric material is *amorphous* [Fig. 8.6(a)]. However, the van der Waals forces are also capable of pulling the chains into ordered, *crystalline* regions: *crystallites*. Most plastics, such as polyethylene and poly(vinyl chloride), consist of amorphous and crystalline regions [Fig. 8.6(b)]. Moreover, there are some, e.g. nylon-6,6, that are almost fully crystalline [Fig. 8.6(c)]. Generally, unbranched, linear polymers are most likely to align into a crystalline order.[8]

The degree of crystallinity can be adjusted (within limits) during the manufacturing process. Letting the molten material cool slowly will result in a more orderly molecular structure, whereas the amorphous structure of the melt may be preserved by quenching.

When a polymeric material (consisting of linear or branched molecules) is heated, the van der Waals bonds gradually break up and the material softens. On cooling, these bonds re-establish themselves; the material regains its strength.

---

6 For example, the nylon yarn has five times the tensile strength of structural steel on a unit mass basis.
7 The average length of the molecular chains, i.e. the average molecular mass, is controlled in the manufacturing process.
    Copolymerization is a process in which two or more kinds of monomer are combined into copolymer molecules. Plastics of random *copolymers* have, as a rule, properties intermediate between those of the individual *homopolymers*. Copolymerization usually produces molecules of reduced symmetry, which are less likely to arrange themselves into crystalline order.
8 For example, high-density polyethylene, which is made up almost entirely of linear molecules, is a highly crystalline material, while low-density polyethylene, which consists of branched molecules, is of much lower crystallinity (see Table 8.2).

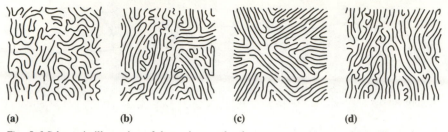

(a)   (b)   (c)   (d)

Fig. 8.6 Schematic illustration of the various molecular structures: (a) amorphous; (b) partly crystalline; (c) highly crystalline; (d) oriented

Plastics that can be turned into a viscous melt and back again into an elastic–plastic solid are referred to as *thermoplastic materials*.

The softening of the material at elevated temperatures can be moderated or prevented by tying the molecular chains together by covalent bonds: *crosslinks* [Fig. 8.5(c)]. Crosslinking is achieved by the application of crosslinking agents (e.g. peroxides) or (as in the case of polyethylene) by irradiation with high-energy electrons. The effect depends on the number of links formed. Lightly crossbonded plastics (those with no more than one crosslink per 50 monomer units), while showing improved strengths, retain some degree of flexibility. Heavily crosslinked plastics, on the other hand, become brittle and remain so on heating.[9] Above some temperature level, the crossbonds between the chains and the bonds between the monomers break irreversibly, and the material decomposes. Plastics of this kind are referred to as *thermosetting* materials.

Melting, a first-order transition, is characterized by enthalpy and volume changes that accompany the conversion of crystalline molecular order into liquid-phase disorder. Clearly, fully amorphous plastics which have a disordered molecular structure to start with do not, in a strict sense, melt, neither do thermosetting materials which decompose on heating. Consequently, the *melting point* (or temperature interval of melting), T_m (K), exists only for crystalline or partly crystalline polymeric materials.

More important than the melting point is the *glass transition temperature*, T_g (K). Glass transition is a second-order phenomenon; it is not accompanied by an abrupt change in the material's enthalpy and volume. The properties that change are the specific heat and the coefficient of thermal expansion. From a practical point of view, however, much more significant are the changes in the material's mechanical properties. Below the glass transition temperature, the material is rigid and glassy; above that temperature, it is rubbery, deformable. Clearly, for applications that require strength and rigidity, the choice is limited to plastics that have their glass transition temperatures well above room temperature. Conversely, whenever toughness is an essential requirement, only materials with low glass transition temperatures can be considered.

---

9   A solid made up of heavily crosslinked molecules is, in fact, one giant molecule.

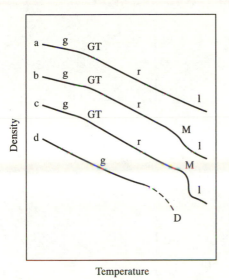

Fig. 8.7 Density versus temperature relation for four kinds of plastics: GT, glass transition; M, melting; D, decomposition; g, glassy; r, rubbery; l, liquid. Curves: *a* amorphous; *b* partly crystalline; *c* highly crystalline; *d* crosslinked

The nature of a plastic material can best be analyzed by studying the variation of its density (or specific volume) with the rise of temperature. Figure 8.7 depicts the density versus temperature relation for four different kinds of plastics. The glass transition temperature is usually less pronounced with highly crystalline materials. It may become hardly noticeable in the case of heavily crosslinked plastics.

The hardness and stiffness of a polymeric material may be reduced by *plasticization*, i.e. by the addition of a monomeric liquid plasticizer compatible with the polymer. The plasticizers are somewhat volatile, and may lose their effectiveness if the material is heated or aged for some length of time.

Certain properties of amorphous and crystalline plastics can be substantially improved by the *orientation* of its molecules or crystallites. Orientation is usually done by stretching the material at a temperature above its glass transition temperature. Figure 8.6(d) displays schematically the oriented structure of a partly crystalline material.

Table 8.1 shows the effect of the chemical and structural makeup of polymeric materials on their mechanical and thermal properties. Some thermal characteristics of a number of common plastics, used as construction materials and as building insulation, are listed in Tables 8.2 and 8.3.

In view of the paucity of reliable information on the properties of plastics, the fire safety practitioner may find the following information useful:

- The ratio of glass transition temperature to melting temperature, T_g/T_m, is usually between 0.5 and 0.67.
- For amorphous plastics below their glass transition temperatures, the (tensile) yield strength (σ_y) is of the order of 55 MPa, and the modulus of elasticity

Table 8.1 Effect of chemical and structural characteristics of polymers on various properties of plastics[a] (Mark 1942; Billmeyer 1971; Deanin 1972; Thorne 1979)

Property	Average molecular mass	Degree of crystallinity	Branching	Crosslinking[b]	Plasticizing	Orientation[c]
Glass transition temperature	+		−	+[d]	−	?
Modulus of elasticity	+	+	?	+	−	+
Yield strength		+				
Ultimate tensile strength	+	?	?	?	−	+[e]
Elongation at rupture	+	−	?	−	+	?
Surface hardness	+	+	?	+	−	+
Resistance to creep	+	+	+	+		+
Resistance to moisture	+	+	−	+		+
Density	?	+			−	
Coefficient of thermal expansion	−	−[f]		−		−
Specific heat	?	?				
Thermal conductivity	+	+[g]				+

[a] Symbols: +, increases; −, decreases; ?, effect unclear or insignificant.
[b] Effect depends on the number of crosslinks.
[c] Some properties may vary the opposite way in the direction perpendicular to orientation.
[d] Heavily crosslinked plastics may not show glass transition.
[e] At moderate orientation.
[f] Above glass transition temperature; below that temperature little effect.
[g] For crystalline plastics it is about twice as high as for amorphous plastics.

Table 8.2 Some thermal characteristics of selected polymeric materials (based on various sources, mainly International Technical Information Institute 1976; Kroschwitz et al. 1985–1989)

Material	ρ (10^3 kg m^{-3})	α (10^{-6} m m^{-1} K^{-1})	c_p (J kg^{-1} K^{-1})	k (W m^{-1} K^{-1})	DC[a] (%)[b]	T_g (°C)[b]	T_m (°C)[bc]
ABS[d]	1.01–1.07	72	1470	0.30	Low	100	
Polycarbonate	1.20–1.22	57	1260	0.20	0	150	225
Polyester	1.10–1.46	78	1050	0.17			
Glass fiber reinforced	1.80–2.30	10–30		0.33–0.50			
Polyethylene							
Low density	0.91–0.93	200	2300	0.33	55	−120	120
High density	0.94–0.98	120	2070	0.52	95		130
Poly(methyl methacrylate)	1.16–1.20	45	1400	0.21	0	105	200
Polypropylene	0.85–0.92	80	1900	0.12	High	−20	160
Polystyrene	1.04–1.08	65	1190	0.14	0	100	240
Poly(vinyl chloride)							
Rigid	1.38–1.41	125	1260	0.21	≈20	85	200
Plasticized	1.19–1.35	160	1670	0.17			
Chlorinated	1.47–1.55	70		0.14			
Phenol–formaldehyde	1.20–1.28	42–68	1470	0.15–0.25			170
Polyurethane	1.21	150	1800	0.21–0.31			
Urea–formaldehyde with cellulose filler	1.47–1.52	27	1670	0.29–0.42			
Nylon-6,6	1.13–1.16	80	1670	0.24	High	110	260

[a] DC = degree of crystallinity.
[b] There is a wide scatter in the reported data; the table values represent averages. Unless the material is highly crystalline, the glass transition and melting may extend over a range of 10–20 °C.
[c] Amorphous plastics have no well defined melting points. What is referred to as melting temperature is a temperature level at which the viscosity of the material attains a value characteristic of liquid plastics.
[d] ABS, poly(acrylonitrile–butadiene–styrene).

Table 8.3 Thermal characteristics of some foamed polymeric materials used as thermal insulation (Kroschwitz *et al.* 1985–1989)

Material	ρ	α	c_p	k	Upper limit of use
	(kg m^{-3})	$(10^{-6}\text{m m}^{-1}\text{K}^{-1})$	$(\text{J kg}^{-1}\text{K}^{-1})$	$(\text{W m}^{-1}\text{K}^{-1})$	$(^{\circ}\text{C})$
Polystyrene	12–48	30–45	1200–1300	0.028–0.036	75
Polyurethane and polyisocyanurate	30–40	30–40	1500	0.020–0.026	120
Urea–formaldehyde	9–18		1500–1800	0.035	210
Phenolic resin	45–60	20–40	2000	0.018	150

(E) is of the order of 3.5×10^3 MPa. In the case of crystalline materials, the spread is quite large. Many polyethylenes have a yield strength below 14 MPa, whereas nylons may have values as high as 83 MPa (Brydson 1975).

- The density (ρ) of amorphous hydrocarbon polymers is usually between 860 and 1050 kg m^{-3}. If large atoms (e.g. chlorine) are present in the polymer, the density may be much higher (Brydson 1975).
- Most polymeric materials have a coefficient of thermal expansion (α) in the range of 20×10^{-6} to 200×10^{-6} m m^{-1} K^{-1}.
- The specific heat (c_p) of most plastics is between 1250 and 2100 J kg^{-1} K^{-1}.
- The values of the thermal conductivity (k) usually range from 0.15 to 0.35 W m^{-1} K^{-1}. The thermal conductivity of plastic foams depends primarily on their porosities. It can be calculated as described in Chapter 5.

Because of the relatively narrow temperature range of their utility, the elevated-temperature properties of plastics are of little consequence. Their specific heats may increase by as much as 3–15 per cent per 10 °C rise of temperature. In contrast, their thermal conductivities show little variation.

Most plastics are not used in pure form, but are mixed with solid powders or fibers (commonly referred to as fillers) before fabrication into products. The fillers are generally incompatible with the plastic matrix, and are present as a separate, discontinuous, solid phase. The effect of the fillers on the thermal properties of the material can be assessed using the applicable form of mixture rule (see, for example, Eqs 6.50, 6.51, and 6.55).

Rogowski (1976) and Malhotra (1977) reviewed the application of plastics in building construction. Table 8.4 presents some of their findings. Although the

Table 8.4 Plastics extensively used in building construction (Rogowski 1976; Malhotra 1977)

Material	Type[a]	Shape/form
ABS[b]	TP	Rigid sheets, pipes
Polycarbonate	TP	Clear sheets
Polyester, glass-reinforced	TS	Rigid sheets, domes, framework, wall panels, ducts
Polyethylene	TP	Sheets, pipes, films
Poly(methyl methacrylate)	TP	Clear sheets, domes, fibers
Polypropylene	TP	Sheets, moldings, fibers
Polystyrene	TP	Sheets, moldings
Poly(vinyl chloride)	TP	Sheets, pipes, ducts, tiles, films
Poly(vinyl chloride), chlorinated	TP	Pipes, ducts
Polystyrene foam	TP	Panels, tiles, core material
Phenol–formaldehyde foam	TS	Panels, laminates
Polyisocyanurate foam	TS	Laminates, core material
Polyurethane foam	TS	Laminates, core material, spray
Urea–formaldehyde foam	TS	Cavity fill

[a] TP, thermoplastic; TS, thermosetting.
[b] ABS, poly(acrylonitrile–butadiene–styrene).

building designer has control only over the use of construction materials, his sphere of interest should include all burnable materials that may be found in buildings.

Taylor (1974) discussed the use of plastics in furnishing items in the context of possible fire hazards. The available data (Thorne 1979) seem to indicate that the amount of plastics brought into the building (as furniture, appliances, housewares, toys, etc.) after its completion exceeds that used in the construction.

Nomenclature

c_p	specific heat (or apparent specific heat) at constant pressure, $J\,kg^{-1}\,K^{-1}$
E	modulus of elasticity (Young's modulus), Pa
k	thermal conductivity, $W\,m^{-1}\,K^{-1}$
l	length, m
Δl	$= l - l_o$, m
m	moisture content of wood (related to oven-dry mass), $kg\,kg^{-1}$
P	porosity, $m^3\,m^{-3}$
T	temperature, K (or °C)
W	mass, kg
α	coefficient of thermal expansion, $m\,m^{-1}\,K^{-1}$
ρ	density (in oven-dry condition), $kg\,m^{-3}$
ρ_t	true density, $kg\,m^{-3}$
σ	strength, Pa

Subscripts

g	of glass transition
m	of melting
o	at the reference (room) temperature
u	ultimate
y	yield

9 The Fire Process

The previous four chapters have dealt with subjects related mainly to material behavior. This chapter is devoted to processes and phenomena, an area referred to in Chapter 4 as 'output-end' fire science. Three decades of research and hundreds of papers relevant to subjects in this sphere of fire science have been reviewed by Drysdale (1985), primarily for the benefit of the academic community. A more recent book by Shields and Silcock (1987) and the *SFPE Handbook* (DiNenno *et al.* 1988) handle the fire-related processes and phenomena in a way that may have an appeal for both the student of fire science and the fire safety practitioner. These three books are recommended reading for those who may find some parts of this book not sufficiently detailed on certain subjects.

The subjects to be discussed in this book have been selected with a view to providing those concerned with the fire safety of buildings with a sound scientific background, necessary for recognizing the common fire risks and devising measures to eliminate or mitigate them. The measures may concern

(a) the layout and dimensioning of the building and its constituent parts,
(b) the provision of safety devices and facilities, and
(c) the selection of construction materials and products.

In providing fire safety, the building designer must keep in mind all the problems that may arise in the course of a fire, i.e. in the course of ignition, burning of single objects, initial fire spread, growth of fire, fully developed burning, and intercompartmental fire spread. He must also address the smoke problem, which may be present at any phase of the fire process.

From a practical point of view, the characteristics of fully developed burning and the nature of intercompartmental fire spread are of principal interest. They will be dealt with in considerable depth in the next two chapters. All the other areas of output-end fire science, including the smoke problem, will be discussed in this chapter.

9.1 Terminology

Defining a few key terms is the unavoidable first step toward any meaningful discussion on the fire process. The rather liberal use of the words 'burning' and 'combustion' has already given rise to a great deal of misunderstanding (Harmathy 1984).

In everyday use, the two words are regarded as synonymous.[1] The noun 'combustion' is thought of as the consumption by fire of any material (solid, liquid, or gas), and is pictured mostly as an oxidation process accompanied by flames. However, flaming can only accompany gas-phase reactions; solids and liquids must convert into the gaseous phase (gasify) before they can undergo flaming combustion. This conversion takes place by

(a) *vaporization* (or sublimation) in the case of liquids (or solids) of simple molecular structure, or

(b) by *pyrolysis* (thermal decomposition) in the case of solids or liquids of complex (polymeric) structure.[2]

Vaporization always takes place at a well-defined temperature, whereas pyrolysis develops over a temperature interval. The energy required to preheat the material to the temperature level where pyrolysis or vaporization takes place is often tacitly included in the *heat of pyrolysis* or *heat of vaporization*.[3].

Vaporization (or sublimation) entails the full conversion of the material into the gaseous phase. Full conversion may also take place in the pyrolysis of polymeric materials. However, with many kinds of polymeric materials the conversion is not complete. In a solid product of high carbon content, char forms concurrently with the *volatile*[4] products. Thus, polymeric materials can be divided into two categories: noncharring and charring materials.

In the case of charring solids, burning is a complex process, consisting of three kinds of dissimilar reactions which, for the most part, develop simultaneously:

(1) *pyrolysis*: a series of (usually) endothermic reactions, occurring at moderately elevated temperatures,

(2) *flaming combustion* of the volatiles: homogeneous (gas with gas), exothermic reactions, and

(3) *char oxidation*: heterogeneous (solid with gas), strongly exothermic reaction.

The following terminology is used consistently in this book

● The noun *burning* is an umbrella term for the overall process that results

---

1   In fact, the verb 'combust' is still not regarded as fully legitimate; its substitution by the verb 'burn' has been almost compulsory.

2   Many plastics melt at temperature levels below the temperature range of pyrolysis. The word *pyrolysis* is often used in a meaning that covers a series of changes, beginning with melting and ending with the completion of the pyrolysis reactions.

3   The word *gasification* is often used to cover both pyrolysis and vaporization (or sublimation). In the case of char-forming materials, however, pyrolysis is not simply a transition from solid to gaseous phase, and therefore the use of the word gasification is misleading. Since fire science is mainly concerned with the burning of solids of polymeric structure, in this book the word *pyrolysis* will often be used as a general term covering both true pyrolysis and vaporization (sublimation).

4   The volatiles may include both true gases, and liquids that are gaseous at the temperature of pyrolysis.

in the consumption of the fuel[5] by fire, irrespective of the reactions involved.

- The noun *combustion* covers the homogeneous (gas-phase), exothermic oxidation reactions that involve the volatile products of vaporization or pyrolysis, and take place in the flame zone.
- The noun *char-oxidation* denotes the heterogeneous (solid-gas), exothermic reaction of the char with oxygen.

The corresponding verbs are *burn* (solids, liquids, gases), *combust* (volatiles), and *oxidize* (char); and the corresponding adjectives are *burnable* (solids, liquids), and *combustible* (volatiles).

Since pyrolysis and vaporization are endothermic reactions,[6] they can proceed only if they are driven, i.e. if heat is continuously supplied to the fuel. The heat may originate from (1) the flame, (2) the oxidizing char, and (3) external sources. In the absence of external heat supply, the vaporization and pyrolysis rely on the heat produced by either of the driving reactions: flaming combustion and char oxidation.

Table 9.1 shows the possible component reactions in a burning process in the absence of external heat supply. Apparently, only with gaseous fuels do the terms 'burning' and 'combustion' have synonymous meanings. If the material is char-forming, three cases are possible. The process may be driven

(a) by both flaming combustion and char oxidation,
(b) by flaming combustion alone, or
(c) by char-oxidation alone.

If char-oxidation alone is the driving reaction, the burning process is called *smoldering*.

It has been customary to regard smoldering and flaming combustion as alternative modes of 'combustion'. Apart from the fact that (in the sense discussed) smoldering is not a combustion process (but pyrolysis driven by char-oxidation),

Table 9.1 Reactions covered by the term 'burning' in the absence of external heat supply (Harmathy 1984)

| Process | Polymeric materials | | | | | |
| | Char-forming | | | Noncharring | Simple liquids and solids | Gases |
	a	b	c			
Pyrolysis/vaporization	×	×	×	×	×	
Flaming combustion	×	×		×	×	×
Char-oxidation	×		×			

5 The word *fuel* is used in fire science as a generic term for any kind of organic material (solid, liquid, gas).
6 See Footnote 1 in Chapter 5 for the extended use of the word *reaction*.

in the absence of external heat sources smoldering can be an alternative to flaming combustion only with char-forming materials.

9.2   Building before Outbreak of Fire

Many important findings of fire research have been developed from studies in which the fire compartment is pictured as a space neatly separated from the rest of the building. Since the distribution and intensity of air currents (drafts) are determinant factors in the spread of fire and smoke, it seems appropriate to start a discussion on the fire process with a survey of the nature of air currents in buildings.

Drafts are caused by

(1) a temperature difference between the building and the outside atmosphere, and
(2) air leakage through the boundaries of building spaces.[7]

Owing to the former, drafts are especially strong during the winter heating season.

Figure 9.1(a) illustrates schematically the direction and intensity of the dominant air currents in winter in a nine-story office building, after the shutdown of the air-handling system. The building is shown to consist of rooms (R), uncompartmented spaces (U), corridors (C), and stairwells and elevator shafts (S). If the arrangement of building spaces is repetitive along the building height, and the leakage characteristic of the outside walls is uniform, the infiltration of cool air takes place below the midheight of the building. After passing through a number of building space boundaries, the warm air enters the shafts, rises to the upper floors, moves toward the outside walls, and exfiltrates to the outside atmosphere.[8] Because of the important role the stack-like shafts play, this phenomenon is often referred to as air movement by 'stack effect'.

The rate of air flow depends on the 'leakiness' of the boundaries of various building spaces, especially on the leakiness of the outside walls. Since the flow through small holes or gaps can be treated as flow through orifices, it is customary to characterize the leakiness of a boundary element by its 'equivalent orifice area', ζ ($m^2\,m^{-2}$), which is the aggregate area of (often invisible) holes, cracks, gaps, etc., per unit area of the element. An analysis of the situation illustrated in Fig. 9.1(a) requires information on three equivalent orifice areas: ζ_w for the outside walls of the building, ζ_r for the room—corridor walls, and ζ_s for the shaft—corridor or shaft—uncompartmented space walls.

The direct causes of air movement in the building are, naturally, the pressure

---

7   It is usually understood that building spaces are those parts of a building that are enclosed by walls, floor, and ceiling, and are most of the time separated from the rest of the building by closed door(s). Building spaces may be placed in four categories: (1) *rooms* or *compartments* (small and medium-sized spaces), (2) *uncompartmented spaces* (large spaces), (3) *corridors*, and (4) vertically extending spaces, such as stairwells and elevator shafts, referred to jointly as *shafts*.
8   The air currents rising from story to story through the ceilings are insignificant and, therefore, will be disregarded in this discussion.

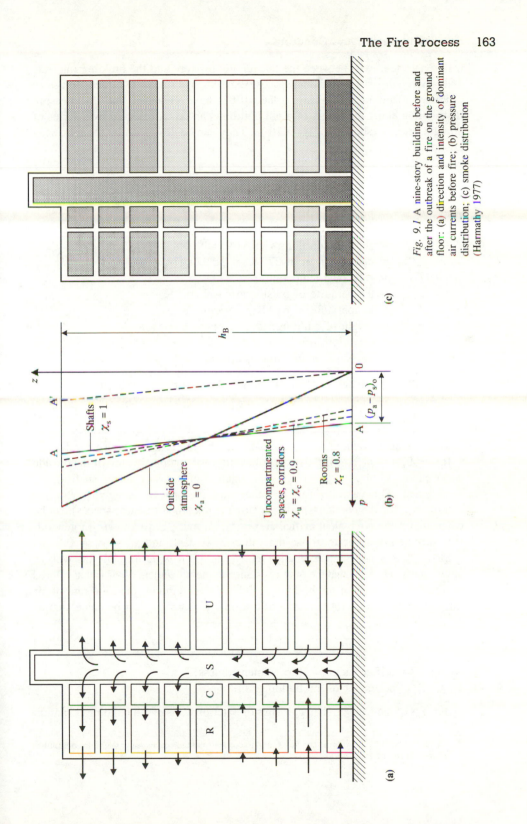

Fig. 9.1 A nine-story building before and after the outbreak of a fire on the ground floor: (a) direction and intensity of dominant air currents before fire; (b) pressure distribution; (c) smoke distribution (Harmathy 1977)

differences that exist between the outside atmosphere and the building interior, and between the various building spaces. Experimental studies of heated multistory buildings (Tamura and Wilson 1966, 1967) have indicated that the pressure distribution along the height of a tall building can be represented by a series of straight lines, as shown in Fig. 9.1(b).[9] They can be described by the following equation:

$$p - p_a = - g(\rho_a)_o T_o \left(\frac{1}{T_a} - \frac{1}{T_i} \right) \left(\frac{h_B}{2} - z \right) \chi \qquad [9.1]$$

where p = pressure, Pa;
p_a = pressure of outside atmosphere, Pa;
g = acceleration due to gravity $\simeq 9.8\,\mathrm{m\,s}^{-1}$;
$(\rho_a)_o$ = density of air at 273 K = $1.292\,\mathrm{kg\,m}^{-3}$
T_o = reference temperature = 273 K;
T_a = temperature of outside atmosphere, K;
T_i = temperature of building interior, K;
h_B = height of building, m;
z = elevation, m;
χ = pressure factor, dimensionless;

and the pressure of the outside atmosphere is

$$p_a = - g(\rho_a)_o T_o \frac{z}{T_a} \qquad [9.2]$$

provided that its value at $z = 0$ is taken as the reference pressure level.

If $\chi = 0$, $p = p_a$, i.e. Eq. 9.1 describes the variation of pressure of the outside atmosphere. With $\chi = 1\ (= \chi_s)$, the variation of pressure in the shaft(s), p_s, is obtained. The values χ for the other building spaces, namely χ_r (for the rooms), χ_c (for the corridors), and χ_u (for the uncompartmented spaces) can be calculated if the equivalent orifice areas, ζ_w, ζ_r, and ζ_s, are known. In general, it is sufficiently accurate to use the values $\chi_r = 0.8$, and $\chi_c = \chi_u = 0.9$.

Since ζ_w is usually much smaller than ζ_r or ζ_s, it is permissible (as well as convenient) to assume that the only resistance to the movement of air is that offered by the outside walls of the building. With this assumption, the total rate of air infiltration (and total rate of air exfiltration), U (kg s^{-1}), can be calculated as

$$U = \frac{\beta \zeta_w P(\rho_a)_o T_o}{3 T_a} \sqrt{g \left(1 - \frac{T_a}{T_i} \right) h_B^3} \qquad [9.3]$$

where β = orifice factor $\simeq 0.6$, dimensionless;
P = perimeter of the building, m.

Surveys of the leakage characteristics of exterior walls of tall buildings (Tamura

---

9   In reality, the lines for rooms, corridors, and uncompartmented spaces show slight discontinuities at the ceiling of each story.

and Wilson 1967; Tamura and Shaw 1976) seem to indicate that, for lack of experimental information, $\zeta_w = 0.5 \times 10^{-3}\,\mathrm{m^2\,m^{-2}}$ is a reasonable selection.

The upward movement of air (as well as smoke) in the shaft(s) can be negated by the 'pressurization' of the building (to be discussed), which is achieved by supplying air to the building interior at a rate sufficient to raise the pressure in the shaft at the ground floor ($z = 0$) to the level of the outside atmosphere. The pressure distribution in the shaft [line A−A in Fig. 9.1(b)] is thereby moved to the right (to a new position 0−A$'$) by an amount

$$(p_a - p_s)_0 = g(\rho_a)_0 T_0 \left(\frac{1}{T_a} - \frac{1}{T_i} \right) \frac{h_B}{2} \qquad [9.4]$$

where p_s (Pa) is the shaft pressure. Equation 9.4 describes the pressure difference against which the supply fan has to work. The required rate of air supply, U_{pr} (kg s^{-1}), is (McGuire and Tamura 1975)

$$U_{pr} = 2^{3/2} U \qquad [9.5]$$

i.e. roughly three times the rate of infiltration of air into the building under normal conditions.

9.3   Pyrolysis

The pyrolysis of polymeric materials has been reviewed by Beyler (1988). The thermal decomposition of plastics may be either straight pyrolysis or oxidative pyrolysis. As the name indicates, oxidative pyrolysis involves oxidation of the polymers during the decomposition process. Concurrent oxidation lowers the heat absorbed in the process.

The following mechanisms may be important in the pyrolysis of polymeric materials:

(1) chain-end scission, in which the monomers at the ends of molecular chains are released in an 'unzipping' reaction which, in essence, is the reverse of chain polymerization,
(2) random chain scission, in which the molecular chains break at random locations,
(3) chain stripping, in which atoms or groups (branches) are cleaved off the polymer's 'backbone', and
(4) crosslinking between the chains or chain fragments.

The decomposition process often involves more than one of these mechanisms.

If all the molecules or molecular groups produced in the process are small enough to be gaseous at the temperature of pyrolysis, the formation of volatiles is completed in one step. If they are not, the larger molecular fragments will remain in the solid or liquid phase until they are broken up by heat into smaller fragments that can vaporize.

Crosslinking occurs most frequently between stripped polymer chains. It is an important mechanism in the formation of char.

Some examples of the pyrolysis process are as follows.

Poly(methyl methacrylate) decomposes entirely by chain-end scission; its pyrolysis product is gaseous (at the temperature of pyrolysis) methyl methacrylate. The decomposition of polystyrene is considerably more complex, even though its monomer yield is fairly high. Ultimately, it will degrade into styrene (40 per cent), toluene (2.4 per cent), and some other products of relatively high molecular masses (Billmeyer 1971). In the case of poly(vinyl chloride), chain stripping and crosslinking are the principal decomposition mechanisms. The material loses hydrogen chloride (HCl) molecules between 230 and 280 °C, and H_2 molecules between 430 and 480 °C. A char-like residue is left behind.

Most polymeric materials that are heavily crosslinked or become crosslinked during the decomposition process tend to yield char on heating.

The heat of pyrolysis, Δh_p ($J\,kg^{-1}$), for a number of polymeric materials is given in Table 9.2.

In spite of the rapidly growing use of plastics, the bulk of the burnable materials in buildings still consist of wood and other cellulosic materials (mainly paper and cotton) manufactured from plant cells.

The pyrolysis of wood[10] is an extremely complex process which is still not fully understood. Wood is a mixture of natural polymers, the most important of which are cellulose (50 per cent), hemicellulose (25 per cent), and lignin (25 per cent) (Madorsky 1964). From the suggested mechanisms of decomposition (reviewed by Drysdale 1985), it appears that the degradation reactions may proceed on alternative paths, and therefore the heat of pyrolysis of wood and the nature and relative quantities of pyrolysis products (volatiles and char) depend on such factors as temperature, rate of heating, and inorganic impurities in the material.

Based on a literature survey, Harmathy (1972) suggested the following chemical formula for a typical wood:

$$CH_{1.455}O_{0.645}\cdot0.233H_2O$$

where the attached 0.233 molecules of water account for the presence of about 15 per cent moisture (referred to overall mass) under normal atmospheric conditions. Since the average charcoal yield of untreated wood is about 15 per cent (referred to oven-dry mass; 12.76 per cent if referred to overall mass), and the chemical formula of charcoal is roughly $CH_{0.2}O_{0.02}$ (Roberts 1964b), the pyrolysis reaction can be written as

$$CH_{1.455}O_{0.645}\cdot0.233H_2O \rightarrow 0.233H_2O + 0.715CH_{1.955}O_{0.894}$$
$$+ 0.285CH_{0.2}O_{0.02} \qquad\qquad [9.6]$$

---

10 For simplicity, fire scientists usually apply the word *wood* to all kinds of cellulosic materials.

Table 9.2 Some information concerning the pyrolysis and burning of selected materials (based on various sources, mainly Tewarson and Pion 1976; Drysdale 1988)

Material	Pyrolysis temperature (°C)	$(dw/dt)_{id}$ (10^{-3} kg m^{-2} s^{-1})	q_L (10^3 W m^{-2})	Δh_p (MJ kg^{-1})	Δh_b (MJ kg^{-1})	Δh_b^0 (MJ kg^{-1})	LOI (%)
Polyacrylonitrile	250–280				−30.8	−13.6	25
Polycarbonate	420–620	25	74.1	2.07	−29.7	−13.1	21
Polyester, glass fiber reinforced		18	16.3	1.39	−23.8		
Polyethylene	335–450	14	26.4	2.32	−43.3	−12.7	17
Poly(methyl methacrylate)	170–300	24	21.3	1.62	−24.9	−13.0	17
Polyoxymethylene		16	13.8	2.43	−15.5	−14.5	
Polypropylene	328–410	14	18.8	2.03	−43.3	−12.7	18
Polystyrene	285–440	35	50.2	1.76	−39.9	−13.0	18
Polyurethane foam							
Rigid		45	57.7	1.52	−26.0		15
Flexible		32	24.3	1.22	−24.6		16
Poly(vinyl chloride)	230–480	13	16.3	2.46	−16.4	−12.8	45
Phenol–formaldehyde				1.64			40
Nylon-6,6	310–380			2.26	−29.6	−12.7	29
Wood	200–500	13–20	23–45	1.5–2.8	−17.0	−14.3	18–21
Carbon (porous)					−32.8	−12.3	55

Notes

(1) When Δh is positive, heat is absorbed in the process; when it is negative, heat is evolved. The sign is important mainly in thermodynamic calculations.

(2) The heat of burning, Δh_b, consists of the heat of combustion of the volatile pyrolysis products, Δh_c, minus the heat of pyrolysis, Δh_p.

(3) Δh_b^0 is heat of burning per unit mass of oxygen. To obtain the heat of burning per unit mass of air, multiply Δh_b^0 by 0.232 (mass fraction of oxygen in air).

(4) For char-forming materials, $(dw/dt)_{id}$ is taken as the peak burning rate.

where the first two terms on the right-hand side represent the volatile decomposition products.[11]

Values reported on the overall heat of pyrolysis of wood range from $-0.3\,\mathrm{MJ\,kg^{-1}}$ (Roberts and Clough 1963) to $+3.2\,\mathrm{MJ\,kg^{-1}}$ (Petrella 1979).[12]

9.4   The Flame

Fire safety practitioners are mainly interested in burning processes accompanied by flames. The flames produced by accidental fires are *diffusion flames*: combustion occurs as air from the ambient atmosphere mixes by entrainment and diffusion with the volatiles rising from the pyrolyzing or vaporizing fuel.

The nature of the flame depends to a large extent on the flow rate of the volatiles that feed the flame. If that flow rate is high, the flame is momentum-dominated; if it is low, the flame is *buoyancy-dominated* (i.e. dominated by the density difference between the hot gases and the ambient atmosphere). A dimensionless parameter, referred to as nondimensional rate of heat release (by the flame), $\bar{R}_\mathrm{F}$, is used to characterize the relative importance of the momentum or buoyancy in the flaming combustion. $\bar{R}_\mathrm{F}$ is defined as

$$\bar{R}_\mathrm{F} = \frac{R_\mathrm{F}}{\rho_a c_a T_a D^2 \sqrt{gD}} \qquad [9.7]$$

where R_F = (average) rate of heat release by the flame (to be calculated from Eq. 9.16), W;

ρ_a = density of ambient air = $1.184\,\mathrm{kg\,m^{-3}}$ at 25 °C;

c_a = specific heat (at constant pressure) of ambient air = $1005\,\mathrm{J\,kg^{-1}\,K^{-1}}$ at 25 °C;

T_a = temperature of ambient air = 298 K (25 °C);

g = acceleration due to gravity $\simeq 9.8\,\mathrm{m\,s^{-2}}$;

D = equivalent diameter of fire source,[13] m.

For room-temperature ambient conditions, substitution of the listed values into Eq. 9.7 yields

$$\bar{R}_\mathrm{F} = 0.9 \times 10^{-6} \frac{R_\mathrm{F}}{D^{5/2}} \qquad [9.7a]$$

The process of flaming combustion is momentum-dominated if $\bar{R}_\mathrm{F}$ is greater than 10 000; it is buoyancy-dominated if $\bar{R}_\mathrm{F}$ is less than about 100. As $\bar{R}_\mathrm{F}$ decreases toward 1, a regular pulsation at the base of the flame grows in amplitude

---

11 In reality, the composition of the volatiles changes and their heat of combustion increases as the pyrolysis progresses (Brenden 1967).

12 A negative value indicates an exothermic process. The sign is important mainly in thermodynamic calculations.

13 The equivalent diameter is defined as $\sqrt{4A_f/\pi}$, where A_f (m^2) is the area of (pool-like) fuel surface.

and the region of intermittent flaming at the top of the flame increases (Zukoski 1986).

In accidental fires, where the gases that feed the flame are generated by pyrolysis or vaporization, the maximum flow rate of gases usually ranges from 0.01 to 0.03 kg m^{-2} s^{-1}, and the maximum velocity of the gases at the fuel surface is between 0.01 and 0.02 m s^{-1}. Under these circumstances, $\bar{R}_F$ is, as a rule, less than 1.

Flames up to a height of about 0.2 m are smooth, and *laminar*; larger flames are *turbulent*.

McCaffrey (1979) studied the structure of natural gas flames produced by a porous refractory burner of 0.3 m square.[14] He described his findings concerning the temperature and vertical velocity of gases on the centerline of the flame with the aid of a parameter, Z (m W$^{-2/5}$), defined as

$$Z = \frac{z}{R_F^{2/5}} \qquad [9.8]$$

where z (m) is vertical distance above the burner. He found that the 'fire plume' consisted of three distinct regimes, namely

(1) the near-field, $0 \le Z < 1.27$, where the temperature was constant, approximately 825 °C, and the velocity of flame gases increased with height;
(2) the field of intermittent flaming, $1.27 < Z < 3.17$, where the temperature decreased from 825 °C to about 325 °C, while the velocity of the gases stayed approximately constant; and
(3) the far-field, i.e. the buoyant plume, $Z > 3.17$, where both the temperature and the velocity decreased with height.

McCaffrey (1988) showed that the height of flames from fuels burning in the open can be expressed in terms of the nondimensional heat release rate. Among the several equations available for the calculation of flame height, the one developed by Heskestad (1983) seems to have the widest range of application. In terms of the nondimensional heat release rate, Heskestad's equation is

$$\frac{h_F}{D} = 3.88\ \bar{R}_F^{2/5} - 1.02 \quad \text{for } 0.12 < \bar{R}_F < 12 \times 10^3 \qquad [9.9]$$

where h_F (m) is the flame height.[15] Equation 9.9 is applicable to both liquid and solid fuels burning in pool configuration (i.e. plane surface facing upward). In the case of solid fuels burning in pile configuration, it is applicable only if the combustion of the volatile pyrolysis products takes place essentially above the pile.

---

14 The non-dimensional heat release rate varied from 0.19 to 0.78.
15 The (average) flame height is defined as the distance from the fire source at which flame is present 50 per cent of the time (Zukoski *et al.* 1985). It may also be defined as the distance at which the temperature in the centerline of the 'fire plume' drops to about 500 °C above that of the ambient atmosphere.

According to Eq. 9.9, the flame height would become negative if $\overline{R}_F \leq 0.035$. Heskestad (1988) suggested that at this value of $\overline{R}_F$ the flame breaks up into a number of independent flamelets.

The amount of air entrained into the flame is an important piece of information in assessing the development of flaming combustion in enclosed spaces. Steward (1970) claimed that the ratio of the air entrained up to the flame tip to the air required by stoichiometric combustion is about 5. Based on various sources, Babrauskas (1980b), Zukoski (1986), and Delichatsios (1988) suggested much higher values, ranging from 10 to 20.

If the fuel is located close to a wall, the entrainment of air into the flame will not be axisymmetric. The restricted flow of air on one side of the flame will have two effects:

(1) the flame will deflect toward the wall and possibly hug the wall, and
(2) the flame height will extend, owing to diminished turbulence in the flame and the reduced chance of the volatiles to intermix with air [Fig. 9.2(a)].

If, in an enclosed space, the ceiling height is lower than the flame height, the flame will impinge on the ceiling and spread out horizontally. Babrauskas (1980b) suggested that the horizontal spread, y_F (m) [Fig. 9.2(b)], could be estimated on the assumption that the amount of air needed for the combustion of the volatiles

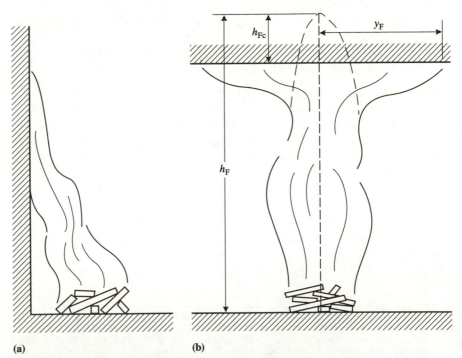

(a)   (b)

Fig. 9.2 Flame in an enclosed space: (a) close to wall; (b) impinging on ceiling

is the same for the ceiling flow as for the plume flow. Yet the rate of entrainment of air into the ceiling flow is greatly reduced, since the substantial difference in the densities of the hot ceiling flow and the cool air hampers the process of intermixing. The graphs presented in Babrauskas's paper make it possible to correlate the 'cut-off height' of the flame, h_{Fc} (m), with the horizontal flame extension. It appears that y_F/h_{Fc} may be as high as 12, or perhaps higher, depending on the rate of heat release by the flame, the width and length of the enclosure, and the location of the fire source with respect to the boundaries of the enclosure.

The heat released in flaming combustion is transmitted to the surroundings by convection and radiation. It is customary, therefore, to divide the total rate of heat release by the flame into convective, R_{FC} (W), and radiative, R_{FR} (W) contributions. The radiant heat flow from a luminous (yellow) flame to nonreflecting surroundings (atmosphere) can be written as

$$R_{FR} = A_F \, \sigma \epsilon \, (T_F^4 - T_a^4) \tag{9.10}$$

where A_F = area of 'flame envelope' (function of base area and height of flame), m^2;

$\quad \sigma$ = Stefan$-$Boltzmann constant = $5.67 \times 10^{-8} \, W \, m^{-2} \, K^{-4}$;

$\quad \epsilon$ = emissivity of (luminous) flame, dimensionless;

$\quad T_F$ = average flame temperature, K;

$\quad T_a$ = temperature of ambient atmosphere, K.

Since $T_a^4 \ll T_F^4$ the second term between the parentheses may be dropped.

The emissivity of the flame can be expressed as

$$\epsilon = 1 - \exp(-al) \tag{9.11}$$

where a = absorption (emission) coefficient for the (luminous) gas, m^{-1};

$\quad l$ = mean beam length[16] ('optical flame thickness'), m.

The coefficient a depends mainly on the luminosity of (concentration of soot particles in) the flame. For flames of liquid hydrocarbon fuels, $a \simeq 1 \, m^{-1}$; therefore it is reasonable to assume that $\epsilon \to 1$ for flames thicker than about 2 m. Flames with emissivities close to 1 radiate like black bodies.

Experimenting with the burning of small samples of solid and liquid fuels, Tewarson (1980) found that the radiative heat release ranged from 18 to 69 per cent of the total heat release. For a few fuels that burn with nonluminous flames, such as methanol or polyoxymethylene, the R_{FR}/R_F ratio may be lower than 0.18. In the case of fires involving wood and furniture, it is usually assumed that R_{FR}/R_F is about 1/3.

The radiant heat flux falling on an object at a distance L (m) from the flame

---

16 The mean beam length may be taken as approximately equal to the equivalent diameter of the flame base.

is approximately $R_{FR}/4\pi L^2$ (W m^{-2}), provided L is much larger than the flame height.

In general, the more luminous the flame, the lower its temperature. Rasbash *et al.* (1956) observed that in some of their experiments the nonluminous (bluish) flames of methanol reached a temperature of 1200 °C, while the luminous flames of benzene reached only 920 °C.

Pool fires of liquid hydrocarbons, such as those resulting from transportation accidents, may create very severe conditions within minutes. A draft ASTM test method describes how these conditions can be simulated. In the proposed test, the specimen is required to be fully engulfed in thick, luminous flames, the average temperature of which is to be maintained at a level between 1010 and 1180 °C after 5 min into the test. The emissive power of the flames (which radiate as a black body; $\epsilon \simeq 1$) is required to be $(158 \pm 8) \times 10^3$ W m^{-2}. The temperature in fires that fulfil these requirements may vary from less than 650 °C near the air-entraining edge of the plume to 1200 °C in a small central core amounting to about 10 per cent of the flame volume.

9.5   Burning

In this section, the burning of solid and liquid fuels in unconfined spaces will be discussed. The subject of burning in building spaces will be dealt with in the next chapter.

Liquid fuels and solids with melting points below their pyrolysis temperatures (such as waxes and many thermoplastic materials) usually burn in pool configuration. In contrast, char-forming materials can burn only for a limited time in pool configuration without aid from external heat sources. Shortly after their ignition, a char layer builds up on their surface which, by blocking thermal feedback from the flame, will quell the pyrolysis and, with it, the supply of volatile decomposition products to the flame.

Provided that the concentration of oxygen in the ambient atmosphere is adequate, all noncharring, nonmelting solids can burn without support from external heat sources in most geometric configurations, and char-forming solids (depending on size) can burn in a variety of configurations other than pool-like.

Figure 9.3 illustrates an object (a small cylindrical bar) made from a char-forming material. The inner core of the object (region f) consists of virgin fuel; the outer 'crazed' region (region c) is the char. Between them lies the zone of pyrolysis (p). The volatile decomposition products emerge from this zone through numerous cracks in the char, and combust mainly above and by the sides of the object. Swept by an air stream induced by the flame, a section of the char layer along the lower surface of the object oxidizes. Part of the heat of char-oxidation (which, per unit mass, is usually much higher than the heat of combustion of the volatile products) penetrates the pyrolysis zone. If this contribution to the heat of pyrolysis is large enough to make up for the attenuation of thermal feedback

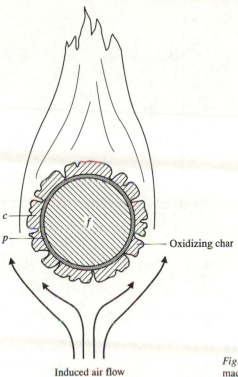

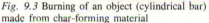
— Oxidizing char

Induced air flow

Fig. 9.3 Burning of an object (cylindrical bar) made from char-forming material

from the flame by the char cover along the top and side surfaces, sustained pyrolysis and burning are ensured.[17]

In the course of burning, the loss of mass by the object, dW (kg), during the time dt (s) consists of two parts: loss of mass by the virgin fuel due to the release of volatile pyrolysis products, αdW_f (kg), and loss of mass by the char due to char oxidation, dW_{ch} (kg). The rate of the mass change is

$$\frac{dW}{dt} = \alpha \frac{dW_f}{dt} + \frac{dW_{ch}}{dt} \qquad [9.12]$$

where α (kg kg^{-1}) is the mass fraction of the volatiles in the pyrolysis products (which consist of volatiles and char).

Figure 9.4(a) shows a typical mass (W) versus time (t) curve for a burning object (or a pile of objects) made from some char-forming material. Figure 9.4(b) is a model of the curve, devised on the assumption that both the pyrolysis of virgin fuel and the oxidation of char proceed at constant rate.

W_o is the original mass of the object. If char-oxidation did not occur, the solid

---

17 Whether sustained burning is possible without support from an external heat source depends on the shape and size of the object.

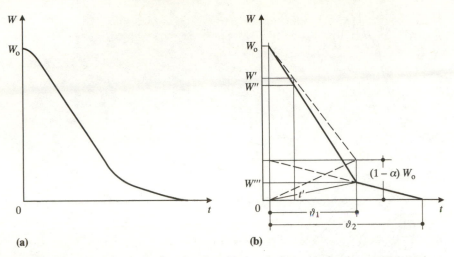

Fig. 9.4 Mass versus time curve for a burning object made from char-forming material: (a) actual curve; (b) model of curve

mass of the object at a time t' would be W'. If char-oxidation does occur, the fuel mass will only amount to W''.

The pyrolysis process is completed at a time of ϑ_1 (s). In the absence of char-oxidation, the mass of solid at that point would be equal to the mass of char formed in the pyrolysis process: $(1 - \alpha)W_o$. However, since char-oxidation has occurred, the mass of the object (as well as the mass of char) will only be W'''. According to the earlier assumption, the char will continue to oxidize beyond ϑ_1 at the same (constant) rate, and the total amount of char will be consumed[18] at a time of ϑ_2 (s).

The model curve can be formulated as follows.

For the period $0 < t \leq \vartheta_1$ (Period I)

$$\left(\frac{W}{W_o}\right)_I = 1 - t\left(\frac{\alpha}{\vartheta_1} + \frac{1 - \alpha}{\vartheta_2}\right) \tag{9.13a}$$

and for the period $\vartheta_1 < t \leq \vartheta_2$ (Period II)

$$\left(\frac{W}{W_o}\right)_{II} = (1 - \alpha)\left(1 - \frac{t}{\vartheta_2}\right) \tag{9.13b}$$

In these equations, α is more or less a constant determined by the pyrolysis stoichiometry; ϑ_1 and ϑ_2, on the other hand, can be looked upon as adjustable parameters, and selected to achieve the best agreement between the experimental W versus t curve and its model. However, as will be pointed out later, ϑ_1 and

---

18   It is assumed that the mass of the solid residue of char-oxidation is negligible.

ϑ_2 cannot be varied independently. The oxidation of char may provide the bulk of the heat consumed in the pyrolysis process, and therefore ϑ_1 and ϑ_2 are interrelated (see, for example, Eq. 9.31).

The slope of the W versus t curve, i.e, the *rate of burning*, is also of interest.

For the period $0 < t \leq \vartheta_1$ (Period I)

$$\frac{1}{W_o}\left(\frac{dW}{dt}\right)_{\text{I}} = -\left(\frac{\alpha}{\vartheta_1} + \frac{1-\alpha}{\vartheta_2}\right) \qquad [9.14a]$$

and for the period $\vartheta_1 < t < \vartheta_2$ (Period II)

$$\frac{1}{W_o}\left(\frac{dW}{dt}\right)_{\text{II}} \equiv \frac{1}{W_o}\left(\frac{dW_{\text{ch}}}{dt}\right) = -\frac{1-\alpha}{\vartheta_2} \qquad [9.14b]$$

The assumption that the pyrolysis and the oxidation of char proceed at constant rates is usually a fair approximation of reality. It would be unreasonable, however, to claim that for a given fuel the rates of pyrolysis and char-oxidation can be looked upon as universal constants. Yet, the common assertion that the rate of burning of wood proceeds at a constant rate of about $0.6\,\text{mm}\,\text{min}^{-1}$ (perpendicular to the surface) is, in essence, equivalent to such a claim. Emerging from time to time in the literature, that assertion has probably been based on measurements of the depths of char layers on wooden columns and beams exposed to standard test fires. Butler (1971) proved experimentally that the rate of burning of wood is not constant, but depends quite substantially on the irradiance of the object. However, Butler failed to consider other variables that are more significant than the radiant heat flux. The oxygen concentration of the surrounding atmosphere and the flow rate of air past the object, which determine the rate of char-oxidation and, to a large extent, the rate of pyrolysis are no doubt the principal factors in the rate of consumption of virgin wood and in the rate of buildup of surface char.

From the temporal constancy of the rate of burning one must not infer that (at least during Period I) the rate of heat release by the fuel, R (W), is also constant. In fact, R usually varies throughout the burning process, mainly because of the variation of the composition of volatile pyrolysis products. Yet, since allowing for the nonconstancy of R would substantially increase the complexity of dealing with the process, the variability of the heat release rate is routinely overlooked in engineering practice.

The net rate of heat release by the burning object, R, is equal to the rate of heat release in the flaming combustion of volatiles, plus the rate of heat release in the oxidation of char, minus the rate of heat absorption in the pyrolysis process.[19]

---

19 It should be remembered that dW/dt, dW_f/dt, dW_{ch}/dt, dw/dt, Δh_c, and Δh_{ch} are negative quantities. However, to avoid some confusions that might arise from insisting on the correct usage of signs, the sign convention will not be strictly followed. In general, the listed symbols should be looked upon as representing absolute values.

$$R = R_F + R_{ch} - \frac{dW_f}{dt} \Delta h_p \qquad [9.15]$$

where R_F = rate of heat release in flaming combustion, W;

R_{ch} = rate of heat release in char-oxidation, W;

Δh_p = heat of pyrolysis (enthalpy change in converting the virgin fuel into volatiles and char), $J\,kg^{-1}$.

In terms of changes of mass, R_F and R_{ch} can be expressed as follows:[20]

$$R_F = \alpha\, \frac{dW_f}{dt}\, \xi_F \Delta h_c \qquad [9.16]$$

$$R_{ch} = \frac{dW_{ch}}{dt}\, \xi_{ch} \Delta h_{ch} \qquad [9.17]$$

where Δh_c = (lower) heat of combustion of the volatile pyrolysis products, $J\,kg^{-1}$;

Δh_{ch} = heat of oxidation of char, $J\,kg^{-1}$;

ξ_F = factor quantifying incomplete combustion of the volatile pyrolysis products in the flame, dimensionless;

ξ_{ch} = factor quantifying incomplete oxidation of the char, dimensionless.[21]

Combining Eqs 9.15, 9.16, and 9.17, the rate of total heat release is

$$R = \frac{dW_f}{dt}\left(\alpha\xi_F \Delta h_c - \Delta h_p \right) + \frac{dW_{ch}}{dt}\, \xi_{ch} \Delta h_{ch} \qquad [9.18]$$

Since $\Delta h_p \ll \Delta h_c$, it is common practice to ignore Δh_p, or to include it in Δh_c.

Various methods have been developed for studying the total heat release rate. The more or less conventional techniques suggested by Smith (1972) and Parker and Long (1972) have been by now largely superseded by an ingenious new technique called oxygen consumption calorimetry. Since the heat of burning (i.e. heat of flaming combustion plus char-oxidation) per unit mass of oxygen, Δh_b^O [$J\,(kg\,O_2)^{-1}$], is approximately the same, about $(13.1 \pm 0.7)\,MJ\,kg^{-1}$, for most organic materials found in buildings (see Table 9.2),[22] Huggett (1980) suggested that the history of heat release in a burning process be followed up by determining the history of oxygen consumption. It can be achieved by monitoring the oxygen concentration and flow rate of the fire gases.[23] Parker (1982) and Babrauskas

---

20 R_F may be slightly higher than the value calculable from Eq. 9.16, because the oxidation of char produces some carbon monoxide which may end up in the flame.

21 Since the appropriate values of ξ_F and ξ_{ch} are difficult to assess, it is customary to assume that they are both approximately equal to 1.

22 Unfortunately, wood seems to be one of several exceptions.

(1982a) worked out the details of putting Huggett's idea to practical use. Babrauskas's 'cone calorimeter' has since been adopted as an ASTM standard (American Society for Testing and Materials 1990d).

The energy balance for the pyrolysis zone in the object shown in Fig. 9.3 is

$$\frac{dW_f}{dt}\,\Delta h_p + Q_L = Q_F + Q_{ch} + Q_E \qquad\qquad [9.19]$$

where Q_L = rate of heat loss at the boundaries due to (a) radiant and convective losses, (b) conduction of heat into the virgin fuel, and (c) decrease of the mass of solid fuel, W;

Q_F = rate of heat gain from energy released in the flame, W;

Q_{ch} = rate of heat gain from energy released in the oxidation of char, W;

Q_E = rate of heat gain from external sources, W.

Q_F and Q_{ch} depend on two different sets of variables. Q_F depends largely on the rate of heat release in the flaming combustion, and therefore on the rate of volatiles production in the pyrolysis process. It also depends on the magnitude of thermal feedback, which usually takes place by a combination of convection and radiation. Q_{ch}, on the other hand, originates from heat release in a heterogeneous (char—oxygen) reaction whose rate is governed by the rate at which air is transferred to the char—air interface by convection and/or diffusion, and into the char zone by diffusion. Heat transfer from the glowing char to the pyrolysis zone and to the virgin fuel takes place by conduction.

The presence of the term Q_{ch} is conditional on the geometry of the burning object. In a still atmosphere, substantial char-oxidation can only occur if the geometry of the object is such that parts of its surface become exposed to air flow induced by the flame. However, it is almost always possible to bring about char-oxidation by the application of forced convection.[24]

The rate of heat gain from the flame, Q_F, can be expressed as a fraction of the rate of heat release by the flame, R_F.

$$Q_F = f_F R_F \qquad\qquad [9.20]$$

Similarly, the rate of heat gain from the oxidizing char, Q_{ch}, can be expressed as

$$Q_{ch} = f_{ch} R_{ch} \qquad\qquad [9.21]$$

---

23 According to Krause and Gann (1980), the fact that the oxidation processes are usually incomplete does not lead to serious errors. If all the carbon atoms in the organic compounds were converted to CO, the value adopted for Δh_b^O (13.1 MJ per kg oxygen) would prove about 30 per cent too high. However, it would prove 20 to 25 per cent too low if all the carbon appeared as soot. Since the smoke usually contains both CO and soot, the overall error is not expected to be more than about 5 per cent.

24 Everyone who has ever tried to get a wood fire started knows that blowing air at the seat of fire will cause the charred wood to glow and the flames to flare up.

In Eqs 9.20 and 9.21

f_F = that fraction of the heat released in the flaming combustion which is transferred back to the burning object by convection and radiation, dimensionless;

f_{ch} = that fraction of the heat released in the oxidation of surface char which is transferred to the pyrolysis zone by conduction, dimensionless.

After substitutions from Eqs 9.16, 9.17, 9.20, and 9.21, Eq. 9.19 becomes

$$\frac{dW_f}{dt}\left(\alpha f_F\xi_F\Delta h_c - \Delta h_p\right) + \frac{dW_{ch}}{dt}f_{ch}\xi_{ch}\Delta h_{ch} + Q_E - Q_L = 0 \qquad [9.22]$$

In the case of noncharring materials[25] (for which $\alpha = 1$, $dW_{ch} = 0$, and $dW_f \equiv dW$)

$$\frac{dW}{dt}\left(f_F\xi_F\Delta h_c - \Delta h_p\right) + Q_E - Q_L = 0 \qquad [9.23]$$

Although char-forming materials are more abundant in buildings than noncharring materials, it has been common practice among fire researchers to investigate some basic aspects of the burning process using liquids and noncharring solids, and to deal with charring solids as more or less aberrant cases. This practice, while understandable in light of the problems presented by the presence of char, has lead to numerous misconceptions concerning the nature of burning of piles of fuels and the nature of fires in enclosures.

In an effort to develop a technique for quantifying the *flammability*[26] of solid fuels, Tewarson and Pion (1976) studied the rate of burning of a large number of fuels, mainly noncharring plastics. Their test samples, $0.006-0.010 \, m^2$ in area and $30-50 \, mm$ in thickness, were mounted horizontally in a quartz tube of 170 mm inside diameter, and weighed continuously as they burned (with or without aid from external radiant heat sources) in an upward stream of a mixture of nitrogen and oxygen.

Tewarson and Pion (1976) expressed the heat balance in terms of heat fluxes referred to the area of the top surface of the samples. From Eq. 9.19 (in which, for noncharring materials, $Q_{ch} = 0$ and $dW_f \equiv dW$):

$$\frac{dw}{dt}\Delta h_p = q_F + q_E - q_L \qquad [9.24]$$

where q_F, q_E, and q_L ($W \, m^{-2}$) are heat fluxes, corresponding to the heat flow rates Q_F, Q_E, and Q_L in Eq. 9.19, and w ($kg \, m^{-2}$) is mass per unit surface area, corresponding to W.

---

25 For simplicity, Δh_p will be used to denote the heat of pyrolysis as well as the heat of vaporization (sublimation).

26 Materials that are subject to easy ignition and rapid flaming combustion are referred to as flammable materials. Since flammability is a rather vague term, its use is generally discouraged.

Investigating the effect of the oxygen concentration in the gas stream, x (kmol kmol^{-1}), on the rate of burning of the samples, Tewarson and Pion found that the experimental results could be satisfactorily correlated by the following equation:[27]

$$\frac{\mathrm{d}w}{\mathrm{d}t} = \frac{x}{0.21}\left(\frac{\mathrm{d}w}{\mathrm{d}t}\right)_{\mathrm{id}} + \frac{q_{\mathrm{E}} - q_{\mathrm{L}}}{\Delta h_{\mathrm{p}}} \qquad [9.25]$$

where $(\mathrm{d}w/\mathrm{d}t)_{\mathrm{id}}$ was referred to as 'ideal' burning rate: the rate of (quasi-steady-state) burning in a normal atmosphere ($x = 0.21$), if $q_{\mathrm{E}} = q_{\mathrm{L}}$. Tewarson and Pion (1976) claimed that ideal burning rate was a characteristic of the material with respect to its 'flammability'.

Table 9.2 gives some information on the pyrolysis and burning characteristics of selected materials, including the parameters developed by Tewarson and Pion. One way of using this kind of information is to assess if a material is capable of sustained burning under normal atmospheric conditions ($x = 0.21$). The condition of burning without support by an external heat source ($q_{\mathrm{E}} = 0$) is

$$\left(\frac{\mathrm{d}w}{\mathrm{d}t}\right)_{\mathrm{id}} - \frac{q_{\mathrm{L}}}{\Delta h_{\mathrm{p}}} > 0 \qquad [9.26]$$

The oxygen index test, proposed by Fenimore and Martin (1966) and later adopted by the ASTM as a standard test procedure (American Society for Testing and Materials 1987a), is a simple technique for assessing the 'flammability' (burning characteristics) of solid fuels. The test consists of determining the minimum oxygen concentration of the atmosphere in which the flaming combustion of a candle-like specimen of the material can be maintained. This oxygen concentration, usually expressed in moles (or volume) per cent, is called the *limiting oxygen index*, or *LOI*. Clearly, materials with *LOI*s higher than 21 cannot burn under normal atmospheric conditions without support from an external heat source.

The last column in Table 9.2 gives the *LOI*s for the materials listed.

Petrella (1979) pointed out the relation between Tewarson and Pion's ideal burning rate and the limiting oxygen index. He claimed that the *LOI* could be defined as the oxygen content at which (with $q_{\mathrm{E}} = 0$) the rate of heat gain from the flame, q_{F}, is equal to the rate of heat losses from the sample's surface, q_{L}. Thus, from Eqs 9.24 and 9.25,

---

27 Δh_{p} was interpreted to include the heat needed for raising the temperature of the sample into the range of pyrolysis. How Tewarson and Pion interpreted $\mathrm{d}w/\mathrm{d}t$ is not clear from their paper. Because of the small size of the samples, the burning apparently was not a steady-state process. $\mathrm{d}w/\mathrm{d}t$ increased during the test, whereas Δh_{p} undoubtedly decreased. It is possible, however, that the product of the two remained approximately constant.

The rate of heat loss, q_{L}, did not seem to be noticeably affected by changes in the rate of burning brought about by changes in the oxygen concentration of the gas stream.

$$LOI = 100x \simeq \frac{21q_L}{\Delta h_p \left(\dfrac{dw}{dt}\right)_{id}}$$ 　　　　[9.27]

The *LOI* values calculated from this equation seemed to compare favorably with experimental values.

The pitfall in relying on either Tewarson and Pion's method or the oxygen index method in the ranking of various materials is that, because of the small size of the samples, the test results do not necessarily reflect burning behaviors under real-world conditions. If the flame is small (0.1–0.3 m), thermal feedback from the flame to the object takes place mainly by convection. The flames in real-world fires are usually larger than 0.3 m, and radiation is the dominant mode of feedback. Studying the burning of a poly(methyl methacrylate) slab 1.22 m square, Modak and Croce (1977) found that more than 80 per cent of the feedback was by radiation. However, the magnitude of feedback depends not only on the size but also on the luminosity of the flame and, in turn, on the concentration of radiating soot particles.

Researchers have always found it convenient to study the characteristics of fires in real-life dimensions on liquids and noncharring solids burning in pool configuration. In their pioneering work, Blinov and Khudiakov (1957) measured the rate of burning, dw/dt, of various liquid fuels in cylindrical pans, ranging in diameter, D, from 0.0037 to 22.9 m, and estimated the flame heights. Their findings indicated (Hottel 1959) that, as long as D was less than 0.05 m, the flames were laminar and the rate of burning decreased with the increase of pan diameter. As D was increased from 0.05 to 1.0 m, the flames grew more and more turbulent, and the rates of burning remained low. For $D > 1.0$ m, the flames were fully turbulent, and the rates of burning became roughly independent of the pan diameter. The flame height to pan diameter ratio, h_F/D, seemed to decrease steadily from about 10 to 1.5 as the pan diameter was increased from 0.0037 to 22.9 m.

If D is large (larger than, say, 1.0 m), one may assume that thermal feedback from the flame takes place predominantly by radiation, and that the heat loss from the pool is negligible (i.e. $q_L \simeq 0$). If there is no heat supply from external sources, $q_E = 0$ also. Thus, from Eqs 9.24, 9.10, and 9.11, the rate of burning is

$$\frac{dw}{dt} \simeq \frac{1}{\Delta h_p} \sigma T_F^4 [1 - \exp(-al)]$$ 　　　　[9.28]

where σT_F^4 (W m^{-2}) is the black-body emissive power of the flame.[28] The term between the brackets represents the flame's emissivity.

The empirical equation recommended by Burgess *et al.* (1961) is of a form similar to Eq. 9.28.

---

28 Since the temperature of the pool surface is much lower than the flame temperature, reradiation from the surface can be neglected.

$$\frac{dw}{dt} = \left(\frac{dw}{dt}\right)_\infty [1 - \exp(-\psi D)] \qquad\qquad [9.29]$$

where $(dw/dt)_\infty$ is the burning rate for $D \to \infty$, and ψ (m^{-1}) is an empirical constant.

When it comes to practical aspects of fire safety, knowing the rate of burning of liquids and noncharring solids in pool configuration is of rather limited value. Ever since society became conscious of the fire problem, the majority of fire protection measures have been inspired by the fact that wood, a char-forming material, is the most abundant burnable material in buildings. Shaped into furnishing and other household items, wood is usually present in configurations that allow char-oxidation to play a substantial part in the burning process. The glowing char surfaces also serve as external radiant-heat sources for other items burning nearby.

When many pieces of wood burn in a *pile* configuration, a large proportion of the heat radiated by the glowing char surfaces becomes trapped within the cluster. Thus, the rate of burning of a pile of wood is (per unit mass) much higher than that of the constituent pieces when separated from the pile.

The following model is suggested for the burning of a *large* pile.

- The heat loss from the pile as a whole is negligible; consequently, heat loss from one object in the pile becomes heat gain for other objects ($Q_L \simeq Q_E$ in Eq. 9.19).
- Since there is no net heat loss from the individual objects, all the heat released in char-oxidation is consumed by the pyrolysis process ($f_{ch} \simeq 1$ in Eq. 9.21).
- The space within the pile is insufficient for satisfactory intermixing between the volatile pyrolysis products and air; as a result, the volatiles leave the pile largely uncombusted, and oxidize above the pile.
- The combustion of the volatiles above the pile creates an updraft of air within the pile which drives the oxidation of char.
- Because of the large size of the pile, the flames cannot 'see' into its interior, and therefore cannot feed the pyrolysis of the objects inside the pile with radiant heat ($Q_F \simeq 0$ in Eq. 9.19); hence, the pyrolysis is driven mainly by heat released in the oxidation of char.

It appears from the work of McCarter and Broido (1965) that this model is a fair approximation of the actual conditions. In their experiments, they eliminated the flame above the burning wood pile by quenching it with carbon dioxide jets. They found that the presence or absence of flames above burning wood piles had little effect on the rate of burning. 'It is highly unlikely', they wrote, 'that radiation from the flames above the fuel bed is the dominant energy flux regulating propagation of the fire.'

These findings are very important in understanding the true nature of burning of piles of char-forming materials (e.g. furniture) in enclosed spaces (to be discussed in the next chapter). Since the rate of burning (dW/dt) of these piles apparently does not rely appreciably on thermal feedback from the flames, a significant proportion of the volatiles released in the pyrolysis process may combust outside the compartment without noticeably affecting the burning process.

After substitutions of $Q_L = Q_E$, $Q_F = 0$, $f_{ch} = 1$, dW_f/dt from Eq. 9.12 and Q_{ch} from Eqs 9.17 and 9.21, Eq. 9.19 becomes

$$\frac{dW}{dt} = \frac{dW_{ch}}{dt}\left(1 + \alpha\xi_{ch}\frac{\Delta h_{ch}}{\Delta h_p}\right) \qquad [9.30]$$

which is an expression of rate of mass loss (i.e. rate of burning) for the pile.

Equation 9.30 reveals that the rate of burning of large piles of char-forming fuels is controlled by the rate of char-oxidation and, in turn, by the flow rate of air through the pile. The important role the oxidation of char plays in the burning of char-forming solids in pile configurations was demonstrated by Harmathy (1978a). He showed that char-forming materials burn faster if the flow rate of air through the pile is increased, whereas the rate of burning of noncharring materials remains virtually unaffected by the flow rate of air.

Substituting the expression of dW/dt from Eq. 9.14a and the expression of dW_{ch}/dt from Eq. 9.14b into Eq. 9.30 yields

$$\frac{\vartheta_2}{\vartheta_1} = (1 - \alpha)\xi_{ch}\frac{\Delta h_{ch}}{\Delta h_p} \qquad [9.31]$$

This equation, because of the several assumptions used in its derivation, is unlikely to give a handle on assessing the $\Delta h_{ch}/\Delta h_p$ ratio from the shape of an experimental W versus t curve. It suggests, however, that ϑ_2/ϑ_1 is probably constant for a given char-forming fuel under similar conditions of burning.

The following material constants may be used for wood (Roberts 1964a; Harmathy 1972), or cellulosic materials in general:

- $\alpha \simeq 0.87$.
- Heat of combustion of the volatiles, $\Delta h_c \simeq 16.7\,\text{MJ}\,\text{kg}^{-1}$.
- Heat of char oxidation, $\Delta h_{ch} \simeq 33.4\,\text{MJ}\,\text{kg}^{-1}$.
- Heat of burning (heat of combustion of volatiles, Δh_c, plus heat of char oxidation, Δh_{ch}, minus heat of pyrolysis, Δh_p), $\Delta h_b \simeq 17.0\,\text{MJ}$ per kg wood.
- Heat of burning per unit mass of oxygen, $\Delta h_b^0 \simeq 14.3\,\text{MJ}\,\text{kg}^{-1}$.

According to Eq. 9.6, the stoichiometric air requirements are:

- For combustion of the volatiles, $\simeq 4.2\,\text{kg}\,\text{kg}^{-1}$.
- For char-oxidation, $\simeq 11.4\,\text{kg}\,\text{kg}^{-1}$.
- For burning (combustion of volatiles plus char-oxidation), $\simeq 5.1\,\text{kg}\,\text{kg}^{-1}$.

Since the $CO_2/(CO_2 + CO)$ ratio in the products of burning is usually about 0.6, the actual air requirement is only about 80 per cent of the above values.

The heat of pyrolysis for wood is difficult to assess. The average of reported values seems to be $\Delta h_p \simeq 1.8 \, \text{MJ kg}^{-1}$. Of course, all the values listed above depend to some extent on the species of wood and some characteristics of the burning process.

The design of wood cribs for maximum rates of burning involves finding the best compromise between two conflicting requirements:

(1) The crib should be sufficiently large and the sticks should be placed sufficiently close together, so that the radiant-heat loss from the crib is minimal.

(2) The sticks should not be so close together as to severely impede the natural (flame-induced) ventilation of the crib.

Gross's experiments (1962) revealed that with densely packed cross cribs of wood, the burning grows more vigorous as the 'porosity' of the crib (i.e. its ventilation) is increased. Above a certain porosity level, however, the rate of burning becomes roughly independent of porosity, and at very high porosities, presumably because of the large radiant-heat losses, sustained burning is not possible.

As Table 1.4 shows, most fire deaths occur in residential buildings. The objects involved in the fire are furnishing items, usually pieces of upholstered furniture or mattresses, ignited by smokers' materials: cigarettes and matches. The burning process is very often of the smoldering kind, and the cause of death is smoke inhalation. Sumi and D'Souza (1982) reviewed the strategies that had been developed to improve fire safety in residential occupancies, and concluded that fire-related casualties could be significantly reduced by the use of furnishings that resist ignition by cigarettes.

Smoke produced by smoldering is a common cause of deaths and nonfatal casualties. Since, as Table 9.1 shows, only char-forming materials are capable of sustained smoldering in the absence of an external heat source, avoiding the use of cellulosic materials (such as cotton felt) and highly crosslinked, thermosetting plastics in the soft coverings of upholstered furniture and in mattresses would greatly improve fire safety, especially in buildings inhabited or frequented mostly by smokers. The use of thermoplastic materials, which tend to melt when subjected to smoldering cigarettes, seems to be preferable, even though these materials may burn more rapidly when exposed to flame.

Sumi and D'Souza (1982) surveyed a number of organizations in North America and the UK that conduct research into the safety of furnishing items and are involved in writing safety regulations. Two ASTM standards (American Society for Testing and Materials 1990b,c) address the problem of resistance of upholstered furniture to cigarette ignition.

9.6   Ignition and Ignitability

Studying the conditions of ignition, Rasbash (1975) used the following modified version of Eq. 9.23:[29]

---

29 In fact, since char-oxidation plays little role in ignition, Eq. 9.32 may also be regarded as a modified version of Eq. 9.22.

$$\frac{dW}{dt}[(f_{FC} + f_{FR})\xi_F \Delta h_c - \Delta h_p] + Q_E - Q_L = S \qquad [9.32]$$

where f_{FC} = that fraction of the heat released in the flaming combustion which
is transferred back to the burning object by convection,
dimensionless;

f_{FR} = that fraction of the heat released in the flaming combustion which
is transferred back to the burning object by radiation, dimensionless;

S = dummy variable (= 0 for steady burning), W.

As pointed out earlier, once a fire has advanced to the stage of steady burning,
transfer of heat from the flame to the burning object will probably take place
mainly by radiation. However, at the start of flaming combustion f_{FR} is usually
much smaller (owing to the small size of a nascent flame) than f_{FC}. Setting f_{FR}
= 0 and $f_{FC}\xi_F = \varphi$ (dimensionless), and writing $dW/dt = (dW/dt)_{in}$ for the
incipient rate of mass loss by the pyrolysis of the virgin fuel, the following equation
is obtained for the conditions under which sustained burning is possible:

$$\left(\frac{dW}{dt}\right)_{in} (\varphi \Delta h_c - \Delta h_p) + Q_E - Q_L \geq S \qquad [9.33]$$

where $\varphi \simeq 0.3$ for many burnable solids. Values of $(dW/dt)_{in}$ and Δh_p are
available for a number of plastics (see, for example, Drysdale 1985),[30] and Q_E
and Q_L can be calculated from information on fuel geometry, material properties,
and boundary conditions. Thus, Eq. 9.33 can, in principle, be used for assessing
the liability of a burnable object to *piloted ignition*[31] under a given set of
conditions.

Q_L in Eq. 9.33 includes not only the heat loss to the surroundings, but also
the heat transmitted by conduction to the interior of the object. This latter
contribution decreases as the temperature of the object rises and approaches the
firepoint, T_{fp} (K), at which the surface temperature is in the vicinity of the
pyrolysis temperature, and the distribution of subsurface temperature is such as
to allow the evolution of volatile pyrolysis products at a rate sufficient to support
continuous flaming combustion.

In order to bring an object to its firepoint, an external heat source, usually a
radiant energy source, is required. The time the object takes to reach the firepoint
is a function of the magnitude of the radiant heat flux falling on its surface. Lawson
and Simms (1952) investigated the relation between irradiance and time to ignition
for vertically held samples of wood. They expressed their findings for oven-dry
wood by two empirical equations. In the case of piloted ignition,

---

30 Table 9.2 also lists values of Δh_p for several materials.
31 In piloted ignition, the combustion of the volatile pyrolysis products is started by the application
of a small flame or an electric spark.

$$[q_R - (q_R)_{\min}]t_{ig}^{2/3} = 0.6[(\sqrt{k\rho c})^2 + 0.119 \times 10^6] \qquad [9.34]$$

and in the case of spontaneous ignition,

$$[q_R - (q_R)_{\min}]t_{ig}^{4/5} = 1.2[(\sqrt{k\rho c})^2 + 0.061 \times 10^6] \qquad [9.35]$$

where q_R = irradiance of object, $W\,m^{-2}$;
$(q_R)_{\min}$ = minimum irradiance at which (as $t \to \infty$) ignition is possible, $W\,m^{-2}$;
t_{ig} = time to ignition, s;
$\sqrt{k\rho c}$ = thermal absorptivity of wood,[32] $J\,m^{-2}\,s^{-1/2}\,K^{-1}$;

and 0.6, 1.2, 0.119 $\times\, 10^6$, and 0.061 $\times\, 10^6$ are empirical constants of various dimensions. For all species, $(q_R)_{\min} \simeq 13.3 \times 10^3\,W\,m^{-2}$ for piloted ignition, and $(q_R)_{\min} \simeq 25.7\,W\,m^{-2}$ for spontaneous ignition. The term $\sqrt{k\rho c}$ ranges from 210 to 440 $J\,m^{-2}\,s^{-1/2}\,K^{-1}$ for the most common species, in oven-dry condition.

In general, the relation between q_R and t_{ig} may be described by a relation of the form

$$[q_R - (q_R)_{\min}]\, t_{ig}^n = F \qquad [9.36]$$

where the exponent n (dimensionless) is usually close to 1, and F is a quantity characterizing the difficulty of igniting the object.

Quintiere and Harkleroad (1985) suggested that the relation between q_R and t_{ig} can also be expressed as

$$\frac{q_R}{(q_R)_{\min}} = \left(\frac{t^*}{t_{ig}}\right)^{1/2} \qquad [9.37]$$

where t^* (s) is defined as

$$t^* = K(\sqrt{k\rho c})^2 \qquad [9.38]$$

and K ($m^4\,s^2\,K^2\,J^{-2}$) is assumed to be a function of the firepoint. From the fact that $q_R \geq (q_R)_{\min}$, it follows that $t_{ig} \leq t^*$.

Ignitability is a rather loose term. In general, t_{ig} (at any value of q_R) is regarded as an inverse measure of the ignitability of an object. It depends on the orientation and geometry of the object, the nature of heat exchange between the object and its environment, and the oxygen content of the ambient atmosphere, in addition to material properties. Among the variables characterizing the material, the firepoint, T_{fp}, and the thermal absorptivity, $\sqrt{k\rho c}$, are the most significant. The difficulty of ignition increases with both the firepoint and the thermal absorptivity of the material. Among the geometric variables, the thickness of the object is probably the most important. In general, it is more difficult to ignite a thick object than a thin one.

---

32 See Footnote 8 in Chapter 3.

9.7   Flame Spread

The creeping spread of flame along the surface of a single object is, in essence, an advancing ignition process, in which the leading edge of the flame acts both as source of heat and tool for piloted ignition.

The basic equation of steady flame spread is quite simple:

$$\rho v_F (h_{fp} - h_o) = q \qquad [9.39]$$

where ρ = density of the object, $kg\,m^{-3}$;
$\quad v_F$ = flame spread rate, $m\,s^{-1}$;
$\quad h_{fp}$ = enthalpy of material at the firepoint, $J\,kg^{-1}$;
$\quad h_o$ = enthalpy of material at the initial temperature, $J\,kg^{-1}$;
$\quad q$ = rate of heat transfer from the flame to the object, $W\,m^{-2}$.

Yet, finding the explicit form of Eq. 9.39 is a very demanding task. The formulation of the problem involves variables characterizing not only the nature of the object, but also the object's orientation and geometry, and the ambient conditions. Because of the insufficient understanding of the overall process under a great diversity of conditions that may arise, and because of the scarcity of information on the applicable values of material properties, the available formulas for flame spread are of rather limited practical value.

It has been usual to formulate the laws of creeping flame spread according to whether the solid is thermally thin or thick. If the solid is so thin that its temperature can be regarded as approximately uniform across its thickness, the treatment of the problem is somewhat simpler. Among the common burnable materials, those with thicknesses less than 2 mm may be treated as thermally thin, and those with thicknesses more than about 15 mm as thermally thick.

Whether the flame does or does not impinge on the surface ahead of the pyrolysis front is an important factor in the rate of spread. It depends on the orientation of the surface and on the velocity of air flow in the vicinity of the object. If the surface is horizontally or laterally oriented, or vertically oriented with the flame spreading downward, the air flow induced by the flame will oppose the spread and prevent the flame from impinging on the surface. The spread will not be very sensitive to the velocity of the air stream as long as it is less than about $0.3\,m\,s^{-1}$ (which is normally the case in the absence of superimposed air currents).

Mathematical models of various levels of sophistication, applicable to the spread of flame over flat surfaces horizontally, laterally, vertically downward (deRis 1969; Lastrina et al. 1971; Parker 1972; Fernandez-Pello 1977), and vertically upward (Fernandez-Pello 1977; Fernandez-Pello 1978; Hasemi 1986) have been developed. Excellent reviews on the subject have been presented by Williams (1976), Drysdale (1985), and Quintiere (1988c).

According to Quintiere (1988c), the formulas derived by various research workers for the rate of spread of flame, v_F, advancing horizontally, laterally, or vertically downward, and opposed by an air flow of velocity, v_a (m s^{-1}), may be presented in the following general forms. For thermally thin objects,

$$v_F = \frac{1}{\rho c \delta (T_{fp} - T_o)} \phi_1(v_a, x, \Pi, G) \qquad [9.40]$$

and for thermally thick objects,

$$v_F = \frac{1}{[\sqrt{k\rho c}(T_{fp} - T_o)]^2} \phi_2(v_a, x, \Pi, G) \qquad [9.41]$$

where ϕ_1 and ϕ_2 denote unspecified functions, $\sqrt{k\rho c}$ is the thermal absorptivity of the solid, T_{fp} (K) and T_o (K) are the firepoint and the initial temperature of the solid, respectively, x is the oxygen concentration of the air stream, and

ρ = density of the solid, kg m^{-3};
c = specific heat of the solid, J kg^{-1} K^{-1};
δ = thickness of the thermally thin object, m;
Π = material properties not mentioned so far, various dimensions;
G = geometric variables other than thickness, m.

In dealing with flames spreading horizontally, laterally, or vertically downward, it is usually assumed that the transmission of heat from the flame to the fuel takes place mainly by conduction through the air, if the object is thermally thin, and mainly by conduction through the solid, if the object is thermally thick.

If the object is vertically or near-vertically oriented and the spread is upward, the flame will impinge on its surface ahead of the pyrolysis front. Saito *et al.* (1986) investigated the upward spread of flame over thermally thick vertical sheets of poly(methyl methacrylate) and wood. The sheets were ignited and supplied with heat by a line-source burner located at the base. Sustained spread occurred with poly(methyl methacrylate), but did not occur with wood.

The upward propagation of flame is a problem of special significance with clothing articles and furnishing items made from textiles. The induced air flow assists the spread in the case of noncharring materials, and is a 'make-or-break' factor in the case of char-forming materials. Heat transfer from the flame to the virgin fuel takes place mainly by convection and radiation. The rate of flame spread will be roughly proportional to the length of the pyrolysis zone, l_p (m) (Markstein and deRis 1972).

$$v_F = b l_p^n \qquad [9.42]$$

where the dimensional coefficient b is characteristic of the material and some of the object's geometric features (e.g. its thickness), and n (dimensionless) is a constant, its value ranging from 0.5 to 0.75.

After reaching a steady state, the rate of spread may be as high as 0.45 m s^{-1} (Markstein and deRis 1972), which is almost 400 times greater than the rate of spread downward on computer cards, 0.0012 m s^{-1}, measured by Hirano *et al.* (1974).

Experimenting with thin sheets of filter paper (a char-forming material), Loh and Fernandez-Pello (1986) found that concurrent air movement enhanced the rate of flame spread. Under forced-flow conditions, the effect of air flow seemed to become independent of the flow velocity.

Imposed radiation assists the spread of flame. According to Quintiere (1981), the rate of flame spread can be expressed as

$$v_F = \frac{B}{[(q_R)_{min} - q_R]^2} \tag{9.43}$$

where, as in Eq. 9.37, q_R and $(q_R)_{min}$ (W m^{-2}) are, respectively, the irradiance of the object and the minimum irradiance at which piloted ignition is possible. B (J^2 m^{-3} s^{-3}) is a factor that depends primarily on the thermal absorptivity ($\sqrt{k\rho c}$), firepoint, T_{fp}, and geometry of the material.

The rate of spread will be extremely high if $q_R \rightarrow (q_R)_{min}$ and sufficient time is allowed for the object to reach its firepoint. Such a condition arises in a compartment on fire at the time of flashover which, as will be discussed later in this chapter, is claimed to occur when the irradiance of the floor reaches a level of about 20×10^3 W m^{-2}.

Investigating the characteristics of flame spread over various carpets exposed to radiant heat fluxes ranging from 4×10^3 to 11.5×10^3 W m^{-2}, Kashiwagi (1974) found that, for the carpets studied, a minimum heat flux was necessary to sustain spread following piloted ignition. With increasing irradiance, the rate of flame spread increased sharply, and reached several centimeters per second. An increase in pilot ignition delay time also increased the rate of spread.

Since the propensity of an object to propagate flame in the presence of an external heat source is related to its ignitability, it is believed that these two properties may be evaluated jointly from a single test. In North America, the ASTM E 162 test apparatus (American Society for Testing and Materials 1987b) has been looked upon as the appropriate test. The heat source in that test is a vertical, gas-heated panel which radiates at 44.8×10^3 W m^{-2} power. The test specimen is placed at a 30 ° angle from the vertical in front of the panel, and ignited near its upper edge, which is set at 120 mm from the radiant panel. The flame spread index for the material is evaluated from the rate of downward travel of the flame front, and from the heat released by the specimen, as determined from the rise of temperature in a stack located above the specimen.

Modern variants of this method are the ASTM E 1321 (American Society for Testing and Materials 1990a) and BS 476 (British Standards Institution 1987) test methods, and a draft ISO method (International Standards Organization 1991). The E 1321 method was devised by Robertson (1979, 1984), and the technique of evaluating the test results is based on Quintiere's work (1981, 1988a). Measured in this test is the rate of lateral spread of flame along a specimen, 155 mm high and 800 mm long, placed with its long axis at an angle of 15 ° to a radiant panel. The radiant heat flux falling on the specimen decreases from 50×10^3 to

$1.5 \times 10^3 \, \text{W m}^{-2}$ as the distance between it and the radiant panel increases. The specimen is ignited by a pilot flame at the end close to the radiant panel, and the 'flame-spread parameter' for the material is evaluated from the rate of advance of the flame front and from the radiant heat flux at the position where the flame stops.

Creeping flame spread along a single object takes place mainly by conductive or convective heat transfer from the flame to the burning object. In contrast, the spread of large flames and the spread of fire from one object to another occur mainly on account of radiation, and are governed by the rate of heat release by the burning object and the emissivity of the flame. Consequently, the sootiness of the flame, which is to a large extent responsible for its emissivity, plays an important role in the spread. deRis (1985) suggested a bench-scale test, capable of providing for the simultaneous measurement of both the heat release rate and soot formation in the combustion of volatiles produced in the pyrolysis process.

The ASTM E 84 test apparatus (American Society for Testing and Materials 1989a; see Chapter 3) is in North America the primary instrument for determining the propensity of ceiling and wall surfaces for propagating flames of considerable size. Building code regulations often refer to flame spread ratings developed by the E 84 test method (or its Canadian equivalent) when imposing restrictions on the flame spread characteristics of materials. In general, the use of interior finishes having flame spread ratings higher than 150−200 is prohibited in buildings of dense occupancy.[33] Based on data presented by Castino et al. (1975), Benjamin (1977) claimed, however, that the merit sequences derived from the E 84 test are not necessarily valid.

The E 84 test was visualized to simulate the draft-aided spread of well developed flames over ceiling-mounted specimens. Quintiere (1985) found, however, that the rate of heat release by the combustion of volatiles produced in the pyrolysis zone of the specimen, which is apparently the principal factor in the spread of flame, depends at least as much on the conditions specified in the test as on the properties of the material tested.

In order to circumvent the often-contradictory conclusions drawn from the results of the E 84 test, building regulators sometimes require the so-called 'mini-corner' test (American Society for Testing and Materials 1988c) for determining the flame spreading characteristics of various compartment lining materials. The test specimen consists of two walls, 1.63 m high and 1.22 m wide, and a ceiling, 1.22 m by 1.22 m, assembled to form a covered corner. The material is ignited with a gas burner, and the flame spread, temperature rise, and damage are recorded.

Nonstandard 'mini-room' arrangements are also used in exploratory experiments. Belles et al. (1988) conducted a series of tests by means of mini-rooms measuring 2.44 m by 3.66 m by 2.44 m high, to study the fire behavior

---

33 Further restrictions are imposed on the flame spread ratings of lining materials installed in exits.

of ordinary carpets applied to walls and ceilings. They concluded that the performance of such lining materials could not be reliably predicted from the E 84 test results.

Another area where the results of the E 84 test were demonstrated to offer misleading conclusions is the vertical spread of fire over burnable exterior walls. Using a full-scale test facility consisting of a burn room and a section of building facade, Oleszkiewicz (1990) studied the behavior of various types of cladding exposed to flames issuing from the burn room, and measured the radiative and total heat fluxes at the surface of the cladding.

9.8   Preflashover Fire

In this section, the development of a fire in ordinary compartments (rooms) will be discussed. The conclusions may not be applicable to fires in large, uncompartmented spaces and corridors.

A full fire process is pictured as consisting of three stages (Fig. 9.5). They are usually defined in terms of the average temperature of compartment gases, T_g (K). During the *preflashover* or *growth period*, the fire is more or less localized, and the average temperature of the compartment atmosphere is relatively low. With the onset of the *period of fully developed burning* or *postflashover period*, the process of burning extends to the entire compartment, and the temperature becomes very high, typically 800 °C. The third period, the *decay period*, is judged to begin when the temperature of the compartment gases declines to 80 per cent of its peak value. This is approximately the point at which the

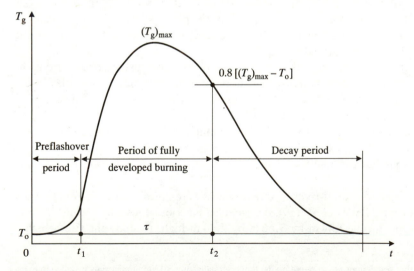

Fig. 9.5 Typical temperature history of compartment gases during a full fire process

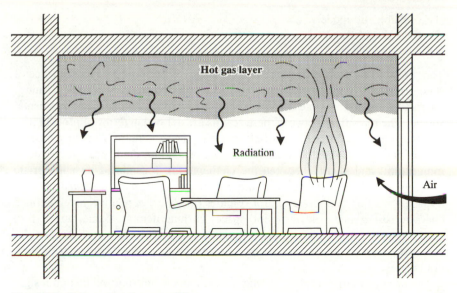

Fig. 9.6 Preflashover development of fire

heat transmission between the gases and the compartment boundaries changes sign; from this time on the boundaries will transfer heat to the gases.[34]

For some time following ignition, the first item ignited burns in approximately the same way as it would in the open. Then, as the flame grows taller and perhaps other items ignite, the process of burning becomes more and more influenced by factors characteristic of the compartment as a whole. A smoky layer of hot combustion and pyrolysis gases builds up below the ceiling. Radiant energy fluxes, originating from the hot ceiling and the adjoining gas layer (Fig. 9.6), gradually heat up the contents of the compartment. If the rate of heat release by the burning item(s) does not start declining soon enough, all burnable objects will ignite in quick succession: *flashover* occurs.

The time to flashover is the maximum length of time that the occupants of a compartment have to escape or be rescued. For this reason, a thorough understanding of the chain of events that connects the ignition of the first object with the flashover has long been one of the major goals of theoretical and experimental fire research. The literature of the mathematical modeling of preflashover fires is quite voluminous; the principles of modeling have been reviewed in many publications (e.g. Roux and Berlin 1979; Pape and Waterman 1979; Emmons 1981; Hägglund 1983; Jones 1983; Mitler 1985; Smith and Green 1987; Quintiere 1988b; Walton and Thomas 1988).

There are three kinds of preflashover fire models:

---

34 The decay period was originally conceived as the stage in which the char left behind by the pyrolyzing wood is consumed by oxidation. Of course, this interpretation of the decay period in not applicable if the fire load consists of noncharring fuels.

(1) probabilistic models,
(2) field models, and
(3) zone models.

The probabilistic models describe the preflashover fire process as a sequence of states (realms), e.g. preburning state, sustained burning, vigorous burning, interactive burning, remote burning, room involvement, and intercompartmental spread. The transitions between these realms are defined in terms of probabilities of occurrence, and time. In the field models, the volume of a compartment is divided into a multitude of finite-volume elements, and the exchanges of mass, momentum, and energy between the elements are formulated by appropriate differential equations. In the zone models, the compartment is looked upon as a space consisting of discrete control volumes (modules) of relatively well-defined characteristics, such as lower (cool) space, burning item(s), flame, upper (hot) space, vent(s), target objects (not yet ignited), and inert surfaces. These modules are then coupled by fluxes of mass, momentum, and energy across their boundaries.

The principal utility of these mathematical models lies in providing an insight into past fire events and, occasionally, into possible future events. Their spheres of application to fire safety design is rather limited, partly because of the large number of essential variables needed for defining a preflashover fire process, and partly because of the dearth of knowledge concerning the most important modules: the burning item(s) and flame. In reality, commonsense reasoning based on simple formulas and intuition still have an edge over mathematical fire modeling, when it comes to predicting the course of preflashover fires.

The criterion of flashover may be described in three alternative ways:

(1) in terms of radiation level in the compartment,
(2) in terms of sustained burning rate (or rather, sustained rate of heat release), and
(3) in terms of ceiling temperature.

Waterman (1968) found that for the most part a heat flux level of about 20×10^3 W m^{-2} at floor level marked the beginning of flashover.[35] He also observed that flashover occurred only if the rate of sustained burning exceeded 0.04 kg s^{-1}. However, much higher threshold values, 0.08 to 0.15 kg s^{-1}, were later reported by Hägglund et al. (1974) and Babrauskas (1979). Finally, the work of Hägglund *et al.* (1974) and Fang (1975) indicated that the attainment of a ceiling temperature of about 600 °C could also be looked upon as a flashover criterion. Of course, these criteria are of little value unless the variables on which the radiation level, rate of burning, or ceiling temperature depend are known.

---

35 This value is higher than the minimum radiant heat flux needed for the piloted ignition of wood, but lower than the minimum flux at which spontaneous ignition can take place (see Eqs 9.34 and 9.35 and the related discussion). It is sufficient for promoting piloted ignition and rapid flame spread along the surface of most burnable materials (Drysdale 1985).

Hägglund *et al.* (1974), Babrauskas (1980a), and McCaffrey *et al.* (1981) developed more detailed criteria for the likelihood of flashover. The criterion recommended by McCaffrey *et al.* (1981) can be presented, for most compartments, in the following form:

$$R t_{fl}^{1/4} \geq 13 \times 10^3 \, (A_t \Phi \sqrt{k\rho c})^{1/2} \qquad [9.44]$$

where R (W) is the rate of heat release by the burning object (Eq. 9.18), $\sqrt{k\rho c}$ ($J\,m^{-2}\,s^{-1/2}\,K^{-1}$) is the thermal absorptivity of the compartment boundaries, 13×10^3 ($J^{1/2}\,K^{1/2}\,kg^{-1/2}$) is a dimensional constant, and

t_{fl} = time to flashover, s;
A_t = total surface area of the compartment boundaries [excluding area of ventilation opening(s)], m^2;

·The so-called *ventilation parameter*,[36] Φ ($kg\,s^{-1}$) is defined as

$$\Phi = \rho_a A_v \sqrt{g h_v} \qquad [9.45]$$

where ρ_a = density of atmospheric air = $1.184\,kg\,m^{-3}$ at 25 °C;
A_v = area of ventilation opening(s), m^2;
h_v = height of ventilation opening(s), m;
g = acceleration due to gravity $\simeq 9.8\,m\,s^{-2}$.

If the exact dimensions of the compartment are not known, A_t may be estimated as

$$A_t = 2A_F + 4h_C \sqrt{A_F} \qquad [9.46]$$

where A_F = floor area of the compartment, m^2;
h_C = height of the compartment, m.

If the boundaries of the compartment are formed by different materials, the thermal absorptivity should be looked upon as a surface-averaged value, to be calculated as

$$\sqrt{k\rho c} = \frac{1}{A_t}(A_1\sqrt{k_1\rho_1c_1} + A_2\sqrt{k_2\rho_2c_2} + A_3\sqrt{k_3\rho_3c_3} + \ldots) \qquad [9.47]$$

where the numerical subscripts refer to the various materials and the surfaces formed by them.

Assuming a tentative value for t_{fl}, Eq. 9.44 may be used for assessing whether the burning of a large object, e.g. an armchair or bed, can lead to flashover. For flashover to occur, it is of course essential that the rate of heat release by the burning object(s) should keep up at a level R for a period of time not less than t_{fl}.

---

36 Some authors refer to the $A_v h_v^{1/2}$ group as the ventilation parameter. The way of calculating the ventilation parameter for compartments that have ventilation openings at various levels was discussed by Magnusson and Thelandersson (1970).

Rate of heat release data for a variety of furnishing items are available from the literature (Quintiere 1977; Babrauskas 1988),[37] or may be determined experimentally, using a furniture calorimeter (Babrauskas *et al.* 1982).

The time to flashover, t_{fl}, is very difficult to assess. In general, it can be assumed that flashover, if it occurs, will follow the flaming ignition of a larger object in 5 to 25 min.

As Eq. 9.44 shows, the thermal absorptivity of the compartment boundaries, $\sqrt{k\rho c}$, is an important factor in the probability of flashover. According to Thomas and Bullen (1979), it is also an important factor in the time to flashover. They suggested that

$$t_{fl} = a + b(\sqrt{k\rho c})^n \qquad\qquad [9.48]$$

where the value of n is typically between 0.25 and 0.5, and a (s) and b (various dimensions) are empirical constants.

The weakness of such flashover criteria is that they all rely on the assumption that the compartment is fully ventilated during the preflashover period. It is reasonable to assume that full ventilation will only start after flashover, as the flames break the windows or pressure differences force a door to open. Restricted ventilation during the growth period is likely to accelerate the buildup of fire toward flashover, owing to low convected heat losses. However, it may also hinder the fire development, especially if the compartment is small and relatively airtight.

Whether the flame will spread from the item first ignited to other items is very important in the rapidity of fire buildup. A few fire scenarios of practical interest were surveyed by Benjamin (1977). He pointed out that burnable wall and ceiling linings may or may not play an essential role in the chain of events that lead to flashover, depending on the fire load (total mass of burnable items), the nature and distribution of the burnable objects in the compartment, and the location and size of the first item ignited. Experiments by Bruce (1959) seem to indicate that the nature of the wall lining materials has little effect on the time to flashover if no piece of furniture is placed closer than about 0.45 m to the walls. This finding has, in essence, been corroborated by Babrauskas (1982b), who claimed that, even with rapidly burning pieces of furniture, radiant heat fluxes exceeding 20×10^3 W m^{-2} could not be measured beyond a distance of 0.88 m.

Since the ceiling of the fire compartment is first exposed to the hot gases produced by the burning item, the nature of ceiling lining materials is more critical from the point of view of preflashover buildup of fire than the nature of the wall coverings. For example, expanded polystyrene ceiling tiles may rain burning droplets on the surfaces below, and thereby greatly accelerate the fire spread.

Fire safety in a building can be greatly increased by circumspect design. The designer knows the intended use of the building and, therefore, has at least a rough idea of the types of objects that may be placed in the various compartments.

---

37 It appears that the time-averaged rate of heat release by major items of furniture is from 0.2 to 1.5 MW, and the heat is usually released over a period of 600–1200 s.

He can add valuable minutes to the time to flashover by not specifying burnable linings in rooms that are most likely to be furnished with upholstered items, or in which clothing articles are kept or stored. He can further improve the level of fire safety by providing closets or built-in cabinets for the storage of clothing articles or paper products. In the design of theaters, lecture rooms, atria, lounges, etc. he can specify slightly elevated or recessed walkways, or built-in planters along the walls that are to be lined with burnable materials, and thus prevent the occupants or the interior decorator from placing upholstered furniture close to those walls.

The probability that a fire will reach the flashover stage can be markedly reduced by the installation of detecting and suppressing devices. Photoelectric and ionization smoke detectors are more suitable than thermal detectors for residential buildings. With the use of smoke detectors, the human losses in residential buildings may be reduced by as much as 40–50 per cent (Gomberg *et al.* 1982; Ruegg and Fuller 1984).

Except in high-rise buildings and buildings with large open floor arrangements (uncompartmented spaces), the installation of a sprinkler system is usually an optional fire safety measure, which may be rewarded by the reduction of certain building code requirements and by lower insurance premiums. However, mandating the use of sprinklers in single family dwellings cannot be justified on economic grounds (Harmathy 1988).

The design of detecting and suppressing devices and systems is beyond the scope of this book. Detailed discussions on these subjects are available in the *SFPE Handbook* (Schifiliti 1988; Fleming 1988). The optimum locations for the alarm-sounding devices have been discussed by Sultan and Halliwell (1990).

9.9   The Smoke Problem

Smoke is the effluent from burning organic materials. It consists of visible components (solid and/or liquid particles) and invisible components (various gaseous combustion products). In building fires, the harmful effect of the particulate phase is twofold: it obscures the vision of the building's occupants, and acts as an irritant. The effect of the gaseous phase is, in general, also twofold: some of its components may act as narcotics (asphyxiants), others as irritants.

The particulate phase may originate from the pyrolysis of the fuel or from the incomplete combustion of the gaseous pyrolysis products. Pyrolysis without flame (smoldering) usually produces volatiles of various molecular masses, of which the high-molecular-mass fraction condenses on contact with cool air into minute droplets of less than 1.0×10^{-6} m mean diameter (as in cigarette smoke). The smoke produced by incomplete combustion of the pyrolysis products contains somewhat smaller particles, made up almost entirely of solid compounds of high carbon content (soot). As the smoke 'ages', the solid and liquid particles tend to coalesce into much larger particles.

The yield of smoke particulates is usually expressed in terms of optical density, D (dimensionless), defined as

$$D = - \frac{1}{2.303} \ln\left(\frac{I}{I_o}\right) \qquad [9.49]$$

where I and I_o (lx) are the luminous fluxes incident on a photocell in the presence and absence of smoke, respectively. According to Beer's law, the relation between I and I_o is

$$\frac{I}{I_o} = \exp(-\gamma Cl) \qquad [9.50]$$

where γ = light absorption (extinction) coefficient, m^{-1};
$\quad\quad C$ = concentration of solid or liquid particles, $m^3\,m^{-3}$;
$\quad\quad l$ = path length of optical beam, m.

From Eqs 9.49 and 9.50

$$\frac{D}{l} = \frac{\gamma C}{2.303} \qquad [9.51]$$

This equation indicates that D/l is proportional to the concentration of the particulate phase. Unfortunately, the proportionality is not perfect, because the absorption coefficient, γ, is not constant; it depends on the density of the particles and on their average size and size distribution (Seader and Chien 1975). In fact, γ cannot be expected to stay constant even for a given smoke; it will decrease as the average particle size increases with aging.

D/l is a measure of the obscuration of light over a unit distance and, therefore, it may be referred to as the *obscuration index*. The finding that the greatest distance at which objects are visible in smoke, L_v (m), is inversely proportional to D/l comes as no surprise.

$$L_v \simeq \frac{\mu}{D/l} \qquad [9.52]$$

where μ (dimensionless) is a factor with a value of the order of 1.

The visibility of exit signs in smoke is an important aspect of fire safety. According to Rea *et al.* (1985), it depends not only on D/l, but also on the brightness of the sign, on the ambient illumination, and on the acuity of vision and spectral sensitivity of the observer. Ambient illumination appears to be an especially important factor.[38]

Rea *et al.* (1985) recommended that the exit signs

(1) should be as bright as possible, and made with large, green colored lettering,

---

38 Using mist generated by the evaporation of cosmetic oil, Rea *et al.* (1985) found that without ambient illumination the exit signs were, on an average, detectable at $D/l \simeq 3.2$ and readable at $D/l \simeq 2.7$. With ambient illumination, the corresponding values were 1.8 and 1.5.

(2) should be designed to enhance the luminous contrast of the text and to minimize the sources of scattered light.

There has been considerable pressure by sectors of local and national governments for a test method that allows the ranking of materials according to their smoke production potentials. Unfortunately, a perfect test may never be found, because the smoke yield is not strictly a material property; it depends to a great extent on such characteristics of the fire environment as irradiance, ambient temperature, ventilation, and oxygen concentration.

The apparatus most widely used for the evaluation of obscuration index is the NBS smoke chamber (Gross *et al.* 1967; American Society for Testing and Materials 1983), which is, in fact, an improved version of the Rohm and Haas XP2 apparatus (American Society for Testing and Materials 1988b). Both the NBS and Rohm and Haas methods were developed mainly with a view to quantifying the potentials of surface lining materials for producing solid or liquid particulates. Measured in the NBS test is the optical density of smoke generated by a vertically mounted specimen, exposed to a radiant heat flux of 25×10^3 W m^{-2}. The generation of smoke may be either by smoldering (i.e. flameless pyrolysis) or by flaming combustion. If the latter mode of burning is selected, the sample is ignited along its lower edge by a multiple-flamelet burner.

In designing the NBS test method, it was apparently assumed that the obscuration index would be proportional to the area of burning surface, and inversely proportional to the volume of building space into which the smoke is fed. In order to produce test results that are applicable to a variety of conditions, Gross *et al.* (1967) described the obscuration index in terms of a 'specific optical density', D_s (dimensionless), defined by the equation

$$\frac{D}{l} = D_s \frac{A_b}{V_S} \qquad [9.53]$$

where A_b = area of the burning surface, m^2;

V_S = volume of the space into which smoke is fed, m$_1$.

The primary objective of the test is to determine the maximum of the specific optical density, $(D_s)_{max}$ (dimensionless). Presumably, it was visualized by the designers of the method that D_s would reach a maximum during the test, either because the burning of the lining material comes to an end,[39] or because after some time into the test the optical density stops increasing, owing to the settling, coalescence, or condensation of the particles. From Eqs 9.49 and 9.53, the expression for $(D_s)_{max}$ is

$$(D_s)_{max} = -\frac{V_{ap}}{A_{ap}l_{ap}} \frac{1}{2.303} \ln\left(\frac{I}{I_o}\right) \qquad [9.54]$$

---

39 In the case of char-forming materials, this may indeed happen.

where V_{ap} = volume of the test apparatus = $0.51\,m^3$;
  A_{ap} = exposed surface area of the specimen in the apparatus = $4.2 \times 10^{-3}\,m^2$;
  l_{ap} = path length of the optical beam in the apparatus = $0.92\,m$.

With the above values, $V_{ap}/(A_{ap}l_{ap})$ = 132, and therefore

$$(D_s)_{max} = -\frac{132}{2.303}\ln\left(\frac{I}{I_o}\right)$$ [9.54a]

Some of the assumptions used by Gross *et al.* (1967) are clearly untenable. Robertson (1973) noted that $(D_s)_{max}$ increases with the thickness of the test specimen, at least up to a thickness of 6.4 mm. Consequently, the test results are meaningful only if they include information on the specimen thickness. Table 9.3 gives typical values of $(D_s)_{max}$ for selected materials.

Rasbash and Phillips (1978) suggested that the obscuration of vision by smoke be related to the mass of volatiles released in the pyrolysis process, ΔW (kg), rather than to the surface of the burning materials. They described the obscuration index in terms of a 'smoke potential', SP ($m^2\,kg^{-1}$), defined by the equation

$$\frac{D}{l} = SP\,\frac{\Delta W}{V_s}$$ [9.55]

A number of researchers (Gaskill 1970; Brenden 1970; Hilado 1970; Grubits 1970) modified the Rohm and Haas and NBS apparatuses to make them suitable for determining the effect of irradiance level, ventilation, and oxygen concentration on the production of smoke particulates. An entirely different technique was devised by Tsuchiya and Sumi (1974). In their test, the specimen is installed in a vertical furnace operating at a preset temperature, and its mass is continuously recorded. A mixture of air and nitrogen is introduced at the bottom of the furnace, and the gaseous mixture leaving the furnace is exposed to electric sparks. The mixture may ignite or pass unignited, depending on the prevailing conditions. The optical density of the combustion or pyrolysis products is measured at top of the test assembly.

Tsuchiya and Sumi conducted tests on 17 materials, and expressed the results in terms of a 'smoke generation coefficient', defined as the ratio of the 'amount of smoke' to the initial mass of the test specimen. They found that for wood the production of smoke particulates decreased with increasing temperature and decreasing oxygen concentration. Most polymeric materials produced more smoke particulates at higher temperatures, irrespective of the oxygen content of the ambient gas.

A number of other techniques are available for assessing the production of smoke by various fuels, such as the ASTM D 4100 (Arapahoe), ASTM E 84 (Steiner tunnel), ASTM E 906 (OSU calorimeter) tests, and an ISO smoke chamber test. They have been reviewed by Tsuchiya and Sumi (1974), Hilado (1982), and Seader and Einhorn (1977).

Table 9.3 Maximum specific optical density of smokes generated by selected materials (Gross *et al.* 1967; Robertson 1973)

Material	Thickness (mm)	Density (kg m^{-3})	Maximum specific optical density $(D_s)_{max}$	
			Flaming combustion mode	Smoldering mode
Wall coverings				
Asbestos—cement board	4.6	2000	0	0
Acoustic tile	19.1	370	20	20
Fiberboard, unfinished	12.7	260	150	380
Gypsum board	9.5	820	15	40
Hardboard				
cherry	6.4	900	70	470
oak	6.4	900	160	350
Spruce	19.1	340	310	420
Floor coverings				
Floor tile	2.4	1490	220	300
Red oak	19.8	660	120	660
Acrilan rug	7.6		160	320
Nylon rug	7.6		270	320
Polypropylene rug	4.6		110	460
Wool rug	7.6		180	220
Plastics				
ABS	1.2	1040	>660	70
Acrylic	5.6	1200	110	160
Polycarbonate	6.4	1200	170	10
Polyester				
glass-reinforced	4.0	1280	400	350
flame-retardant	4.0	1390	620	360
Polystyrene	6.4	1060	>660	370
Poly(vinyl chloride)	6.4	1410	>660	300
Plastic foams				
Polyethylene,				
chlorinated	25.4	32	30	20
Polystyrene	25.4	29	390	30
Polyurethane	15.2	77	230	160
fire-retardant	25.4	30	30	320
Natural rubber	19.1	96	>660	240

The results of these tests offer some guidance on material selection, whenever choices are at all possible. Yet, selecting materials of low smoke-producing propensities is only one aspect of fire safety design. In general, the high rate of fire development, rather than the chemical makeup of the burning materials, is responsible for poor visibility in some, possibly vital, areas of the building. Consequently, the problem of visibility in smoke (as well as the toxic effects of smoke) cannot be considered separately from problems related to the rate of spread of fire within and between compartments.

As pointed out in Chapter 1, inhalation of smoke is the most frequent cause of fatal and nonfatal casualties in fires. During the early stages of fire, people

Table 9.4 Narcotic and irritant substances in the pyrolysis and combustion products of materials found in buildings, based on information collected by Purser (1988). Reprinted with permission from *SFPE Handbook of Fire Protection Engineering*, copyright © 1988, Society of Fire Protection Engineers, Boston, MA 02109

Compound	Molecular mass (kg kmol^{-1})	Lethal dose, LD_{50} (m^3 min m^{-3})
Narcotic products		
CO	28.01	0.075 to 0.120
CO_2	44.01	>2.7
HCN	27.03	0.005 to 0.007
O_2 deficiency[a]	32.00	−4.2 to −4.5
Irritant products		
Acrolein	56.06	0.004 to 0.005
Acrylonitrile	53.06	0.12 to 0.14
Formaldehyde	30.03	0.020 to 0.025
Phenol	94.11	0.012 to 0.020
Styrene	104.14	0.3 to 2.4
Cl_2	70.91	≈0.003
HBr	80.92	0.05 to 0.18
HCl	36.47	0.05 to 0.18
HF	20.01	0.03 to 0.11
NH_3	17.03	0.04 to 0.24
NO_2	46.01	0.002 to 0.008
SO_2	64.06	0.009 to 0.015

[a] Oxygen deficiency is to be interpreted as concentration (volume fraction) of O_2 minus 0.21. Strictly speaking, oxygen deficiency cannot be given in terms of 'toxic dose'. The tabulated value is valid for a 30 min exposure to an atmosphere of 6−7 per cent O_2 content, which is equivalent to 0.14−0.15 O_2 deficiency. An atmosphere of less than 5 per cent O_2 may be lethal in 5−10 min.

who are not fully alert, or are handicapped by age, sickness, or some disability, are most commonly the victims of fire smoke. In their injuries or deaths, obscuration of vision by smoke is not necessarily an important factor. However, after the fire has developed beyond the stage of flashover in one or more of the building spaces, fully alert, agile adults may also become trapped in dense smoke and fall victim to toxic gases.

The toxicity of a smoky atmosphere may be dominated by narcotic (asphyxiant) gases or by irritants. Table 9.4 lists a number of narcotic and irritant products of pyrolysis or combustion for materials commonly found in buildings.

The most important narcotic gas is carbon monoxide, CO, which is produced in the incomplete combustion of the volatile pyrolysis products of any organic material, as well as in incomplete char oxidation. It is, to all intents, the only narcotic gas worth considering in fires of cellulosic materials, and in fires of plastics composed only of carbon and hydrogen atoms (such as polyethylene, polypropylene, or polystyrene). If nitrogen is present in the material's chemical structure (wool, acrylic fiber, nylon, urea−formaldehyde, polyurethane), nitrogen-containing toxicants will also be formed, such as ammonia (NH_3) and hydrogen cyanide (HCN). Furthermore, the presence of halogens will produce a yield of hydrochloric, hydrofluoric, and similar acids.

In the case of narcotic products, the toxic effect is due to high concentration of the toxicant and/or low concentration of oxygen in the cerebral blood supply and in the brain itself. The severity of the effect depends on the dose accumulated by inhalation (to be discussed). Incapacitation or death usually occurs during the time of exposure to the toxic atmosphere. Irritants, on the other hand, become effective through their presence in the lining of the nose, throat, or lung. The symptoms, such as running nose, burning sensation in the nose, mouth, and throat, inflammation of eyes, and violent coughing, occur immediately on exposure. They depend on the concentration and potency of the irritant, rather than on the accumulated dose. The aftereffects, however, do depend on the dose. There seems to be a threshold above which severe respiratory difficulties and even death may occur, most often within 6−24 h following the exposure (Purser 1988).

The presence of irritants at low and moderate concentrations does not significantly impair the person's ability to think and move about. By providing a strong stimulus to escape, it may even be beneficial.

The concentration of the toxicants, C, is, as a rule, quoted as volume fraction, namely parts per million (ppm, which is equal to $10^{-6}\,m^3\,m^{-3}$). Sometimes it is quoted as mass per unit volume, $mg\,m^{-3}$ ($= 10^{-6}\,kg\,m^{-3}$) or $mg\,L^{-1}$ (i.e. $mg\,liter^{-1}$).[40]

The potency of a toxicant is usually characterized by its rodent LC_{50} number ($m^3\,m^{-3}$). It denotes the (constant) concentration of the toxicant in an atmosphere which causes the death of 50 per cent of exposed mice or rats in a specified period of time, τ (min). Only those deaths are counted that occur during exposure or within 14 days after the exposure. The duration of exposure is normally 30 min.

Haber's rule is based on the assumption that the toxic effect depends on the dose, D ($m^3\,min\,m^{-3}$), inhaled (accumulated) during the time of exposure. If the concentration of the toxicant, C, varies over the interval $0 \leq t \leq \tau$, D is to be calculated (Hartzell *et al.* 1985) as

$$D = \int_0^\tau C\,dt \qquad\qquad [9.56]$$

where t is time (min).

The lethal dose, LD_{50} ($m^3\,min\,m^{-3}$), is the dose that leads to the death of 50 per cent of exposed rodents. Table 9.4 includes some information on the lethal doses for the narcotic and irritant compounds listed.

It is convenient to quantify the severity of exposure to a toxicant by the ratio of accumulated dose to lethal dose, D/LD_{50}, which will be referred to as the *harm coefficient*. Clearly, the condition of lethality is $D/LD_{50} \geq 1$.

It is usually assumed that the harm resulting from an exposure to a mixture of toxicants can be determined additively from the harms attributable to the

---

40 To convert from ppm to $mg\,m^{-3}$, multiply the value by $M/22.41$, where M ($kg\,kmol^{-1}$) is the molecular mass of the toxicant, and 22.41 ($m^3\,kmol^{-1}$) is the specific volume of (perfect) gases at 273 K and atmospheric pressure.

components.[41] Thus, the harm coefficient for the mixture, $(D/LD_{50})_m$, is

$$\left(\frac{D}{LD_{50}}\right)_m = \frac{D_1}{(LD_{50})_1} + \frac{D_2}{(LD_{50})_2} + \frac{D_3}{(LD_{50})_3} + \ldots \qquad [9.57]$$

where the subscripts 1, 2, and 3, refer to the various components of the mixture. Again, the condition of lethality is $(D/LD_{50})_m \geq 1$.

The writers of fire safety regulations have long been concerned with the relative potencies of smokes produced in the burning of furnishing items and building materials. In the United States, at least ten states have introduced legislation concerning the use of commercial materials on the basis of the toxicity of smoke they may produce (Alexeeff and Packham 1984). Similar legislative measures have been planned, or already implemented, in some countries in Europe and in Japan.

Alexeeff and Packham listed seven test methods that may be used for obtaining information on the propensities of various materials and products to produce toxic pyrolysis and/or combustion products. These methods have been discussed in more detail by Kaplan *et al.* (1983).

There are three methods that have gained general acceptance in North America and Europe: the National Bureau of Standards (NBS) method, the University of Pittsburgh (UP) method, and the German DIN 53436 method. Although they are quite different with respect to the procedures employed, they are similar with respect to the underlying assumptions, which are:

- The condition of lethality established for rodents is directly applicable to humans.
- The toxic potency of the products of burning (pyrolysis and/or flaming combustion) is a characteristic of the material, and therefore there is no practical need to identify the toxicants.
- A toxic atmosphere can be described in terms of the mass of the material consumed in, or subjected to, the burning process, divided by the volume into which the products are dispersed (i.e. as $kg\,m^{-3}$ or $mg\,L^{-1}$).

The first assumption is tacitly accepted in all branches of toxicology. The other two, however, are hardly justifiable. The nature and quantity of toxicants are known to depend not only on the material but, perhaps even more significantly, also on the fire scenario. Thus, a multitude of tests (and subjecting a multitude of rodents to immense suffering) would, in principle, be required to describe fully the toxic hazards associated with the use of any material. There is a move, therefore, towards reducing the need for animal tests by identifying and quantifying

---

41 A situation referred to as synergism may occur if the toxic potency of a mixture is greater than the sum of the potencies of the components. According to Tsuchiya and Sumi (1973), synergistic interactions are rarely significant. However, Levin *et al.* (1987) found that with mixtures of CO and CO_2 synergism may not be negligible.

the principal toxicants generated by the various materials in simulated fire situations, and making predictions on the expected hazards on the basis of known information on the potency of the component toxicants.

The move dates back to the 1960s, when Tsuchiya and Sumi (1967) introduced the 'toxicity index' concept. They determined the toxicity index by dispersing the volatiles evolved from the burning of 1 g of material in 1 m^3 of air, and relating the concentration of the principal toxicants to the appropriate LC_{50} numbers taken from the literature. Sumi and Tsuchiya (1975) suggested later that every material be studied under a variety of realistic conditions, and labeled with the maximum toxicity index resulting from the study. Still later, Tsuchiya (1981) redefined the toxicity index by relating it to the rate of evolution of products of burning from a unit area of the material.

The draft of a proposal ASTM standard also reflects the move toward reducing the need for animal experiments in the assessment of toxic hazard. In the proposed test, a radiant heat flux of 50×10^3 W m^{-2} is applied to the upper surface of a horizontally placed sample, 76 mm by 127 mm and a thickness not greater than 25 mm. A spark igniter is positioned at 25 mm above the sample's surface. The concentrations of the principal toxicants evolving from the burning sample are monitored in a closed chamber of 0.2 m^3 volume,[42] and the harm coefficient for the gaseous mixture, $(D/LD_{50})_m$, is calculated periodically. The time at which $(D/LD_{50})_m$ reaches unity is referred to as 'irradiation time 50', or IT_{50}, and looked upon as a material characteristic. The test may be repeated with six animals exposed to the gases in the chamber, if there seems to be a need for confirming the calculated IT_{50} value, or if safety regulations so require.

The proposed ASTM test has been visualized to provide information on the toxic hazard created by various materials when exposed to conditions that may arise in real-world fires. However, there is already strong opposition to the proposed draft — in fact to any move to have the problem of toxicity of the pyrolysis and combustion products of plastics brought into the limelight. The underlying fear is that the public and the writers of safety codes may attach an exaggerated significance to the test results. Indeed, available data (Alexeeff and Packham 1984) seem to indicate that the toxic potencies of the smokes generated by a unit mass of most materials fall into a relatively narrow range[43] of approximately 1.9 orders of magnitude. It is widely believed that, in the design of fire-safe buildings, the rate of production of smoke, rather than the smoke's toxic potency, is the problem that requires primary attention.

The distribution and intensity of air currents in winter in a nine-story building

---

42 A small plastic bag attached to the chamber allows for gas expansion during the test.

43 Purser (1988) suggested that 0.3 kg min m^{-3} may be taken as a representative value of the lethal dose for the pyrolysis and/or combustion products evolved from any fuel. Note that this value is based on smoke concentrations expressed in kg m^{-3} rather than m^3 m^{-3}. Since the density of smoke is roughly equal to that of air (1.2 kg m^{-3}), the above value is equivalent to $LD_{50} \simeq 0.25$ m^3 min m^{-3}.

were illustrated earlier [Fig. 9.1(a)]. Figure 9.1(c) shows how these air currents would distribute smoke on the various levels of the building within 10−15 min following the outbreak of fire on the ground floor.

If the door of the fire compartment remains closed, it will serve as an effective barrier not only against the spread of fire, but also against the spread of smoke. A numerical example worked out for a 20-story building, with fire occurring on the first floor, indicated that the rate of spread of dense smoke could be reduced by a factor of at least 30 by making sure that the door of the fire compartment is closed throughout the fire. In tall buildings, using self-closing doors is probably the best and least expensive fire safety measure. Further reduction in the rate of smoke spread can be achieved by applying a strip of intumescent material to the edges of the doors.

It is essential to have smoke or thermal detectors installed on the exhaust side of the ventilation and air-conditioning system. Surveillance of this kind will enable specific actions to be taken, such as shutting down the system and closing the dampers (Anonymous 1987).

In milder climates, where the stack effect (discussed earlier in this chapter) may not be significant in the dispersion of smoke, the technique of diluting the smoke is often employed in certain vital areas of the building, such as lobbies and stairwells. It has been suggested (McGuire et al. 1970) that dilution with fresh air in a 100:1 proportion will ensure relatively safe conditions with respect to both visibility and smoke toxicity. The information needed to design a smoke dilution system includes the equivalent orifice area, ζ, for the boundaries of the spaces to be supplied with air, and the rate of smoke generation by the fire.[44]

The time of evacuating a tall building may be longer than the average duration of a compartment fire (Pauls 1988). Consequently, complete evacuation of a building taller than 10 to 15 stories may not be practical. The danger of exposing the occupants to smoke can be greatly reduced by providing refuge areas (preferably in the vicinity of stairwells), where the occupants can stay in relative safety for the duration of a fire. The required rate of air supply is not likely to be determined by the leakage characteristics of the boundaries, but rather by the need for maintaining tolerable conditions for the assembled crowd. The recommended rate of air supply is about $0.45 \text{ m}^3 \text{ min}^{-1}$ per person.

The most effective way of preventing the spread of smoke in tall buildings is to pressurize the building or some major parts of it. As pointed out earlier in this chapter, the travel of smoke through the vertical shafts (stairwells, elevator shafts, etc.) to the upper floors can be negated by raising the pressure in the shafts by an amount given in Eq. 9.4, and thereby moving the line representing the shaft pressure from the A−A position to the 0−A′ position [Fig. 9.1(b)]. In buildings that are not too tall, pressurization can be conveniently achieved by injecting outside air into all shafts at the top of the building.

---

44 The rate of generation of smoke in fully developed fires will be discussed in Chapter 10.

The discussion of smoke control measures has been restricted here to high-rise buildings with uniform compartmentation and with shafts that run the full height of the building. A more detailed discussion on the subject has been presented by Klote (1988). A *Supplement to the National Building Code of Canada* (Associate Committee on the National Building Code 1973) contains an exhaustive survey of measures for safety from smoke in high buildings. Some of these measures are just commonsense solutions and impose very little restriction on the building design.

Nomenclature

a	absorption (emission) coefficient, m^{-1}; empirical factor, s
A	area; surface area, m^2
b	empirical factor, various dimensions
B	empirical coefficient, $J^2\,m^{-3}\,s^{-3}$
c	specific heat at constant pressure, $J\,kg^{-1}\,K^{-1}$
C	concentration of particulate matter; concentration of gases, $m^3\,m^{-3}$
D	(equivalent) diameter of fire source, m; optical density, dimensionless; toxic dose, $m^3\,min\,m^{-3}$
f	fraction of heat transferred back to the burning object, dimensionless
F	empirical factor, various dimensions
g	acceleration due to gravity $\simeq 9.8\,m\,s^{-2}$
G	geometric variables, m
h	height, m; enthalpy, $J\,kg^{-1}$
Δh	latent heat, $J\,kg^{-1}$
Δh^O	latent heat, referred to the mass of reacting oxygen, $J\,kg^{-1}$
I	illuminance of photocell, lx
IT_{50}	time at which $(D/LD_{50})_m$ reaches unity, min
$\sqrt{k\rho c}$	thermal absorptivity, $J\,m^{-2}\,s^{-1/2}\,K^{-1}$
K	empirical coefficient, $m^4\,s^2\,K^2\,J^{-2}$
l	mean beam length; path length of optical beam; length, m
L	distance, m
LC_{50}	lethal concentration, $m^3\,m^{-3}$
LD_{50}	lethal dose, $m^3\,min\,m^{-3}$
LOI	limiting oxygen index, volume (or mole) per cent
M	molecular mass, $kg\,kmol^{-1}$
n	exponent, dimensionless
p	pressure, Pa
P	perimeter of building, m
q	heat flux, $W\,m^{-2}$
Q	rate of heat flow, W
R	rate of heat release, W
$\overline{R}$	nondimensional rate of heat release, dimensionless
S	dummy variable, W
SP	smoke potential, $m^2\,kg^{-1}$
t	time, s, min
t^*	parameter defined by Eq. 9.38, s

T	temperature, K (°C)
U	rate of air flow (infiltration or exfiltration), kg s^{-1}
v	rate of spread; velocity, m s^{-1}
V	volume, m^3
w	mass per unit area, kg m^{-2}
W	mass, kg
ΔW	mass of volatiles released in the pyrolysis process, kg
x	oxygen concentration, kmol kmol^{-1}, or m^3 m^{-3}
y	horizontal spread, m
z	elevation; vertical distance, m
Z	parameter, defined by Eq. 9.8, m W$^{-2/5}$
α	mass fraction of volatiles in the pyrolysis products, kg kg^{-1}
β	orifice factor $\simeq 0.6$, dimensionless
γ	light absorption (extinction) coefficient, m^{-1}
δ	thickness, m
ϵ	emissivity, dimensionless
ζ	equivalent orifice area, m^2 m^{-2}
ϑ	period of time, s
μ	factor related to visibility in smoke, dimensionless
ξ	factor quantifying incomplete oxidation, dimensionless
Π	material properties, various dimensions
ρ	density, kg m^{-3}
σ	Stefan–Boltzmann constant $= 5.67 \times 10^{-8}$ W m^{-2} K^{-4}
τ	duration of the period of fully developed burning, s; time of exposure to toxicants, min
φ	factor $= f_{FC}\xi_F$, dimensionless
ϕ	function
Φ	ventilation parameter, kg s^{-1}
χ	pressure factor, dimensionless
ψ	empirical constant m^{-1}

Subscripts

a	of outside atmosphere; of ambient air
ap	of/in apparatus; of specimen in apparatus
b	of burning; of burning surface
B	of building
c	for corridors; of (gas-phase) combustion
ch	of/by/from char
C	by convection; of compartment
E	from external sources
f	of fuel; of virgin fuel
fl	of flashover
fp	of/at firepoint
F	of/by/in/from flame; of flame envelope; of floor
Fc	cut-off (of flame)
g	of compartment gases

i	of building interior
id	ideal
ig	of ignition
in	incipient
L	loss
m	for mixture of toxicants
max	maximum
min	minimum
o	in reference condition; at reference temperature (273 or 298 K); initial; at $t = 0$; in the absence of smoke
p	of pyrolysis (or vaporization, sublimation); of pyrolysis zone
pr	required for pressurization
r	for room–corridor walls; for rooms
R	by radiation; radiant
s	for shaft–corridor and for shaft–uncompartmented space walls; for/in shaft; specific
S	of building space; of space into which smoke is fed
t	total (surface area)
u	for uncompartmented spaces
v	of ventilation opening; at visibility limit
w	for outside walls
0	at $z = 0$
∞	for $D \to \infty$

10 Fully Developed Fires

For the fire safety practitioner, the *period of fully developed burning* of compartment fires is of primary interest. The propensity of fires to spread to other compartments depends largely on their characteristics during this period.

As discussed in connection with Fig. 9.5, the various stages of a fire (unattended by the fire department) are defined in terms of the temperature of the compartment gases, T_g (K). The period of fully developed burning is usually defined as the time interval between flashover (signified by a sharp rise in the value of T_g) and the time at which T_g declines to about 80 per cent of its peak value. The duration of this period, the end points of which are marked with t_1 and t_2 in Fig. 9.5, will be denoted by τ (s).

There seems to be no practical reason for dealing separately with the final stage of fire, the *decay period*. By the beginning of this stage, the flaming combustion is more or less completed, and the compartment boundaries gradually release the sensible heat absorbed during the earlier periods. Hence, the fire's potential to spread by destruction or convection is negligible.

10.1   Characteristics of Fuel

The aggregate mass of conventional burnable objects (usually made up mainly of cellulosic materials, wood in particular) is called fire load. The *specific fire load* is the mass of burnable materials per unit floor area.[1]

$$L = \frac{W_o}{A_F}$$ [10.1]

where L = specific fire load, $\mathrm{kg\,m^{-2}}$;

---

1   The term 'fire load density' is perhaps more widely used than 'specific fire load'. However, the latter term is believed to be more in line with the usage accepted in physics and engineering. It is also customary, and perhaps logically more justified, to define the specific fire load as the total heat that the burnable contents per unit floor area of the compartment are capable of releasing in a complete burn-out. To convert from $\mathrm{kg\,m^{-2}}$ to $\mathrm{J\,m^{-2}}$, multiply the value of L by 17×10^6.

W_o = mass of (cellulosic) fuel in the compartment before fire (or rather, at the beginning of fully developed burning[2]), kg;

A_F = floor area of compartment, m^2.

Fire load surveys indicate that wood is still the most common among the burnable materials to be found in buildings. If the amount of noncellulosic materials is appreciable, their mass must be converted to the calorifically equivalent mass of wood by the application of the multiplier $(\Delta h_b)_n/(\Delta h_b)_w$, where $(\Delta h_b)_n$ ($J\ kg^{-1}$) is the heat of burning[3] for the noncellulosic material, and $(\Delta h_b)_w$ ($J\ kg^{-1}$) is the heat of burning of wood (about $17\ MJ\ kg^{-1}$).

The specific fire load is probably the most important among the input data needed in fire safety design. Unfortunately, it is also the most troublesome. The problem is not the paucity of information, but rather the enormous scatter of available data. The discrepancies in the values obtained from various sources may be attributed to a number of factors, such as inherent national or regional differences, differences in the techniques of sampling and evaluation, and estimation of that part of the fire load which is not likely to be consumed by a fire.

All burnable objects (movable or fixed) and materials used in the surface finish of the compartment boundaries are, in general, contributors to the specific fire load. However, the mass of bulky objects or burnable items encased in nonburnable drawers, cupboards, etc., are not to be fully (or at all) included in the fire load.

Figure 10.1 illustrates the customary presentation of information on specific fire load. In some countries, the 80th percentile, L_{80}, is used as design value.

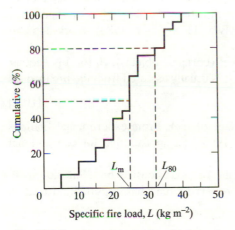

Fig. 10.1 Customary presentation of information on specific fire load

---

2   It is usually assumed that the loss of mass by the fuel during the preflashover period is negligible.
3   As discussed in the previous chapter, the heat of burning is equal to the heat of combustion of the volatile pyrolysis products, plus (if applicable) the heat of char-oxidation, minus the heat of pyrolysis.

Table 10.1 Information on specific fire load
(Pettersson *et al.* 1976)[a]

Occupancy	L_m (kg m^{-2})	σ_L (kg m^{-2})
Dwelling	30.1	4.4
Office	24.8	8.6
School	17.5	5.1
Hospital	25.1	7.8
Hotel	14.6	4.2

[a] The tabulated values are, on the whole,
representative of the occupancies shown.
However, large discrepancies exist in values
acquired from various sources. See text.

The figure explains the meaning of the mean and the 80th percentile of the specific fire load: L_m and L_{80} (kg m^{-2}). In general, it is assumed that the frequency distribution of the specific fire load follows the normal distribution curve, and thus can be quantified by the mean, L_m, and the standard deviation, σ_L (kg m^{-2}).

The occupancy of the building and the use of the compartment are the principal factors that affect the value of L. Based on Swedish data (Pettersson *et al.* 1976), some information is presented in Table 10.1 on the mean specific fire load, L_m, and the standard deviation, σ_L. More detailed information for a number of occupancies may be found in several publications, such as those by Baldwin *et al.* (1970), Law and Arnault (1972), Culver (1978), Issen (1980), Workshop CIB W14 (1986), and Bush *et al.* (1991).

The aggregate surface area of the fuel (burnable objects), A_f (m^2) is another important piece of design information. It is usually assessed from the fire load, as

$$A_f = L A_F \varphi \qquad\qquad [10.2]$$

where φ (m^2 kg^{-1}) is the specific surface of fuel. From experimental findings and from the work of Butcher *et al.* (1968b), it appears that for conventional furniture $\varphi \simeq 0.13$ m^2 kg^{-1}.

As will be discussed later, whether the fire load consists of cellulosic (or, in general, char-forming) materials or noncharring fuels has important bearing on the fire safety design.

10.2   Process Variables

The fire resistance test procedure, the central feature of which is a unique temperature–time curve, is no doubt responsible for the fact that fire safety practitioners look upon the temperature history of the fire compartment as the embodiment of the fire's destructive potential. The area under the temperature–

time curve above some arbitrarily selected temperature level is still widely regarded as the descriptor of the severity of fires, standard test fires or real-world fires.

Kawagoe and Sekine (1963) and Ödeen (1963) were the first to calculate the course of 'compartment temperature' (i.e. average temperature of compartment gases) in fully developed, real-world fires. They applied heat balance equations to the burning compartment, and used numerical techniques to follow-up the flow of heat into the compartment boundaries. Based on similar calculations, Magnusson and Thelandersson (1970) and Lie (1974) developed families of temperature—time curves which, in a simplified way, show the influence of a few important parameters on the temperature history of the fire (see Section 10.10). More recently Wickström (1985) described a technique that, with the aid of scaling factors, allows the derivation of temperature—time curves from a single curve for certain kinds of post flashover fires.

Harmathy (1972) warned, however, that the temperature history of the compartment gases does not, in itself, allow a complete insight into the nature of fire,[4] and claimed that the true nature of a fire can be discerned only if the process variables that characterize the fire are known. He also submitted that, owing to the poor definability of several input variables, employing complex follow-up techniques in the theoretical simulation of fully developed fires gives only a false impression of accuracy, and that a fire can be adequately characterized by means of the time-averaged values of the process variables.

The most important process variables for a fully developed fire are as follows:

U_a = rate of flow of air into the fire compartment, $\mathrm{kg\,s^{-1}}$;
T_g = temperature of compartment gases, K;
R_i = rate of heat release by the fuel inside the compartment, W;
q = heat flux penetrating the compartment boundaries, $\mathrm{W\,m^{-2}}$.

These symbols will usually be regarded as representing the values of the process variables averaged over the period of fully developed burning, τ.

10.3   Flow Rates of Air and Fire Gases

Kawagoe (1958) was the first to formulate the flow of air and fire gases through the ventilation opening(s) [broken window(s), or open or burned-down door(s)]. He assumed that the fire compartment was isolated from the rest of the building, and only communicated with (i.e. received fresh air from and discharged hot fire gases to) the outside atmosphere through the ventilation opening(s), as shown in Fig. 10.2(a).

---

4   The concept of regarding the temperature—time relation for the compartment gases as the principal descriptor of the nature of fire is in error on two grounds. First, the temperature of the gases is determined by a strong and complex interaction between the gases and the compartment boundaries. Second, since the heat transfer from the gases to the boundaries takes place mainly by radiation (which depends on the fourth power of the temperature) a simple gas temperature versus time relation has no relevance to the fire's destructive potential.

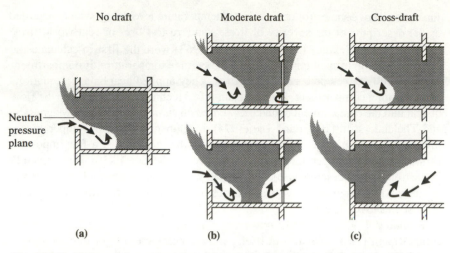

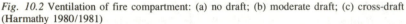

Fig. 10.2 Ventilation of fire compartment: (a) no draft; (b) moderate draft; (c) cross-draft (Harmathy 1980/1981)

In real-world fires, the compartment almost always has some additional routes of communication with one or more secondary environments. For example, when the primary communication is with the outside atmosphere through a broken window, secondary communications may develop with the neighboring compartments through gaps around a door, or through ducts, service holes, etc. The ventilation of the fire compartment thus becomes draft-modified, as shown in Fig. 10.2(b).

Calculations indicate that in high-rise buildings, during the heating season, even small areas of secondary communication allow a significant exchange of air and fire gases between the fire compartment and the secondary environments. If one of these areas is substantial, e.g. if a window is broken and a door is left open by the fleeing tenants, a cross-draft may develop [Fig. 10.2(c)]. The entire window area may be taken up by the inflowing air and the entire door area by the departing gases, or vice versa. In a cross-draft the ventilation may increase by a factor of 5 or more, compared with that in a no-draft situation. Consequently, the flow rate of air calculated from Kawagoe's model represents just about the minimum that can occur in real-world fires.

Kawagoe (1958) (and a host of researchers following him) assumed that the entry of fresh air into, and the departure of fire gases from, the compartment are brought about by a hydrostatic pressure field induced by the temperature difference between the hot fire gases in the compartment and the cool outside atmosphere. As Fig. 10.2(a) shows, the fire gases are driven through the upper section of the window area by the above-atmospheric pressure created by the temperature difference, and the outside air enters through the lower (roughly) one-third of the window area, where the pressure is below the atmospheric level. Between the two sections lies the neutral pressure plane.

Studies conducted by Thomas *et al.* (1967), Prahl and Emmons (1975), and Steckler *et al.* (1982) gave some support to Kawagoe's hydrostatic pressure model. Harmathy (1980b) pointed out, however, that the fire plume acts as a pump, the effect of which may not be disregarded in calculating the flow rates of air and hot fire gases.

Harmathy (1979b) showed that in a no-draft situation, when the fire compartment communicates only with the outside atmosphere, the rate of flow of fresh air into the compartment can be approximated by the following simple expression:[5]

$$U_a^* = 0.138 \chi^{1/2} \Phi - 0.53 \frac{dW}{dt} \qquad [10.3]$$

where U_a^* = rate of flow of air into the fire compartment under no-draft conditions, $kg\,s^{-1}$;

χ = factor quantifying the 'pumping' action of the fire plume on the streams of air and fire gases, dimensionless $(\chi^{1/2} \simeq 1.15)$[6];

Φ = ventilation parameter (defined by Eq. 9.45), $kg\,s^{-1}$;

W = mass of fuel (at some time t) in the compartment, kg;

t = time, s.

The factors 0.138 and 0.53 are dimensionless constants; dW/dt $1(kg\,s^{-1})$ is the rate of loss of mass by the fuel, due to the departure of volatile pyrolysis products and (in the case of char-forming materials) to char-oxidation. It is usually referred to as *rate of burning*.

Since the second term on the right-hand side of Eq. 10.3 is usually much smaller than the first, the rate of air flow into the compartment in a no-draft situation is determined essentially by the ventilation parameter, Φ, and depends mainly on the dimensions of the ventilation opening(s) (see Eq. 9.45).

Under drafty conditions

$$U_a \geq U_a^* \qquad [10.4]$$

because, as mentioned earlier, the air flow rate under no-draft conditions, U_a^*, is just about the minimum that can arise in real-world fires.

The rate of flow of fire gases out of the compartment, U_g $(kg\,s^{-1})$, is, in a way, also a process variable. An expression for U_g can be obtained from a material balance for the compartment

$$U_g = U_a + \frac{dW}{dt} \qquad [10.5]$$

---

5   The reader is reminded again that, although dW/dt, dW_f/dt, dW_{ch}/dt, Δh_c, and Δh_{ch} are negative quantities, to avoid possible confusion the sign convention will not be followed. The listed symbols should be looked upon as representing absolute values. Equation 10.3 is applicable if $250 < T_g < 1000\ °C$; its accuracy is better than 4 per cent for the practically important range of $0 < (dW/dt)/U_a^* < 6$ (Harmathy 1979b).

6   The value of χ is usually between 1 and 2, closer to 1 for large ventilation openings.

Like the rate of inflow of air, the flow rate of fire gases also depends largely on the ventilation parameter and, in turn, on the dimensions of the ventilation opening(s).

U_g can be recognized as the rate of smoke generation by the fire. As such, it may serve as input information in the design of smoke control measures (Section 9.9 in Chapter 9).

10.4   Temperature of Compartment Gases

Because of the high emissivity and high temperature of fire gases, the heat exchange between the fire gases and the compartment boundaries can be assumed to take place essentially by radiation. Setting out from this assumption, Harmathy (1972) developed the following equation for the time-averaged value of T_g:

$$T_g = \left\{ \frac{q}{\sigma \eta_1} + \left(T_a + \frac{q}{\sqrt{k\rho c}} \sqrt{\frac{2\tau}{\pi}} \right)^4 \right\}^{1/4} \qquad [10.6]$$

where τ = duration of fully developed fire, s;

T_a = temperature of atmospheric air, K;

σ = Stefan–Boltzmann constant = $5.67 \times 10^{-8}\,\mathrm{W\,m^{-2}\,K^{-4}}$;

η_1 = empirical factor, dependent largely on the emissivities of compartment boundaries and fire gases, dimensionless ($\simeq$ 0.9);[7]

$\sqrt{k\rho c}$ = thermal absorptivity of the compartment boundaries (see Eq. 9.47), $\mathrm{J\,m^{-2}\,s^{-1/2}\,K^{-1}}$.

In the above equation, the temperature of atmospheric air, T_a, has been used, instead of the initial temperature of the compartment boundaries, for convenience.

In deriving expressions for two of the process variables, U_a and T_g, the question of nature of the burning fuel has not arisen. To be able to develop information on the other process variables, it is necessary now to inquire about the fuel. Previous discussions on burning (Section 9.5 in Chapter 9) have revealed that significant differences exist between the behavior of cellulosic materials (or char-forming materials, in general) and the behavior of noncharring fuels.

10.5   Mechanism of Burning of Cellulosic Fuels

The rate of burning of cellulosic fuels in compartments has long been the principal target of fire research. In the earlier days, it was customary in compartment burn experiments to use, instead of furniture, loosely-packed wood piles (cross cribs) laid out on the floor. Research conducted in the late 1950s and in the 1960s revealed that, under no-draft conditions, the rate of burning of these piles increased in proportion to the ventilation parameter, Φ, and that the ratio of $(dW/dt)/\Phi$ was approximately equal to 0.024. It was also found, however, that this simple proportional relation broke down at higher values of Φ.

---

7   Typical values of the empirical factors are given in parentheses.

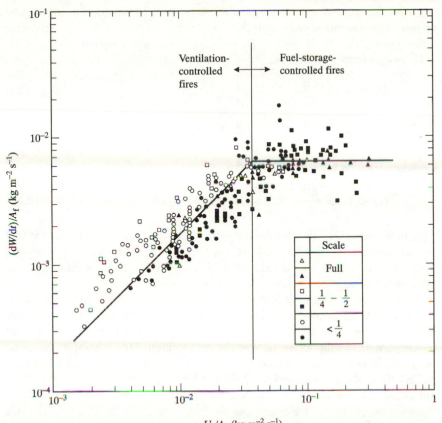

Fig. 10.3 Correlation of experimental data on the rate of burning of cellulosic fuels in compartments (Harmathy 1972). Reprinted with permission from *Fire Technology* (Issue 8), copyright © 1972, National Fire Protection Association, Quincy, MA 02269

Based on an analysis of a multitude of experimental data (Kawagoe 1958; Webster *et al.* 1959: Webster and Raftery 1959; Simms *et al.* 1960; Heselden 1961; Webster *et al.* 1961; Webster and Smith 1964; Gross and Robertson 1965; Butcher *et al.* 1966, 1968a), Harmathy (1972) showed (see plot in Fig. 10.3) that the rate of burning of cellulosic fuel could be represented by the following two equations:

$$\frac{dW}{dt} = 0.0236\Phi \qquad \text{if} \quad \frac{\Phi}{A_f} < 0.263 \qquad\qquad [10.7a]$$

$$\frac{dW}{dt} = 0.0062A_f \qquad \text{if} \quad \frac{\Phi}{A_f} \geq 0.263 \qquad\qquad [10.7b]$$

where 0.0236 (dimensionless), 0.0062 (kg m^{-2} s^{-1}), and 0.263 (kg m^{-2} s^{-1}) are empirical constants. The $0 < \Phi/A_f < 0.263$ interval is referred to as the regime of *ventilation controlled burning*, and the $\Phi/A_f \geq 0.263$ interval as the regime of *fuel-surface-controlled burning*.[8]

Combining Eqs 10.7a and 10.7b with Eq. 10.3, and assuming that $\chi^{1/2} \simeq 1.15$, one obtains

$$U_a^* = 6.19 \, \frac{dW}{dt} \qquad \text{if} \quad \frac{\Phi}{A_f} < 0.263 \qquad\qquad [10.8a]$$

$$U_a^* \geq 6.19 \, \frac{dW}{dt} \qquad \text{if} \quad \frac{\Phi}{A_f} \geq 0.263 \qquad\qquad [10.8b]$$

Consequently, on account of Eq. 10.4,

$$U_a \geq 6.19 \, \frac{dW}{dt} \qquad\qquad [10.8c]$$

At first, the explanation for the proportionality between U_a^* and dW/dt at low ventilations, as expressed by Eq. 10.8a, seemed straightforward. The limited availability of air, it was argued (Tsuchiya and Sumi 1971; Babrauskas and Williamson 1975), puts a limit on the rate of burning.

As discussed in Section 9.5 of Chapter 9, the stoichiometric air requirement for burning (combustion of volatiles plus char-oxidation) is 5.1 kg air per kg wood. Under real-world conditions, however, the oxidation reactions do not proceed to completion. About 40 per cent of the carbon atoms are generally converted to carbon monoxide, and thus the actual air requirement is only about 4.1. Since, as Eq. 10.8c shows, $U_a^*/(dW/dt)$ is always larger than 6.19, shortage of air is not likely to occur at any time during the fire. One must conclude, therefore, that the rate of flow of fresh air *controls*, not limits, the rate of burning; more exactly, it controls the rate-determining link in the burning process.

The popular belief that, in the course of a fully developed fire, the compartment can be modeled as a 'well-stirred reactor' reveals a long-standing misconception about the mechanism of burning, for which the careless use of the words *burning* and *combustion* has, no doubt, been partly responsible. The rate of burning, which is determined from measurements of the loss of mass by the fuel, reflects the rates of pyrolysis and char-oxidation, but it does not reflect the progress of gas-phase combustion (see Eq. 9.12). In fact, the combustion of the volatile pyrolysis products may take place largely outside the compartment boundaries without affecting the measured 'rate of burning'.

The rate-determining factor in the burning process is the oxidation of char, which drives the pyrolysis of fuel. The central role of char-oxidation in the burning of piles of wood was discussed in some detail in the last chapter. Based on the model of pile-burning of char-forming fuels, Harmathy (1978b) developed a model

---

8   Some fire load surveys included valuable information on the size of compartment windows. Judging from the fire load and potential ventilation, Baldwin *et al.* (1970) claimed that in 95 per cent of office compartments a fire would be controlled by the surface area of the burnable materials.

for the mechanism of burning of fully developed compartment fires, the principal features of which are as follows:

- The pyrolysis process is driven mainly by the heat evolving from the oxidation of char. Consequently, the rate of release of volatile pyrolysis products does not rely appreciably on thermal feedback from the hot fire gases and compartment boundaries.
- The flow rate of fresh air into the compartment controls the rate of burning (more precisely, the rate of pyrolysis) by way of controlling the rate of char-oxidation.
- If the rate of inflow of air is small compared with the amount of fuel in the compartment, a *zone of* (char-oxidation-induced) *vigorous pyrolysis* forms after flashover near the ventilation opening. The extent of this zone is approximately proportional to the air flow rate. As the fuel is gradually consumed, this zone (roughly unchanged in size) retreats toward the rear of the compartment. Thus, the rate of burning (i.e. the rates of pyrolysis and char-oxidation) in the zone of vigorous pyrolysis remains more or less constant.
- If the air flow rate is sufficiently high (or the amount of fuel is small), the zone of vigorous pyrolysis will encompass the entire compartment. Hence, the surface area of the fuel (or rather, the area of oxidizing surface char) becomes the principal factor that controls the rate of burning.

Harmathy (1980/1981) demonstrated experimentally that the outlined model of zonal pyrolysis is on the whole correct.

Figure 10.4(a) illustrates the mechanism of burning at low ventilation levels. The furniture is modeled as a loosely packed wood pile uniformly distributed on the floor. The lower regions of the compartment consist of three zones. Zone s1 contains the glowing remains of a section of the fuel network. Zone s2 is the

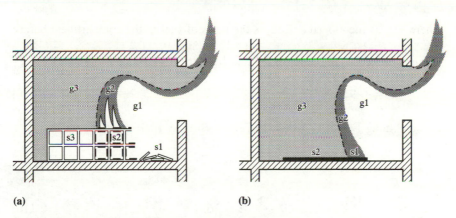

(a)   (b)

Fig. 10.4 Modeling of fully developed compartment fires: (a) fires of cellulosic materials; (b) pool-like fires of noncharring materials (Harmathy 1979b)

zone of vigorous pyrolysis. Here the surface of the pyrolyzing fuel is covered with oxidizing char (shown in heavier lines). Finally, zone s3 contains the rest of the fuel network, which is at an early stage of pyrolysis.

The gas-filled space above the fuel network is divided into three zones. Zone g1 is occupied by air. Zone g3 consists mainly of a mixture of volatile pyrolysis products and combustion gases. Between these two lies the flame zone, g2.

Air passes by zone s1 (where the pile has already been reduced to embers) without significant depletion of its oxygen content. Then, as it reaches zone s2, it separates into two streams. One penetrates the fuel network and feeds the oxidation of the surface char which, in turn, drives the pyrolysis process. The other stream rises with the volatiles released by the fuel, and the bulk of it becomes entrained into the flame in zone g2. Zone s3 receives scarcely any air. For lack of char-oxidation, the pyrolysis of fuel in this zone relies solely on thermal feedback (mainly by radiation) from the fire gases and the hot compartment boundaries and, therefore, proceeds at a relatively low rate.

Butcher *et al.* (1968a, 1968b), experimenting with different fuel arrangements and actual furniture, observed that the configuration of the wood piles had little effect on the rate of burning (i.e. rate of mass loss). Thomas and Nilsson (1973), however, noted some effect, and attributed it to the porosity (i.e. density of packing) of the fuel network.

An approximate expression for the rate of burning (rate of mass loss), dW/dt (kg s^{-1}), can be derived from Eq. 9.14a, by means of the following assumptions:

- The length of the period of fully developed burning, τ (s), is equal to the duration of the pyrolysis process, ϑ_1 in Fig. 9.4(b).
- The duration of char-oxidation, ϑ_2, is equal to $2\vartheta_1$.

With $\alpha \simeq 0.87$ (see previous chapter), Eq. 9.14a yields

$$\frac{dW}{dt} = 0.935 \frac{W_o}{\tau} \qquad [10.9]$$

where W_o, as mentioned before, is the mass of fuel in the compartment before fire.

From Eq. 10.9, the duration of fully developed burning is obtained as

$$\tau = 0.935 \frac{W_o}{dW/dt} \qquad [10.10]$$

By combining Eqs 10.1, 10.2, 10.7a, 10.7b, and 10.10, one obtains

$$\tau = 39.6 \frac{A_F L}{\Phi} \qquad \text{if} \quad \frac{\Phi}{A_f} < 0.263 \qquad [10.11a]$$

$$\tau = \frac{151}{\varphi} \qquad \text{if} \quad \frac{\Phi}{A_f} \geq 0.263 \qquad [10.11b]$$

The second of these equations is especially interesting. It suggests that the duration of fuel-surface-controlled burning depends solely on the specific surface of the fuel. Since, as mentioned in connection with Eq. 10.2, for conventional furniture $\varphi \simeq 0.13 \, \text{m}^2 \, \text{kg}^{-1}$, one may conclude that the expected duration of a fully developed, fuel-surface-controlled fire is 1160 s, or about 19 min.

This finding has important implications in light of the findings of Baldwin *et al.* (1970), according to which the fire load and ventilation conditions prevailing in modern buildings are conducive to fuel-surface-controlled fires. Fully developed compartment fires in modern buildings are not expected to last longer than 15–30 min. In fact, by substituting realistic values of A_F, L, and Φ in Eq. 10.11a, one will find that even ventilation-controlled compartment fires are unlikely to go on longer than 1.5 h (Harmathy 1979a). *Building fires burning vigorously for more than 1 h are almost certain indications of fire spread beyond the compartment where the fire started.*

As discussed in the previous chapter (see Eq. 9.15), the rate of heat release by a char-forming fuel, R (W), consists of three terms: rate of heat evolution from the flaming combustion, rate of heat evolution from the oxidation of char, and rate of heat absorption by the pyrolyzing fuel. If the burning process takes place in a compartment, heat from the flaming combustion may partly evolve outside the compartment boundaries. Thus, the rate of heat released by a char-forming fuel inside the compartment, R_i (W), can be written as

$$R_i = \delta R_F + R_{ch} - \frac{dW_f}{dt} \Delta h_p \qquad [10.12]$$

where R_F = rate of heat release in flaming combustion, W;
δ = fraction of heat of flaming combustion evolved inside the compartment, dimensionless;
R_{ch} = rate of heat release by char-oxidation, W;
W_f = mass of virgin fuel present (at some time t) in the compartment, kg;
t = time, s;
Δh_p = heat of pyrolysis (heat of converting the virgin fuel into volatiles and char), $\text{J} \, \text{kg}^{-1}$.

By expressing R_F and R_{ch} in terms of mass losses (see Eqs 9.16 and 9.17), the following equation is obtained:

$$R_i = \frac{dW_f}{dt} (\alpha \delta \xi_F \Delta h_c - \Delta h_p) + \frac{dW_{ch}}{dt} \xi_{ch} \Delta h_{ch} \qquad [10.13]$$

where α = mass fraction of volatiles in the pyrolysis products (which consist of volatiles and char), $\text{kg} \, \text{kg}^{-1}$;
Δh_c = (lower) heat of combustion of the volatile pyrolysis products, $\text{J} \, \text{kg}^{-1}$;

Δh_{ch} = heat of char-oxidation of char, J kg^{-1};

W_{ch} = mass of charred fuel present (at some time t) in the compartment, kg;

ξ_{F} = factor quantifying incomplete combustion of volatile pyrolysis products in the flame, dimensionless ($\simeq 0.8$);

ξ_{ch} = factor quantifying incomplete oxidation of char, dimensionless ($\simeq 0.8$).

Inspecting Eq. 9.14a, one will see that, with $\alpha = 0.87$ and $\vartheta_2/\vartheta_1 = 2$, the rate of formation of volatile pyrolysis products is $0.87 W_{\text{o}}/\tau$, and the rate of char-oxidation is $\frac{1}{2}(1 - 0.87)W_{\text{o}}/\tau$. Consequently, one may rewrite Eq. 10.13 in the form

$$R_{\text{i}} = \frac{W_{\text{o}}}{\tau}(0.87\delta\xi_{\text{F}}\Delta h_{\text{c}} - \Delta h_{\text{p}} + 0.065\xi_{\text{ch}}\Delta h_{\text{ch}}) \qquad [10.14]$$

or, by neglecting the term Δh_{p} and assuming that $\xi_{\text{F}} \simeq \xi_{\text{ch}} = \xi$ ($\simeq 0.8$), in the form

$$R_{\text{i}} = \frac{W_{\text{o}}}{\tau}\xi(0.87\delta\Delta h_{\text{c}} + 0.065\Delta h_{\text{ch}}) \qquad [10.15]$$

From a large number of published experimental data, Harmathy (1972) developed the following empirical equations for δ (fraction of the heat evolved from flaming combustion inside the compartment):

$$\delta = 1 \qquad \text{if } h_{\text{F}} \leq h_{\text{C}} \qquad [10.16a]$$

$$\delta = \left(\frac{h_{\text{C}}}{h_{\text{F}}}\right)^{3/2} \quad \text{if } h_{\text{F}} > h_{\text{C}} \qquad [10.16b]$$

where h_{C} (m) is the height of the compartment and h_{F} (m) is a hypothetical flame height for compartment fires. Based on Thomas's work (1963) and some observations by Webster and Smith (1964), Harmathy (1972) recommended the following equations for the calculation of h_{F}:

$$h_{\text{F}} = 1.17\Phi^{1/3} \qquad \text{if } \frac{\Phi}{A_{\text{f}}} < 0.263 \qquad [10.17a]$$

$$h_{\text{F}} = 0.75A_{\text{f}}^{1/3} \qquad \text{if } \frac{\Phi}{A_{\text{f}}} \geq 0.263 \qquad [10.17b]$$

where 1.17 (m s$^{1/3}$ kg$^{-1/3}$) and 0.75 (m$^{1/3}$) are dimensional constants.

Harmathy (1972, 1979b) expressed the heat flux penetrating the compartment boundaries, q, from a heat balance over the ventilation opening, as follows:

$$q = \frac{1}{A_{\text{t}}}\left\{R_{\text{i}} - \left(U_{\text{a}} + \frac{dW}{dt}\right)c_{\text{g}}(\nu T_{\text{g}} - T_{\text{a}}) - A_{\text{v}}\sigma\left(T_{\text{g}}^4 - T_{\text{a}}^4\right)\right\} \qquad [10.18]$$

where, as discussed before, T_a is the temperature of atmospheric air, as well as the initial temperature of the compartment boundaries, and

A_t = total surface area of compartment boundaries [which does not include the area of ventilation opening(s)], m^2;

c_g = specific heat (at constant pressure) of fire gases, $J\,kg^{-1}\,K^{-1}$ ($\simeq$ 1150 $J\,kg^{-1}\,K^{-1}$ for the usual ranges of T_g);

ν = factor characterizing the excess temperature in the flame zone (g2) with respect to the average temperature of compartment gases, dimensionless ($\simeq$ 1.05);

A_v = area of ventilation opening(s), m^2.

The first term within the braces represents the rate of heat evolution within the compartment, as expressed by Eq. 10.15. The second represents the convected heat losses, and the third the radiated heat losses through the ventilation opening(s).

No information is available on fully developed compartment fires involving fuels of char-forming plastics. There seems to be no reason for questioning the validity of Eqs 10.16a to 10.18 for char-forming fuels other than cellulosics. However, Eqs 10.11a, 10.11b, and 10.15 are applicable only to cellulosic fuels. For char-forming plastics, the expressions for τ and R_i are expected to be similar in form to these three equations, but to contain constants specific to the fuel.

10.6   Mechanism of Burning of Noncharring Fuels

Noncharring fuels include all liquid fuels and many of the common plastics. With liquids, the fuel is expected to be found on the floor in the form of a pool. For many common noncharring plastics, the temperature range of melting is lower than the temperature range of pyrolysis. If such plastics are present in quantities substantial enough to survive beyond flashover, they are also expected to settle in a pool on the floor.

Because of the practical significance of pool burning and the considerable difficulties of analyzing pile-like burning, studies of fully developed fires of noncharring materials have so far been conducted mostly on fuels in a pool-like arrangement (Bullen 1977; Quintiere *et al.* 1979; Bullen and Thomas 1979; Babrauskas and Wickström 1979).

The most essential difference in the burning of charring and noncharring fuels is that the latter fuels do not have a 'built-in' heat source (oxidizing char) to feed the pyrolysis[9] process, and therefore the process must rely on thermal feedback from the fire gases and (to a lesser extent) from the hot compartment boundaries.

A simple model for the pool-like burning of noncharring fuels[10] is illustrated

---

9   In the following discussion, the word pyrolysis will be interpreted to cover pyrolysis as well as vaporization (sublimation).

10 More elaborate models have also been devised. None of the existing models can claim more than qualitative agreement with the available, rather scanty, experimental findings.

in Fig. 10.4(b). The gas-filled space is again divided into three zones. Zone g1 is occupied by air, and zone g3 by a mixture of combustion gases and volatiles formed in the pyrolysis of the fuel. Between g1 and g3 lies the flame zone, g2.

The segment of the pool surface immediately under the flame zone receives much of the radiant and convective heat feedback from the flame, and therefore this zone, s1, is the zone of vigorous pyrolysis. Yet, the pyrolysis is not negligible in zone s2 (the rest of the pool surface) either, since this zone also receives thermal feedback from the flame zone, g2, from the gases in zone g3, and from the hot compartment boundaries. The result is a fairly high rate of overall volatile formation, especially if the heat of pyrolyzation is low. Since the rate of entry of air is restricted, there is a good probability that a substantial portion of the volatiles will leave uncombusted, and undergo combustion outside the compartment.

Although, in regard to the potential of fire to spread by the destruction of the compartment boundaries, the shifting of a major part of combustion to the neighboring spaces is advantageous, it is extremely dangerous and undesirable for the performance of the building as a whole.

Harmathy (1979b, 1983a) used the outlined model to develop expressions for dW/dt, R_i, q, and τ. Although these expressions do not have sufficient experimental support, they do provide some insight into the basic differences between fires of cellulosic materials and noncharring fuels.

The rate of loss of mass by the pool due to pyrolysis can be expressed as

$$\frac{dW}{dt} = \frac{\sigma\eta_2 A_P}{\Delta h_p}\{\beta[(\psi T_g)^4 - T_p^4] + (1 - \beta)[T_g^4 - T_p^4]\} \qquad [10.19]$$

where A_P = surface area of the pool, m^2;

T_p = temperature of pyrolysis (assumed to be a definite value), K;

Δh_p = heat of pyrolysis, $J\,kg^{-1}$;

η_2 = empirical factor, dependent largely on the emissivities of the pool and the fire gases, dimensionless ($\simeq 0.9$);

β = fraction of area of pool surface that lies in the zone of vigorous pyrolysis [zone s1 in Fig. 10.4(b)],[11] dimensionless;

ψ = factor characterizing excess temperature of fire gases in the flame zone [zone g2 in Fig. 10.4(b)], with respect to average temperature of compartment gases, dimensionless ($\simeq 1.1$).

Clearly, Eq. 10.19 has been formulated on the assumptions that radiation is the dominant mode of heat exchange in the fire compartment, and that by the start of fully developed burning the fuel surface has reached the temperature of pyrolysis. In a simplified way, the equation includes all terms that need to be considered in formulating the rate of mass loss by pyrolysis. The product βA_P is the area of pool surface that lies in the zone of vigorous pyrolysis [zone s1

---

11 βA_P may be estimated roughly as a 1 m deep strip next to the ventilation opening(s).

in Fig. 10.4(b)]. It is supplied with radiant heat by gases in the flame zone, g2, where the temperature is ψT_g, somewhat higher than the temperature of the bulk of compartment gases, T_g. In the less vigorously pyrolyzing zone, zone s2, the pool surface [whose area is $(1 - \beta)A_P$] exchanges radiant heat mainly with the bulk of the gases at temperature T_g.

With liquid fuels or melted solids, there is a good chance that the flame envelope will remain stationary in the vicinity of the ventilation opening, and therefore A_P and β may be regarded as constants. Although for other situations A_P and β are, in a strict sense, variable quantities ($A_P \rightarrow 0$ and $\beta \rightarrow 1$ as the fire progresses), at present there is no experimental foundation for treating them as true variables.

If, as usual, the rate of production of volatiles is substantial, the rate of combustion of the volatiles and, in turn, the rate of heat evolution within the confines of the compartment, R_i, is controlled directly by the rate of entrainment of air into the flame zone, and indirectly by the rate of entry of air through the ventilation opening. Nevertheless, one must also consider the possibility, however unlikely, that there is an abundant supply of air, and therefore the rate of heat evolution within the compartment is controlled by the rate of production of volatiles (i.e. by dW/dt). Thus, one or the other of the following equations will apply:

$$R_i = \begin{cases} \gamma U_a \xi_F \Delta h_c / r \\ \\ \gamma \dfrac{dW}{dt} \xi_F \Delta h_c \end{cases} \quad \text{whichever is less} \qquad [10.20]$$

where Δh_c = (lower) heat of combustion of pyrolysis products, J kg^{-1};
$\quad r$ = mass ratio of air to volatiles in stoichiometric combustion of the volatiles, kg kg^{-1};
$\quad \gamma$ = factor quantifying combustion deficiency due to the incomplete mixing of the pyrolysis products with air inside the compartment, dimensionless ($\simeq 0.8$);
$\quad \xi_F$ = factor quantifying the incomplete combustion of pyrolysis products (owing to formation of carbon monoxide), dimensionless ($\simeq 0.8$).

The heat flux penetrating the compartment boundaries, q, can now be expressed as

$$q = \frac{1}{A_t - A_P} \left\{ R_i - \left(U_a + \frac{dW}{dt} \right) c_g \left(T_g - \frac{T_a + T_p}{2} \right) \right.$$
$$\left. - A_v \sigma \left(T_g^4 - T_a^4 \right) - \frac{dW}{dt} \Delta h_p \right\} \qquad [10.21]$$

The first, third, and fourth terms within the braces represent, respectively, the rate of heat evolution in the compartment (expressed by Eq. 10.20), the rate of heat loss by radiation through the ventilation opening, and the rate of heat absorption by the pyrolyzing pool of fuel. The second term represents the convected heat loss by the compartment through the ventilation opening(s), based on the assumption that the reference level for the enthalpy of gases is roughly halfway between the inlet air temperature and the temperature of pyrolysis of the fuel.

Since the gases leaving through the ventilation opening probably contain, in addition to combustion products, uncombusted volatiles in initially unknown quantities, the assessment of the specific heat of fire gases, c_g, is rather difficult. Although the ratio of combustion products to uncombusted volatiles at the ventilation opening(s) is obtained as a byproduct of the first round of trial-and-error calculations (to be discussed), it is sufficient, in view of the approximate nature of the model, to select for c_g an all-purpose value in the range $1150-1500 \, \text{J kg}^{-1} \, \text{K}^{-1}$.

The duration of the fully developed fire, τ, is equal to $W_0/(dW/dt)$ (where W_0 is the original fuel mass). Thus, making use of Eq. 10.1, the expression for τ is

$$\tau = \frac{A_F L}{dW/dt} \qquad [10.22]$$

As mentioned earlier, the expressions developed for U_a and T_g (Eqs 10.3, 10.4 and 10.6) are applicable to any fully developed fire, irrespective of the nature of fire load.

10.7   Normalized Heat Load

For some time, it was believed that the knowledge of an ensemble of at least three parameters, namely the duration of fully developed burning, τ, and two of the process variables, T_g, q, was necessary for unambiguously characterizing the severity of fully developed compartment fires. Harmathy (1972) referred to these three variables as 'fire severity parameters'. It was later realized that one of these three parameters, T_g, was inconsequential, and that the other two, q and τ, could be combined into a single parameter. Harmathy (1980a) argued that in the last analysis it is the total heat absorbed by the compartment boundaries per unit surface area (as represented by the $q\tau$ product) that effects their structural or thermal failure. He referred to the $q\tau$ product as *heat load*.

The next step in understanding of the nature of compartment fires was the realization (Harmathy and Mehaffey 1982; Harmathy 1983a) that by 'normalizing' the heat load (i.e. dividing it by the thermal absorptivity, $\sqrt{k\rho c}$, of the boundaries) a parameter is obtained that offers significant benefits in fire safety design. The definition of *normalized heat load*,[12] H ($\text{s}^{1/2} \, \text{K}$), is

$$H = \frac{1}{\sqrt{k\rho c}} \int_0^\tau q \, dt \qquad [10.23]$$

or, with a time-averaged value of q,

---

12 The reader is reminded that the normalized heat load concept was already touched upon in Chapter 3.

$$H = \frac{q\tau}{\sqrt{k\rho c}} \qquad\qquad [10.24]$$

The normalized heat load has three important characteristics (Harmathy 1990/1991):

- *Characteristic 1*: If the boundaries of a compartment are surfaced with different building materials, H is approximately the same for all surfaces and, in addition, for the compartment as a whole. In mathematical formulation:[13]

$$H = \frac{1}{\sqrt{k\rho c}}\int_0^\tau q\,dt = \frac{1}{\sqrt{k_1\rho_1 c_1}}\int_0^\tau q_1\,dt = \frac{1}{\sqrt{k_2\rho_2 c_2}}\int_0^\tau q_2\,dt$$

$$= \frac{1}{\sqrt{k_3\rho_3 c_3}}\int_0^\tau q_3\,dt = \cdots \qquad\qquad [10.25]$$

where the numerical subscripts refer to various boundary surfaces and the materials forming them. For boundary elements consisting of layers of different materials, the layer exposed to fire is of primary significance (Harmathy and Mehaffey 1987).[14]

- *Characteristic 2*: For any boundary element of the enclosure, H is an approximate measure of the *maximum* temperature rise at some critical depth from the surface where, in conventional compartment boundaries, the important load-bearing components are usually located (Harmathy 1980a). It has been proved (Harmathy and Mehaffey 1987) that this claim is valid irrespective of the surface geometry of the boundary element, although (at identical values of H) the value of the maximum is higher at locations below rectangular corners than at locations below flat surfaces.
- *Characteristic 3*: The normalized heat load does not depend significantly on the temporal variation of the heat flux that penetrates the compartment boundary (Harmathy 1980a; Harmathy and Mehaffey 1987). Consequently, two fires (one may be a real-world fire, the other a simulation of a standard test fire) that yield the same values for H will create at the critical depths the same maximum temperatures in the compartment's boundary elements; in other words, the two fires will be of equal severity.

Characteristic 3 makes it possible:

(1) to regard the normalized heat load as a quantifier of the *potential of fire to spread by destruction*[15] (i.e. fire severity) with respect to a given compartment; and

---

13 Equation 10.25 is usually referred to as the theorem of uniformity of normalized heat load.
14 It has been shown (Harmathy and Mehaffey 1987) that, under practical conditions, the normalized heat load does not depend on the thickness of the layers that form the compartment boundaries.
15 A study by Schwartz and Lie (1985) has revealed that with the present-day construction practices fire spread by excessive heat conduction through the compartment boundaries is unlikely to occur. In this book, the expression *spread by destruction* is usually meant to include the illusory case of spread by excessive heat transmission.

(2) to relate the performance of (i.e. the severity endured by) a building in a real-world fire to its performance in a standard test fire.

On account of (2), fire resistance values developed from standard fire tests can indeed be looked upon as legitimate performance quantifiers.

From the way these three characteristics have been worded, it is clear that they are not of unrestricted validity. Some of the restrictions are of little significance, and therefore will not be considered here.[16] The only important restriction concerns the materials of the compartment's boundary elements. As discussed in connection with Fig. 3.7, the normalized heat load concept is not applicable to compartment boundaries made from or coated with metallic materials.

10.8   The μ-factor

The most common cause of fire spread is the spilling of combusting volatiles from the compartment on fire through the ventilation opening(s). Irrespective of the nature of burnable materials in the surrounding spaces, the fuels and conditions that are conducive to the massive release of volatiles in the fire compartment present a very real danger for fire spread.

A factor has been introduced to quantify the *potential of fire to spread by convection* (Harmathy 1983a). It is referred to as the *μ-factor* (dimensionless), and defined as

$$\mu = \frac{R - R_i}{R} \qquad [10.26]$$

where, as discussed before, R is the rate of heat release by the fuel, and R_i is the portion of that heat released inside the compartment on fire.

Utilizing Eqs 10.15 and 10.20, the following expressions are obtained:

●   For fires with cellulosic fire load:

$$\mu = \frac{0.87(1 - \delta)\Delta h_c}{0.87\Delta h_c + 0.065\Delta h_{ch}} \qquad [10.27]$$

●   For fires with a fire load of noncharring plastics:

$$\mu = \frac{\dfrac{dW}{dt} - \dfrac{U_a}{r}}{\dfrac{dW}{dt}} \quad \text{or} \quad \mu = 0 \quad \text{whichever is larger} \qquad [10.28]$$

The normalized heat load and the μ-factor are two important design parameters. Since they convey in a succinct form all the information that the ensemble of process variables is capable of conveying, these parameters greatly simplify the

---

16 The reader will find more details in the referenced papers.

design procedure. Their role in the design of buildings against spread of fire will be discussed in more detail in the next chapter.

10.9   Calculation of Process Variables and Design Parameters

In the preceding sections, equations have been derived for the calculation of the time-averaged values of the process variables, the duration of fully developed burning, and the two design parameters.

The first step toward performing the calculations is assembling the input variables. It is clear from the nature of the relations presented that much of the input data must be based on rough estimates, at least in the first stage of design. Even in the later stages, there is nothing wrong with aiming at less than perfection. Thus, for example, the specific fire load may be taken as uniform throughout the building; the surface areas of the compartment boundaries may be assessed from Eq. 9.46; the generic information presented in Table 10.2 may be used for assessing the thermal properties of the compartment boundaries; and so on.[17]

As to the ventilation of the compartments, the input variables provide for the calculation of the ventilation parameter, Φ, and, in turn, the rate of flow of air into the fire compartment in a no-draft situation (Eq. 10.3), U_a^*. The actual air flow rate, U_a, which is always larger than U_a^* (Eq. 10.4), is almost impossible to assess. Fortunately, as will be discussed later, the destructive potential of compartment fires decreases with increasing ventilation. Hence, using in fire safety design the ventilation parameter, Φ, to quantify the ventilation conditions is not expected to lead to unsafe conclusions.

Table 10.2 Typical values of the thermal properties of common construction materials (in moistureless condition) for a temperature level appropriate for a 20-min fire exposure

Material	Thermal conductivity k ($W\,m^{-1}\,K^{-1}$)	Density ρ ($kg\,m^{-3}$)	Specific heat c ($J\,kg^{-1}\,K^{-1}$)	Thermal diffusivity κ ($10^{-6}\,m^2\,s^{-1}$)	Thermal absorptivity $\sqrt{k\rho c}$ ($J\,m^{-2}\,s^{-1/2}\,K^{-1}$)
Steel	44.0	7800	460	12.3	12600
Marble	2.0	2650	1000	0.755	2300
Normal-weight concrete	1.7	2250	1200	0.630	2140
Brick	1.0	2100	900	0.529	1370
Lightweight concrete	0.50	1450	1000	0.345	850
Plaster board	0.25	750	2500	0.133	680
Wood	0.15	550	1800	0.152	390
Mineral wool	0.04	160	1150	0.217	86

17 The error introduced by estimated input data can be compensated for by using a reliability-based design procedure, to be discussed in the next chapter.

Once the input variables are assembled, the calculations can be performed on a programmable calculator, using the trial-and-error scheme illustrated

- in Fig. 10.5, for cellulosic materials,
- in Fig. 10.6, for noncharring plastics.

As the figures show, assuming a plausible value for T_g is the starting point. The calculations will yield all the process variables, as well as the two parameters: the normalized heat load and the μ-factor.

Making use of the results of 22 full-scale compartment burn experiments performed by Butcher et al. (1966, 1968a), Harmathy (1972) showed that the expressions and calculation procedures introduced in this chapter are capable of yielding (at least for fires of cellulosic materials) sufficiently accurate information on the (time-averaged) values of the process variables and on the duration of fully developed fire.

To explore the differences in the characteristics of compartment fires involving the two types of fuel, Harmathy (1983a) performed two series of calculations, employing the iterative techniques illustrated in Figs 10.5 and 10.6. In the first series, the fire load was taken to consist of cellulosic materials. The dimensions of the fire compartment were chosen to duplicate those of the compartment used in the full-scale burn experiments conducted by Butcher et al. (1966, 1968a). The three levels of specific fire load selected for the study, $L = 60$, 30, and $15 \, \text{kg m}^{-2}$, also duplicated those used in the experiments.

In the second series, it was assumed that a noncharring fuel was present in pool configuration, and its properties were similar to those of poly(methyl methacrylate) (PMMA).

The results of the calculations are presented in Figs 10.7(a) and 10.7(b), in the form of H versus Φ and μ versus Φ plots. As pointed out earlier, H is a measure of the fire's potential to spread by destruction, μ is a measure of the fire's potential to spread by convection, and Φ quantifies the compartment's ventilation under no-draft conditions.

The following conclusions can be drawn:[18]

- From the point of view of spread by destruction, fires of cellulosic materials are more dangerous than fires of noncharring fuels; from the point of view of spread convection, the opposite is true.
- For both types of fuel, the fire's potential for spread by destruction decreases with increasing ventilation.
- The fire's potential for spread by convection increases (in general) with increasing ventilation if the fire load consists of cellulosic materials, and decreases with increasing ventilation if the fire load consists of noncharring fuels.

---

18 For cellulosic fuels, these conclusions are well supported by dozens of experimental data, among them those published by Butcher et al. (1966, 1968a). As to noncharring fuels, the conclusions are of a somewhat speculative nature.

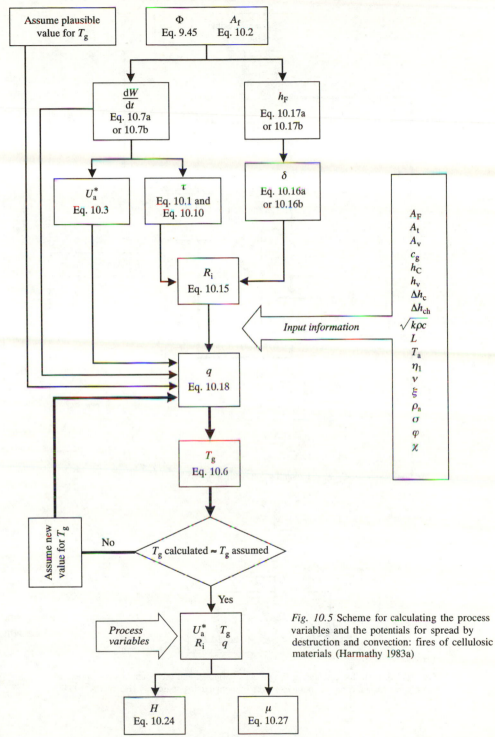

Fig. 10.5 Scheme for calculating the process variables and the potentials for spread by destruction and convection: fires of cellulosic materials (Harmathy 1983a)

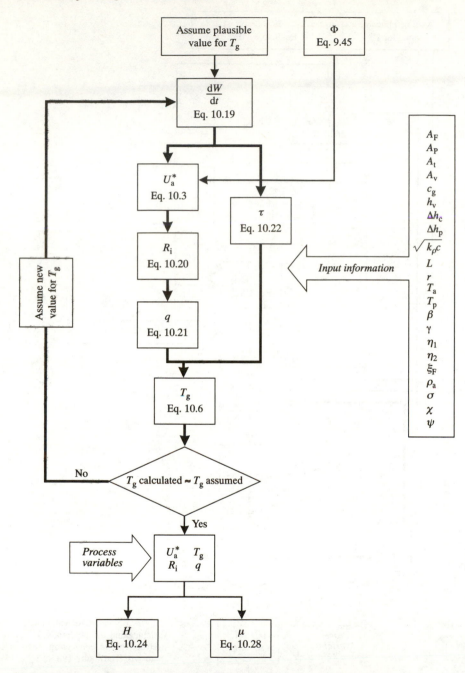

Fig. 10.6 Scheme for calculating the process variables and the potentials for spread by destruction and convection: pool-like fires of noncharring materials (Harmathy 1983a)

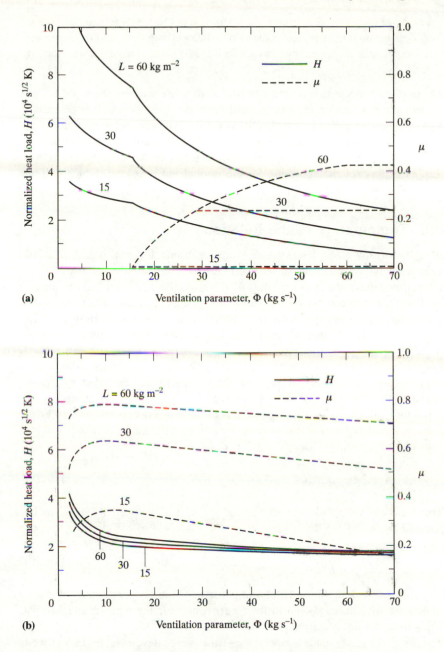

Fig. 10.7 Potentials of fires to spread by destruction and convection: (a) fires of cellulosic materials; (b) pool-like fires of noncharring materials (Harmathy 1983a)

- The level of fire load is important (a) in respect to the potential for spread by destruction if the fire load consists of cellulosic materials, and (b) in respect to the potential for spread by convection if the fire load consists of noncharring fuels.

The popular concept that (at a given fire load) a fire may be short and hot or long and less hot but, on the whole, will have the same destructive potential, has probably grown out of Ingberg's (1928) work. That concept has no foundation at all. In reality, short (well ventilated) fires are less destructive than long (poorly ventilated) fires. This fact will appear obvious to the reader after an inspection of Eq. 11.3, an explicit expression for the normalized heat load, which will be introduced in the next chapter.

10.10   Temperature–Time Relations

Among the ready-made temperature–time relations for real-world fires of cellulosic fuels, those suggested by Lie (1974) are the simplest to use.

Lie claimed that most fires are controlled by ventilation,[19] and that the nature of fuel-surface-controlled fires, for all intents, is unpredictable. Since, from the point of view of fire safety design, any error resulting from the assumption that all fires are ventilation-controlled is on the safe side, he felt that only ventilation-controlled fires deserve consideration.

Employing the calculation technique used by Kawagoe and Sekine (1963), Lie developed two families of curves: one for compartments bounded by heavy materials ($\rho \geq 1600 \, \mathrm{kg \, m^{-3}}$), and another for compartments bounded by lightweight materials ($\rho < 1600 \, \mathrm{kg \, m^{-3}}$). As shown in Figs 10.8(a) and 10.8(b), the curves give the history of the temperature of compartment gases, using a single parameter, the *opening factor*, F ($\mathrm{m^{1/2}}$), defined as

$$F = \frac{A_v h_v^{1/2}}{A_t} \qquad [10.29]$$

These curves are valid over the period of fully developed burning, to be calculated from

$$\tau = 10.9 \, \frac{A_F L}{A_v h_v^{1/2}} \qquad [10.30]$$

where 10.9 ($\mathrm{m^{3/2} \, s \, kg^{-1}}$) is an empirical constant.[20] Lie also presented empirical formulas for the curves shown in the figures, and for the temperature history of fires over the decay period.

The use of the ready-made temperature–time curves developed by Lie, as well as those developed by other research workers, may lead to erroneous conclusions if applied to fuel-surface-controlled fires.

---

19 This claim seems to conflict with the findings of Baldwin *et al.* (1970).
20 τ may also be calculated from Eq. 10.11a.

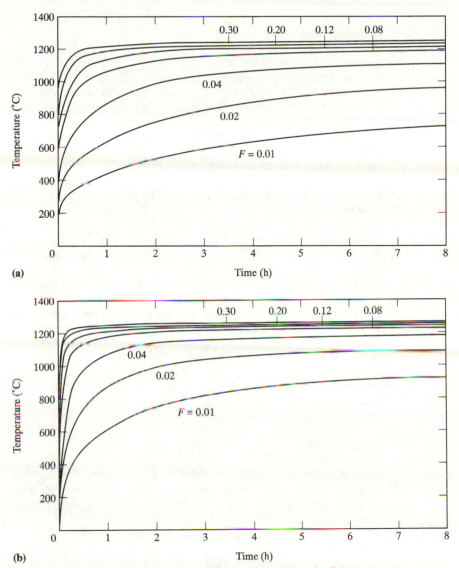

(a)

(b)

Fig. 10.8 Temperature−time curve for ventilation-controlled fires of cellulosic materials: (a) compartments bounded by heavy materials ($\rho \geq$ 1600 kg m^{-3}); (b) compartments bounded by lightweight materials ($\rho <$ 1600 kg m^{-3}) (Lie 1974). Reprinted with permission from *Fire Technology* (Vol. 10, No. 4), copyright © 1974, National Fire Protection Association, Quincy, MA 02269

Nomenclature

A	area, surface area, m^2
c	specific heat at constant pressure, J kg^{-1} K^{-1}
F	opening factor, m$^{1/2}$

h	height, m
H	normalized heat load, $s^{1/2} K$
Δh	heat (of reaction), $J\,kg^{-1}$
k	thermal conductivity, $W\,m^{-1}\,K^{-1}$
$\sqrt{k\rho c}$	thermal absorptivity of compartment boundaries, $J\,m^{-2}\,s^{-1/2}\,K^{-1}$
L	specific fire load (mass of burnable items per unit floor area), $kg\,m^{-2}$
q	heat flux (usually time-averaged) penetrating the compartment boundaries, $W\,m^{-2}$
r	mass ratio of air to volatiles in stoichiometric combustion of volatiles, dimensionless
R	(time-averaged) heat release rate by the fuel, W
t	time, s
T	temperature; time- and space-averaged temperature, K
U	flow rate, $kg\,s^{-1}$
$U*$	flow rate under no-draft conditions, $kg\,s^{-1}$
W	mass of fuel (at some time t), kg
α	mass fraction of volatiles in the pyrolysis products of char-forming materials, $kg\,kg^{-1}$
β	fraction of area of pool surface in the zone of vigorous pyrolysis (vaporization): zone s1 [see Fig. 10.4(b)], dimensionless
γ	factor quantifying combustion deficiency due to incomplete mixing of the products of pyrolysis (vaporization) with air inside the compartment, dimensionless
δ	fraction of heat evolved from flaming combustion inside the compartment, dimensionless
η_1	empirical factor, dependent largely on the emissivities of compartment boundaries and fire gases, dimensionless
η_2	empirical factor, dependent largely on the emissivities of the pool and the fire gases, dimensionless
ϑ	period of time, s
κ	thermal diffusivity, $m^2\,s^{-1}$
μ	μ-factor: quantifier of spread potential of fire by convection, dimensionless
ν	factor characterizing the excess temperature in the flame zone, g2 [Fig. 10.4(a)], with respect to the average temperature of compartment gases, dimensionless
ξ	factor quantifying incomplete oxidation, dimensionless
ρ	density, $kg\,m^{-3}$
σ	Stefan–Boltzmann constant $= 5.67 \times 10^{-8}\,W\,m^{-2}\,K^{-4}$; standard deviation, $kg\,m^{-2}$
τ	duration of fully developed burning (see Fig. 9.5), s
φ	specific surface of fuel, $m^2\,kg^{-1}$
Φ	ventilation parameter, $kg\,s^{-1}$
χ	factor quantifying the 'pumping' action by the fire plume on the streams of air and fire gases, dimensionless
ψ	factor characterizing the excess temperature of fire gases in the flame zone [zone g2 in Fig.10.4(b)], with respect to the average temperature of compartment gases, dimensionless

Subscripts

a	of (atmospheric) air
b	of burning (i.e. the process consisting of pyrolysis, combustion of volatile pyrolysis products, and — in the case of char-forming materials — char-oxidation)
c	of combustion of volatile products of pyrolysis (vaporization)
C	of compartment
ch	of/by/in char-oxidation
f	of fuel; of virgin fuel
F	of floor; of/by/in flame
g	of gases in the fire compartment
i	inside the fire compartment
L	of specific fire load
m	mean
n	of noncellulosic fuels
o	initial (at $t = 0$)
p	of pyrolysis (vaporization)
P	of pool surface
t	total (for compartment boundaries)
v	of ventilation opening(s)
w	of wood
80	pertaining to 80th percentile

11 Design of Buildings against Fire Spread

The basic philosophy of protecting buildings against fire evolved about the turn of the century. Central to that philosophy is the assumption that fires spread by the successive destruction of, or conduction of heat through, compartment boundaries [Fig. 11.1(a)]. Observations and statistical data seem to indicate that, with the present construction practices, spread by destruction or heat conduction is rare. Building fires usually spread by convection: the advance of flames and hot gases through doors (left open by fleeing tenants), windows (broken by the fire), and other openings between the compartment on fire and the surrounding spaces [Fig. 11.1(b)]. Yet, since the building code regulations are still deeply rooted in the age-old philosophy of fire protection, the problem of how to counter the spread of fire by destruction or conduction deserves a great deal of attention.

11.1 Countering Spread of Fire by Destruction

In general, the provision of fire safety consists of setting *performance requirements* and establishing *performances*.

The basic measure in countering the spread of fire by destruction or heat conduction is fire-resistant compartmentation. A fire-resistant building is, in a strict sense, one in which the *fire resistance* of all building elements (compartment boundaries, as a rule) is equal to or greater than that required to prevent the spread of potential fires. In practice, a fire-resistant building is one built in compliance with the requirements prescribed for fire-resistant buildings in the applicable building code. In the United States, for example, the required fire resistances are often 3 or 4 h for load bearing walls and principal structural members, and 2 or 3 h for secondary structural members and nonbearing partitions enclosing stairwells and other vertically extending spaces (shafts). Other nonbearing partitions must be constructed from nonburnable materials.[1]

---

1 In Canada, the building code prescribes the fire resistance of the building's structural components on the basis of height and occupancy of the building, floor area per story, and access to the building by the fire department. The code recognizes two types of construction: combustible (burnable) and noncombustible (nonburnable).

An excellent review of fire resistance requirements prescribed by the four model codes recognized in the United States (Basic Building Code, Standard Building Code, Uniform Building Code, and National Building Code) and by the National Building Code of Canada has been presented in a handbook published by the CRSI (Gustafson 1980).

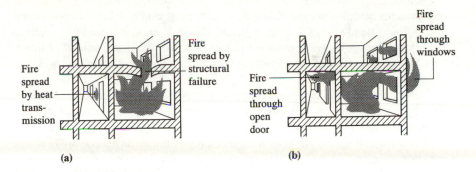

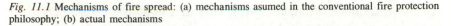

Fig. 11.1 Mechanisms of fire spread: (a) mechanisms asumed in the conventional fire protection philosophy; (b) actual mechanisms

Fire resistance is a mark of performance. It is usually determined by subjecting specimens of building elements (walls, floors, beams, columns) to standard *fire resistance tests* (see Chapter 3). In the test, the specimen is exposed from one direction (except column specimens which are exposed on all sides) to the atmosphere of a furnace whose temperature is controlled to follow a prescribed course. The fire resistance of a building element is the time at which the representative specimen fails, i.e. the time at which, owing to structural weakness or excessive heat transmission, it ceases (according to the test criteria) to be an effective barrier against the spread of fire.[2] For a great many types of building elements, however, information on fire resistance can also be developed by computer-executed calculations that simulate the standard fire test.

Strictly speaking, the fire resistance of a building element only quantifies the element's performance under a specific set of conditions prescribed in the standard fire resistance test. However, by virtue of Characteristic 3 of the normalized heat load (see Section 10.7 in Chapter 10), it is possible to relate the performance of a building element in the standard test fire to its performance in real-world fires.

11.2   Setting the Requirements

The responsibility for setting fire resistance requirements has traditionally been assigned to the writers of building codes. Originally, code requirements were determined on the basis of the expected severity (or destructive potential) of potential fires, and severity was usually judged by the specific fire load typical of the building's occupancy.[3]

---

2   The fire test standards usually refer (1) to major destructive failures as loss of load-bearing capacity, (2) to minor destructive failures (but substantial enough to allow the passage of flames through a building element) as loss of integrity, and (3) to failures by excessive heat conduction as loss of insulating capability (see Chapter 3). In the fire test standards, a temperature rise of 140 °C above the original temperature at the unexposed (reverse) surface of the test specimen is the criterion of failure by excessive heat transmission.

3   The fire resistance requirements now in force are only distantly related to the original requirements. In general, the requirements include an inordinate degree of safety.

Following the *building code approach* consists of complying with the requirements specified in the building code. However, using alternative approaches is possible, thanks to the 'equivalence clauses' in many building codes. Unfortunately, the building designer rarely opts for taking advantage of those clauses, owing to the numerous practical difficulties to which shunning the building code approach may lead.

The practice of assigning fire resistance requirements, or *equivalent times of fire exposure*,[4] based on estimated values of specific fire load originated from an experimental study conducted in the 1920s by Ingberg (1928). Further studies, experimental and semitheoretical, have made it possible to develop more realistic methods for assessing these equivalent times. Best known among them are Law's method (1971), Pettersson's method (1975), and the DIN 18230 method (Deutsches Institut für Normung 1978). They have been reviewed and evaluated in the light of experimental information by Harmathy (1987).

Law's method formulates fire resistance requirements in terms of four variables:

- specific fire load, L (kg m^{-2})
- floor area of compartment, A_F (m^2);
- total surface area of compartment boundaries A_t (m^2);
- area of ventilation opening(s), A_v (m^2):

Pettersson has improved on Law's method by taking into account another variable, the height of the ventilation opening(s), h_v (m), and, in an implicit way, the thermal absorptivity of the compartment boundaries, $\sqrt{k\rho c}$ (J m^{-2}s$^{-1/2}$K^{-1}).

The DIN 18230 method formulates the fire resistance requirements in terms of a few rather arbitrarily defined factors, namely the 'burning factor', the 'ventilation factor', and the 'thermal insulation factor'.

11.3   Fire Engineering Design

The trend in Europe is toward abandoning the more or less conventional approaches discussed thus far, and using a so-called *fire engineering design* procedure which, in its most advanced form, does not rely on the knowledge of fire resistance values, assigned either by tests or developed by calculations. The design procedure consists of three steps (Witteveen 1982; Workshop CIB W14 1986; Pettersson 1986):

(1) Calculation of the temperature history of fire gases in the burning compartment, taking account of all essential variables on which the fire's severity may depend.
(2) Calculation of the temperature rise in the compartment boundaries (including columns) in response to exposure to the fire gases.

---

4   Equivalence is judged in regard to the severity of the standard test fire and that of a potential real-world fire.

(3) Calculation of the decline of load-bearing capacity of compartment boundaries in response to rise of temperature, up to the point of structural failure.

The first step may be bypassed using all-purpose temperature–time curves for the fire gases developed by Magnusson and Thelandersson (1970) and Lie (1988), or employing modified forms of the temperature–time curve of the test standard, as suggested by Wickström (1985). In the design procedure, the random nature of some design variables is usually taken into account by the use of safety factors.

The advantage of this design procedure is that, in principle, it can be applied not only to simple building elements, but also to complex structures composed of two or three elements (walls, floors, beams, etc.). This advantage does not seem so great, however, if one considers that for a complex structure the experimental validation of the theoretical model would be prohibitively expensive.

The disadvantage of the fire engineering design is twofold:

(1) The calculation procedure is involved and, therefore, the range of its applicability is limited;
(2) Repeated application of the procedure is likely to be needed before the final decision can be reached.

11.4   Applying the Normalized Heat Load Concept

The normalized heat load, in its role as quantifier of the potential of fires to spread by destruction, was discussed in the previous chapter. Applying the normalized heat load concept to fire safety design consists, in essence, of comparing the normalized heat load endured (without failure) in a standard test fire by (a specimen of) the building element under consideration, H'' ($s^{1/2}$ K), to that to which the building element is expected to be exposed in real-world fire, H' ($s^{1/2}$ K). The condition of acceptability of the building element is

$$H'' \geq H' \qquad\qquad\qquad [11.1]$$

Because of the uniqueness of the standard fire exposure conditions, the normalized heat load in standard fire tests, H'', is expected to depend essentially on a single variable: the duration of test fire, τ (h). Indeed, as Fig. 3.7 shows, for common building materials the curves representing the H'' versus τ relation (Curves 1–6) lie in a close cluster. However, some deviations from the cluster may occur on account of differences in the operational characteristics of the various fire test furnaces. The effect of differences in the emissivity of furnace gases may be rather substantial. Figure 11.2 shows the H'' versus τ relation (average for materials within the cluster) for two kinds of test furnace: a hypothetical test furnace heated with 'black' (highly emissive) gases (Curve 1), and a well designed test furnace (Curve 2).

The following equation is an approximation of Curve 2 (Mehaffey and Harmathy 1984):

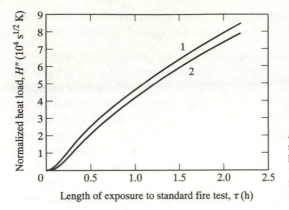

Fig. 11.2 H'' versus τ relations for standard fire tests; Curves: 1, hypothetical furnace, heated with 'black' gases; 2, well designed test furnace (Mehaffey and Harmathy 1984)

$$H'' = 10^4 \times (\sqrt{76.92\tau + 29.41} - 6.15) \qquad [11.2]$$

where 76.92, 29.41, and 6.15 are empirical constants of various dimensions, and the dimension of τ is h.

The iterative techniques for calculating the normalized heat load under real-world conditions, H' ($s^{1/2}$ K), were discussed in the previous chapter, and illustrated in Figs 10.5 and 10.6. Although the calculations are quite straightforward, they may become tedious if the random nature of some input variables and the inaccuracies arising from a number of assumptions used in the derivation of the formulas are also taken into account.

In order to eliminate the need for iteration, Mehaffey and Harmathy (1981) developed, based on the results of hundreds of iterative calculations, an approximate formula for the normalized heat load under real-world fire conditions. The equation, which in a strict sense is applicable only to fires of cellulosic materials, is as follows:

$$H' = 1.06 \times 10^6 \frac{(11.0\delta + 1.6)A_F L_m}{A_t\sqrt{k\rho c} + 935\sqrt{\Phi A_F L_m}} \qquad [11.3]$$

with[5]

$$\delta = \begin{cases} 0.79\sqrt{h_C^3/\Phi} \\ 1 \end{cases} \quad \text{whichever is less} \qquad [11.4]$$

where 1.06×10^6 (J kg^{-1}), 935 (J kg^{-1} K^{-1}), and 0.79 (kg$^{1/2}$ m$^{-3/2}$ s$^{-1/2}$) are dimensional constants, and

δ = fraction of heat evolved from flaming combustion inside the fire compartment, dimensionless;

5   Note that Eq. 11.4 is somewhat different from the expressions derivable from Eqs 10.16a–10.17b.

A_F = floor area of the compartment, m^2;

L_m = mean specific fire load for a given occupancy, $kg\,m^{-2}$;

A_t = total surface area of compartment boundaries, m^2;

$\sqrt{k\rho c}$ = thermal absorptivity of compartment boundaries (Eq. 9.47), $J\,m^{-2}\,s^{-1/2}\,K^{-1}$;

Φ = ventilation parameter (Eq. 9.45), $kg\,s^{-1}$;

h_C = height of the compartment, m.

Substituting realistic values of the variables that appear in Eqs 11.3 and 11.4, one will find that range of H' is from 1×10^4 to $9 \times 10^4\,s^{1/2}\,K$, and therefore, according to Fig. 11.2, the range of realistic fire resistance requirements is from 0.25 to about 2.25 h.

As pointed out in the previous chapter, the two most important factors on which H' depends, the fire load and the rate of flow of air into the compartment, are random variables. Apparently, Eq. 11.3 is valid only for the mean value of the applicable specific fire load. The air flow rate in the equation is represented by the ventilation parameter, Φ, which (as discussed in section 10.3 of Chapter 10) is a measure of minimum ventilation. However, as Eq. 11.3 shows, H' decreases with increasing values of Φ, and therefore using Φ as input information is bound to yield a maximum for H', and thus lead to conservative fire safety design.

A large series of full-scale compartment burnout experiments (Mehaffey and Harmathy 1986) indicated fair agreement between the values of H' calculated from the experiments and those calculated from Eqs 11.3 and 11.4. The coefficient of variation for the error associated with the use of these two equations, Ω_e (dimensionless), was found to be 0.101 (i.e. 10.1 per cent).

Since the assessment of fire resistance requirements for the building elements is one of the main purposes of safety design, it must not be overlooked that, owing to the imperfect repeatability and reproducibility of the fire test results, the fire resistance values are also random quantities. The coefficient of variation, Ω_τ (dimensionless), is somewhere between 0.01 and 0.15 (i.e. between 1 and 15 per cent) (American Society for Testing and Materials 1982); $\Omega_\tau \simeq 0.1$ may be regarded as an all-purpose value.

Taking into consideration the random nature of the specific fire load, L, and the fire test results, τ, as well as the possible error associated with the use of Eqs 11.3 and 11.4 in calculating H', and employing well-known reliability-based procedures (Cornell 1969; Zahn 1977), Mehaffey and Harmathy (1984) derived the following expressions for the assessment of the fire resistance requirements:[6]

$$\tau_d = 0.11 + 16.0 \times 10^{-6}\,H_d'' + 0.13 \times 10^{-9}\,(H_d'')^2 \qquad [11.5]$$

where 0.11, 16.0×10^{-6}, and 0.13×10^{-9} are dimensional constants, and

$$H_d'' = H' \exp\left(\beta\sqrt{\Omega_1^2 + \Omega_2^2 + \Omega_e^2}\right) \qquad [11.6]$$

---

6 Eq. 11.5 is the inverse of Eq. 11.2.

in which

$$\Omega_1 = \frac{\sigma_L}{L_m} \frac{A_t\sqrt{k\rho c} + 468\sqrt{\Phi A_F L_m}}{A_t\sqrt{k\rho c} + 935\sqrt{\Phi A_F L_m}} \qquad [11.7]$$

$$\Omega_2 = 0.9\Omega_r \qquad [11.8]$$

In Eqs 11.6 to 11.8, Ω_e and Ω_r have already been defined. As to the other symbols,

τ_d = design value of fire resistance (i.e. minimum length of successful exposure to standard fire resistance test);

H_d'' = design value of normalized heat load (i.e. normalized heat load pertaining to τ_d), $s^{1/2}$ K;

Ω_1 = coefficient of variation for H', dimensionless;

Ω_2 = coefficient of variation for H'', dimensionless;

β = factor, function of allowed failure probability, P_f, dimensionless.

The allowable failure probability, P_f (dimensionless), is to be interpreted as conditional on the occurrence of flashover in the compartment.[7] The β versus P_f relation is shown in Fig. 11.3.

A worked example of calculating the design value of fire resistance is given in Section 11.7.

Since the 1970s, there has been a move toward extending the philosophy of limit state design of structures to the design of buildings and building components for fire safety (Magnusson and Pettersson 1980/1981, Comité Euro-International du Béton 1982; Kersken-Bradley 1982; Witteveen 1982; Workshop CIB W14 1986; Pettersson 1986). In structural engineering design, applying the limit state concept consists of making sure, by means of partial safety factors, that the

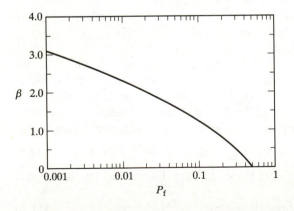

Fig. 11.3 Relation between the factor β and the probability of failure, P_f (Mehaffey and Marmathy 1984)

---

7   A fairly recent review of the statistics of the National Fire Incident Reporting System has indicated that, given ignition, the probability that flashover will follow is 21 per cent (Harmathy *et al.* 1989).

probability of structures becoming unfit for service remains below specified limits. The safety factors take account of the various aspects of design uncertainty, such as possible inaccuracy of the design model, poor definability of the load levels, and scatter of material property values. They are, as a rule, derived on probabilistic considerations, which may be supplemented by engineering judgment.

Applying the limit state concept to fire safety design consists, in essence, of requiring that the experimentally or theoretically derived fire resistance times be equal to or greater than the equivalent times of fire exposure (see Section 11.2). Partial safety factors are used to take care of the various kinds of uncertainties, inherent in both the (experimental or calculated) fire resistance times and the (calculated) equivalent times of fire exposure.

11.5   Countering Spread of Fire by Convection

Unfortunately, the design to counter the spread of fires by convection is not well established. The designer has to rely mainly on commonsense considerations.

Four factors are of major significance for the extent and direction of fire spread by convection:

(1) the μ-factor which quantifies the potential of fire to spread by convection,
(2) the intensity and direction of drafts in the building before the outbreak of fire,
(3) the nature of lining materials along the path of fire spread, and,
(4) the existence of hidden cavities in the building elements, or wall or floor penetrations by drain, waste, and vent (DWV) pipes, by electric or telephone cables, etc.

The presence of burnable overlay in the corridor or along the building facade is not prerequisite to intercompartmental fire spread. The propensity of spread from one compartment to another depends not only on the presence of burnable materials in the path of fire but also, perhaps even more, on the value of the μ-factor for the compartment that feeds flames and hot gases into the surroundings. The designer is well advised to study the variables on which the fire process depends (see Figs 10.5 and 10.6), and possibly adjust them so as to minimize both the normalized heat load and the μ-factor.

As discussed in Section 9.2 of Chapter 9, below the midheight of a tall building (where, during the winter heating season, the air moves toward the building core) the most common path for the spread of fire is a door left open by the fleeing tenants. This is why the simplest and most effective way of countering the spread of fire is to separate the various occupancy units by self-closing doors. Several building codes have already made the use of self-closing doors mandatory in high-rise buildings.

Platt *et al.* (1989) devised a probabilistic model for comparing the effectiveness of various fire safety strategies. They assumed that fire can spread through three alternative paths:

(1) horizontally through open doors,

(2) vertically from window to window, and

(3) horizontally or vertically, by the destruction of building elements, such as walls, closed doors, ceilings.

They found that, for single-story buildings, keeping doors closed is the most effective way of reducing fire losses.

Hinged doors may present some problems. If fire breaks out in a compartment on one of the lower stories, and a window of the compartment is open or broken by the fire, it may be difficult or even impossible to open the corridor door, due to a substantial pressure difference between the compartment and the corridor. On the other hand, if fire occurs on one of the upper stories, the pressure difference may keep a hinged door open even against the force exerted by a closing device. The designer may consider specifying weight-operated sliding doors. If such doors are hung on rollers from a concealed rail (Fig 11.4), opening them at any pressure difference will require less force than that required to open a hinged door equipped with a closing device at no pressure difference. Other suitable solutions have also been suggested.

Vertical and horizontal spread of fire can at times be traced back to the penetration of floors and walls by plastic drain, waste, and vent (DWV) pipes and telephone or electric cables. Studies by McGuire (1973, 1975), and Orals and Quigg (1976) indicated that the problem may be allayed by surrounding these pipes or cables with nonburnable packings housed in thin steel sleeves that extend

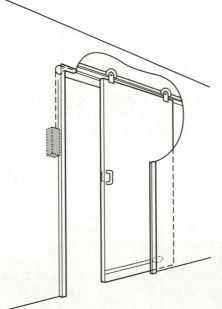

Fig. 11.4 Weight-operated self-closing door (Harmathy 1976). Reprinted with permission from *Fire Technology* (Vol. 12, No. 3), copyright © 1976, National Fire Protection Association, Quincy, MA 02269

beyond the surface of the floor or wall. Further studies by Attwood (1980) and Choi (1987) revealed, however, that sizable delay in the spread of fire through plastic pipe penetrations can only be achieved by the installation of mechanical or chemical (intumescent) fire stops.

The spread of fire is often assisted by the presence of cavities within structural elements and roof spaces. Suitable materials and methods of fabricating and erecting cavity barriers and fire stops have been described in a comprehensive paper by Spiegelhalter (1977).

The practice of using burnable insulation over the external sheathing is rather common with buildings of less than three stories height. In case of fire, flames issuing from the windows may penetrate the gap between the insulation board and the external cladding, ignite the insulation, and spread to the upper stories. Experimenting with polyurethane and polystyrene foam boards, Taylor (1981) found that the danger of fire spread along these hidden cavities depends on the thickness of the air space, the presence or absence of flame barriers, and the nature of the plastic foam and the cladding material. With nonburnable claddings and air spaces less than 25 mm thick, the spread along the boards tends to stop at a distance of 2−3 m above the flaming heat source.

Serious fire safety problems may arise from the use of burnable cladding materials over the building façade. According to Oleszkiewicz (1990), some burnable claddings will support unlimited spread of flames. However, vertical spread of fire from window to window may occur even if the façade is nonburnable, provided the μ-factor for the fire compartment is sufficiently high.

A systematic investigation conducted in Australia (Anonymous 1971) confirmed an earlier British finding (unpublished) that a projection less than 0.6 m wide above a window does not prevent the emerging flames from curling back and igniting the window above. It was also found, however, that a projection wider than 1.2 m is usually very effective in keeping the flames away from the façade, and in reducing radiant heat transmission to the façade above to an acceptable level. Clearly, continuous balconies and open corridors can play useful parts in the overall fire safety design. Unfortunately, they are rarely considered nowadays, even for residential buildings, because they cut down the natural daylight reaching the interior, increase building costs, and may be undesirable from the point of view of security or aesthetics.

Simple flame deflectors (Harmathy 1976) could provide the same degree of protection as continuous balconies and open corridors, without the mentioned drawbacks. They have been visualized as light metal, insulated panels mounted vertically above the windows and held in vertical position with a fusible part (possibly a nut). The width of these panels should obviously be at least 1.2 m, and their length equal to the window breadth plus about 1.2 m. As Fig. 11.5 shows, these flame deflectors would turn down to assume a horizontal position when activated by flames issuing from the window below.

A fire safety system referred to as a 'fire drainage system' was described by Harmathy and Oleszkiewicz (1987). Using the energy of the fire itself, the system

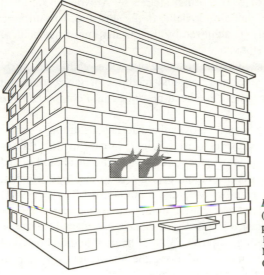

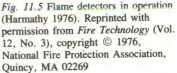

Fig. 11.5 Flame detectors in operation (Harmathy 1976). Reprinted with permission from *Fire Technology* (Vol. 12, No. 3), copyright © 1976, National Fire Protection Association, Quincy, MA 02269

is capable of preventing the convective spread of fires. It confines the fire and smoke to the fire compartment and a small adjacent section of the corridor. The system can be designed to operate without use of water and electricity.

11.6   Countering Spread of Fire Between Buildings

The spread of fire from one building to another by radiation was studied theoretically by Williams-Leir (1966, 1970). With a view to reducing the risk of spread by radiation, the *National Building Code of Canada* (Associate Committee on the National Building Code 1990) specifies the minimum distance of a building from the property line, in terms of (among others) the percentage area of openings (windows, in general) in the building façade, which may become radiation sources in case of fire.

A more refined method for dealing with the problem was suggested by Barnett (1988). In defining the permissible area of openings in the façade of the 'exposing building' (for a given separation between two buildings), Barnett's method takes account of

(a) the expected severity of fire in the exposing building,
(b) the presence or absence of burnable cladding materials over the exposed building, and
(c) the presence or absence of fire windows (wired glass, in general) in the façades of both the exposing and exposed buildings.

By staying in place, the fire windows of the exposing building are assumed to reduce the radiation level and to prevent the projection of flames,[8] and those of the exposed building are assumed to eliminate the possibility of piloted ignition of burnable items located near the openings.

11.7 Worked Example

Design for Fire Resistance

Given
A ten-story office building with a floor area per story of 454 m². There are 16 compartments on each story; they are 2.5 m high, and have an average floor area of 26 m². The plan is to have the compartments lined with conventional materials: the walls and ceiling with plaster board, and the floor with wood parquet. Each compartment will have two windows, 1.52 m tall and 1.02 m wide.

Calculate the fire resistance requirements in such a way as to ensure a failure probability less than 0.05 (5 per cent given flashover, or about 1 per cent given ignition; see Footnote 7).[9]

Input information
Specific fire load for office buildings (Table 10.1):

$$L_m = 24.8 \text{ kg m}^{-2}$$
$$\sigma_L = 8.6 \text{ kg m}^{-2}$$

Surface area of the fuel (Eq. 10.2, with $\varphi = 0.13 \text{ m}^2 \text{ kg}^{-1}$):

$$A_f = 24.8 \times 26.0 \times 0.13 = 83.8 \text{ m}^2$$

Ventilation parameter (Eq. 9.45):

$$\Phi = 1.2 \times (2 \times 1.52 \times 1.02) \times \sqrt{9.8 \times 1.52} = 14.4 \text{ kg s}^{-1}$$

Average surface area of the compartments (Eq. 9.46):

$$A_t = 2 \times 26 + 4 \times 2.5 \times \sqrt{26} = 103 \text{ m}^2$$

Average thermal absorptivity of the compartment boundaries (Eq. 9.47 and Table 10.2):

$$\sqrt{k\rho c} = \frac{1}{103} [(26 + 4 \times 2.5 \times \sqrt{26}) \times 680 + 26 \times 390]$$

$$= 607 \text{ J m}^{-2} \text{ s}^{-1/2} \text{ K}^{-1}$$

δ (Eq. 11.4):

$$\delta = 0.79 \times \sqrt{2.5^3/14.4} = 0.82$$

$$\Phi/A_f = 14.4/83.8 = 0.172$$

---

8   Referring to information published by Butcher and Parnell (1983), Barnett claimed that 2 m can be regarded in the design as the distance of flame projection, and 600 °C as the flame temperature.
9   When deciding on the allowable failure probability, P_f, the expected magnitude of human and property losses that may result from a failure has to be carefully considered.

Therefore the real-world fire is expected to be ventilation-controlled (see Eq. 10.7a); its duration (Eq. 10.11a) is

$$\tau = 39.6 \, \frac{26.0 \times 24.8}{14.4} = 1773 \text{ s} \quad (0.49 \text{ h})$$

Normalized heat load under real-world fire conditions (Eq. 11.3):

$$H' = 1.06 \times 10^6 \, \frac{(11.0 \times 0.82 + 1.6) \times 26 \times 24.8}{103 \times 607 + 935 \times \sqrt{14.4 \times 26 \times 24.8}}$$

$$= 4.76 \times 10^4 \text{ s}^{1/2} \text{ K}$$

Ω_1 (Eq. 11.7):

$$\Omega_1 = \frac{8.6}{24.8} \, \frac{103 \times 607 + 468 \times \sqrt{14.4 \times 26 \times 24.8}}{103 \times 607 + 935 \times \sqrt{14.4 \times 26 \times 24.8}} = 0.245$$

Ω_2 (Eq. 11.8, with $\Omega_r = 0.1$):

$$\Omega_2 = 0.9 \times 0.1 = 0.09$$

β for $P_f = 0.05$ (Fig. 11.3):

$$\beta = 1.64$$

Design value of the normalized heat load (Eq. 11.6, with $\Omega_e = 0.101$):

$$H_d'' = 4.76 \times 10^4 \times \exp (1.64 \times \sqrt{0.245^2 + 0.09^2 + 0.101^2})$$

$$= 7.53 \times 10^4 \text{ s}^{1/2} \text{ K}$$

Fire resistance requirement (Eq. 11.5):

$$\tau_d = 0.11 + 16.0 \times 10^{-6} \times 7.53 \times 10^4 + 0.13 \times 10^{-9} \times (7.53 \times 10^4)^2$$

$$= 2.05 \text{ h}$$

μ-factor (Eq. 10.27, with $\Delta h_c = 16.7 \times 10^6 \text{ J kg}^{-1}$, $\Delta h_{ch} = 33.4 \times 10^6 \text{ J kg}^{-1}$):

$$\mu = \frac{0.87 \times (1 - 0.82) \times 16.7 \times 10^6}{0.87 \times 16.7 \times 10^6 + 0.065 \times 33.4 \times 10^6} = 0.157$$

Conclusions
- The fire's potential for spread by convection will be relatively low.
- Using building elements of fire resistance about 2 h will ensure that the failure probability in fully developed burning will not be more than 5 per cent.

The building designer may select from available listings[10] building elements (walls, floors, etc.) for which the fire resistance is equal to at least 2 h, or may

---

10 Such listings are published periodically by fire testing organizations, e.g. in North America by Underwriters' Laboratories of Canada (1980), and Underwriters' Laboratories Inc. (1981). Przetak's book (1977) also contains information on the fire resistance of a multitude of building elements.

design the compartment boundaries for 2 h fire resistance. (The design principles will be discussed in Chapters 12 to 14.)

Some further calculations were performed (Harmathy 1988b) to illustrate the relative importance of three design parameters: the specific fire load (as reflected by the building's occupancy), ventilation (as reflected by the window height), and permissible failure probability. The calculated values of fire resistance ranged from 0.84 h (for hotels; large windows, 2 per cent permissible failure probability, given ignition) to 3.57 h (office buildings; small windows, 0.2 per cent permissible failure probability).

Nomenclature

A	area; surface area, m^2
h	height, m
Δh	heat of reaction (combustion or char-oxidation), J kg^{-1}
H'	normalized heat load under real-world fire conditions, s$^{1/2}$ K
H''	normalized heat load under standard fire test conditions, s$^{1/2}$ K
$\sqrt{k\rho c}$	thermal absorptivity, J m^{-2} s$^{-1/2}$ K^{-1}
L	specific fire load, kg m^{-2}
P	allowable probability, dimensionless
β	factor, function of P_f, dimensionless
δ	fractional heat evolution within the fire compartment, dimensionless
μ	μ-factor, dimensionless
σ	standard deviation, kg m^{-2}
σ_L	standard deviation for specific fire load, kg m^{-2}
τ	length of exposure to fire resistance test; fire resistance time; duration of real-world fire, h (s)
φ	specific surface of fuel, m^2 kg^{-1}
Φ	ventilation parameter, kg s^{-1}
Ω	coefficient of variation, dimensionless

Subscripts

c	of combustion (of pyrolysis gases)
ch	of char oxidation
C	of compartment
d	design value
e	for error associated with the use of Eqs 11.3 and 11.4
f	of failure, conditional on flashover; of fuel
F	of floor
L	for specific fire load
m	mean
t	total for the compartment
v	of the ventilation opening
τ	for fire resistance test results

12 Fire Resistance

Although fire-resistant compartmentation is apparently not as important an aspect of a well-thought-out fire safety design as was believed earlier, to satisfy the building regulations the designer must still have reliable information on the fire performance of all the elements of a planned building.

The fire performance of building elements is evaluated from tests conducted according to an applicable fire resistance test standard (e.g. according to ISO 834; see Chapter 3). The development of fire test standards started toward the end of the nineteenth century, with a view to allaying the danger of catastrophic fire losses. At that time, the thought that information on the fire performance of building elements could ever be developed by calculation likely did not arise. Yet, thanks to the advent of the computer and to a remarkable progress in materials science, the fire performance of a large proportion of building elements can now be studied without the need for fire resistance tests.

Although the capability of theoretically simulating the fire test and predicting the test result has improved dramatically during the past 30 years, there is still some resistance by the writers of building codes and by the insurance companies to granting general recognition to calculated fire performance information. However, the realization that science works in the interest of society as a whole will, no doubt, soften that resistance.

The analytical and numerical techniques that may be used for determining the performance of building elements in fire (by simulating either fire tests or real-world fires) will be dealt with in some detail in the next two chapters. In this chapter, some general aspects of the fire resistance concept will be discussed, and shortcut methods will be presented which may render the estimation of fire resistance possible without a thorough insight into the heat flow process and the behavior of materials. Further information on the subject is available from the following sources, among others: in a book by Malhotra (1982), in a paper by Harmathy (1979a), in the ACI and ASCE Guides (ACI Committee 216 1989; Lie 1992), in the *Supplement to the National Building Code of Canada* (Associate Committee on the National Building Code 1990), and in two draft Canadian standards (Underwriters' Laboratories of Canada 1988, 1989).

12.1   General Considerations and Terminology

Calculations aimed at predicting the fire resistance of building elements usually consist of two steps:

(1) mathematical simulation of the temperature history of the element (or some crucial component of it) and, based on the results of these calculations,
(2) assessment of the element's structural performance up to the point of failure.

Clearly, carrying out the calculations in two steps is permissible only if the thermal and structural responses are uncoupled, i.e. if the temperature history of the element is not affected by its stress−deformation history, at least until the point of failure is closely approximated. This condition is not fulfilled if the element undergoes substantial dimensional changes (e.g. due to burning, spalling, or partial disintegration) as a result of heating or excessive stresses.

In certain situations, it is possible to forgo the second step calculations, and to predict the outcome of the fire test solely from a simulation of the element's temperature history. Such situations arise whenever

(a) it is well established (either from experience or from theoretical studies) that the type of element in question will fail by excessive heat transmission, and
(b) the fire test standard allows the point of structural failure to be interpreted as the attainment of a critical level by the temperature of an important load-bearing steel component.

The time to failure by excessive heat transmission is often referred to as *thermal fire resistance*, and the time to failure by destruction or by the attainment of a critical temperature level by a key steel component, as *structural fire resistance*.

Design considerations related to thermal and structural fire resistances are not necessarily parallel. For example, adding a layer of insulation to that side of a building element which is likely to be exposed to fire increases the element's thermal fire resistance, as well as its structural fire resistance. On the other hand, adding a layer of insulation to the element's opposite side, while advantageous for its thermal fire resistance, may adversely affect its structural fire resistance, owing to increased temperature buildup near the exposed side where the crucial load-bearing steel components are usually located.

Selecting the *cover thickness*, i.e. the thickness of concrete or other insulating material over crucial load-bearing steel components, is an important aspect of fire safety design.

The *concept of critical temperature*, though very popular in the assessment of fire resistance, is based on an oversimplification of the actual conditions. Clearly, the temperature at which the failure of the crucial steel components occurs cannot be looked upon as a constant value; it depends on the loading conditions and on the kind of steel. As Fig. 12.1 shows, the temperature at failure is a function of the stress level allowed in the design, and is higher for structural steel than for concrete prestressing steel.

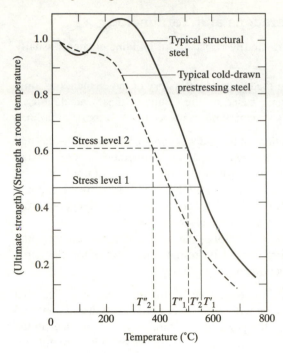

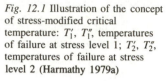

Fig. 12.1 Illustration of the concept of stress-modified critical temperature: T_1', T_1'', temperatures of failure at stress level 1; T_2', T_2'', temperatures of failure at stress level 2 (Harmathy 1979a)

The concept of evaluating the point of failure by means of the strength—temperature relation for the crucial steel component is often referred to as the *concept of stress-modified critical temperature*.

Reflecting on the problem, one will find that the latter concept is not strictly accurate either. As discussed in Chapter 7, the strength of steel is not a unique function of the temperature. Owing to creep, failure may occur earlier than one would expect on the basis of the strength—temperature relation. Hence, to arrive at reasonably accurate predictions of the time of structural failure, the designer may have to resort to the *creep concept*.

The performance of horizontal separating elements (floors, roofs) and flexural elements (girders, beams, trusses, joists), consisting of two or more parallel components, depends on whether there is *composite action* between the components. The absence and presence of composite action is illustrated in Figs 12.2(a) and 12.2(b), which show two ceiling—floor assemblies, both consisting of a protected steel beam and a concrete slab.

The absence or presence of *restraint* is also an important factor in fire performance, especially in the case of flexural elements. As Fig. 12.3(a) shows, restraint means the action of thermal thrust by the surrounding structures, which keeps the element's thermal expansion and the rotation of its ends (edges) within limits. With ceiling—floor assemblies, roofs, and walls, restraint may be applied to two edges or all four edges of the test specimen.

Restraint, if it does not induce spalling, generally imparts an extra degree of

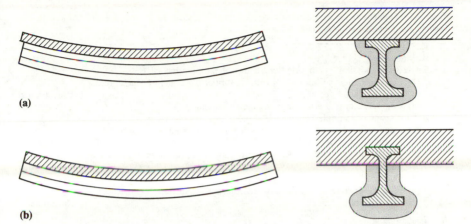

Fig. 12.2 Illustration of composite action on ceiling–floor assemblies consisting of a protected steel beam and concrete floor slab: (a) no composite action; (b) composite action

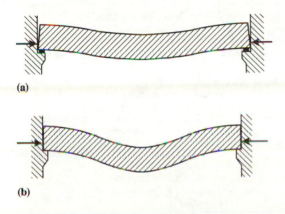

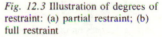

Fig. 12.3 Illustration of degrees of restraint: (a) partial restraint; (b) full restraint

stability to the building element, and therefore it is beneficial for performance in fire.

Figure 12.3(b) illustrates *full restraint*, which does not allow any thermal expansion and end (edge) rotation. Only a few test furnaces are known to be equipped with special loading devices capable of imposing on the test specimen the axial forces and moments required to achieve full restraint. Most testing agencies use heavy restraining frames which impose *partial restraint* of some unknown magnitude. However, the degree of restraint does not seem to influence significantly the performance of flexural elements in fire (Lin and Abrams 1983).

Perfect restraint rarely arises in real-world fires. ASTM E 119 (American Society for Testing and Materials 1988a) gives some guidelines on how to decide whether the fire resistance of a construction should be determined with or without application of restraint. They are reproduced in Table 12.1.

Table 12.1 Guidelines for assessing if a construction is restrained or unrestrained (American Society for Testing and Materials 1988a)

I. Wall bearing:	
Single span and simply supported end spans of multiple bays:[a]	
(1) Open-web steel joists or steel beams, supporting concrete slab, precast units, or metal decking	Unrestrained
(2) Concrete slabs, precast units, or metal decking	Unrestrained
Interior spans of multiple bays:	
(1) Open-web steel joists, steel beams, or metal decking, supporting continuous concrete slab	Restrained
(2) Open-web steel joists or steel beams, supporting precast units or metal decking	Unrestrained
(3) Cast-in-place concrete slab systems	Restrained
(4) Precast concrete where the potential thermal expansion is resisted by adjacent construction[b]	Restrained
II. Steel framing:	
(1) Steel beams welded, riveted, or bolted to the framing members	Restrained
(2) All types of cast-in-place floor and roof systems (such as beam-and-slabs, flat slabs, pan joists, and waffle slabs) where the floor or roof system is secured to the framing members	Restrained
(3) All types of prefabricated floor or roof systems where the structural members are secured to the framing members, and the potential thermal expansion of the floor or roof system is resisted by the framing system or the adjoining floor or roof construction[b]	Restrained
III. Concrete framing:	
(1) Beams securely fastened to the framing members	Restrained
(2) All types of cast-in-place floor or roof systems (such as beam-and-slabs, flat slabs, pan joists, and waffle slabs) where the floor system is cast with the framing members	Restrained
(3) Interior and exterior spans of precast systems with cast-in-place joints resulting in restraint equivalent to that which would exist in Condition III (1)	Restrained
(4) All types of prefabricated floor or roof systems where the structural members are secured to such systems, and the potential thermal expansion of the floor or roof systems is resisted by the framing system or the adjoining floor or roof construction[b]	Restrained
IV. Wood construction:	
All types	Unrestrained

[a] Floor and roof systems can be considered restrained when they are tied into walls with or without tie beams, the walls being designed and detailed to resist thermal thrust from the floor or roof system.

[b] For example, resistance to potential thermal expansion is considered to be achieved when:
   (1) continuous structural concrete topping is used,
   (2) the space between the ends of precast units, or between the ends of units and the vertical face of supports is filled with concrete or mortar,
   (3) the space between the ends of precast units and the vertical faces of supports, or between the ends of solid or hollow-core slab units, does not exceed 0.25 per cent of the length for normal-weight concrete members or 0.1 per cent of the length for structural lightweight concrete members.

The consequences of the failure of various building elements in real-world fires depend on the role the elements play in the structural and functional soundness of the building. It is customary to call some elements *key elements* and others *separating elements*. Those in the latter group are not parts of the principal structural network (even though they may carry substantial load); the failure of one of them allows only a one-step fire spread: from one building space to another. Once the fire has reached the reverse side of the element (either by loss of integrity or by convective means), its total destruction is of little import for the further course of the fire. Consequently, determining the fire performance of separating elements by means of the existing practice of fire testing, i.e. by exposing them (or in calculations, simulating their exposure) to fire from one direction or the other, is correct.

The structural soundness of the building depends on the key elements which form the load-bearing network. The destructive failure of any of them may result in the spread of fire at once to several building spaces. It is essential, therefore, that their load-bearing capacity be maintained throughout a fire exposure, whether it occurs on one side, or two sides, or all sides (as in the case of columns). Examining their performance in fire exposures from one direction only, as is done in the standard fire tests (except for column tests), is clearly insufficient.

The fire resistance test standards require that the test specimen be truly representative of the construction of which the performance is to be determined, with respect to dimensions, materials, workmanship, and conditions of loading and restraint. However, fully duplicating an actual building element is difficult, if not impossible. It is usually necessary to make certain compromises, beginning with the fact that in most cases the specimen is smaller than the element it represents. It is imperative, therefore, that the personnel of the testing laboratory clearly see which are the important factors that must be duplicated.

Not too far removed from the decisions on how to devise the fire tests are the deliberations concerning the possible application of knowledge gained from the test of one building element to other building elements. Once the role of the various factors in the test result is understood, assessing the performance of other, somewhat similar, building elements is also possible.

Theoretical considerations may allow further extension of the knowledge gained from a test to building elements that are not quite similar to the one tested. The theoretical insight may be presented in the form of simple *rules*, or may be stated in a more quantitative manner, in the form of *extension formulas*.

To help the fire safety practitioner, fire researchers have performed a multitude of experiments on a variety of building elements and conducted (mostly by means of computer) thousands of calculations. The results of the experimental and theoretical studies have been condensed into *semiempirical* and *empirical formulas*, *graphs* and *tables*.

12.2   Rules and Extension Formulas

The value of the numerous pieces of information obtained from standard fire tests is very limited without some kind of theory which is capable of cementing the pieces into a consistent framework. In the 1960s Harmathy (1965b) developed a set of rules with a view to giving the testing agencies guidance on how to plan fire tests, how to interpret test results, and how to apply the results to other building elements. At that time the rules (commonly referred to as Harmathy's ten rules) were looked upon as representing a new insight into the art of fire testing. Although at the present level of understanding some of them may sound commonplace, they are still widely used by fire safety practitioners; therefore they will be reproduced here in their entirety. Figure 12.4 serves as illustration of the ten rules.

Rule 1   The thermal fire resistance of a building element consisting of a number

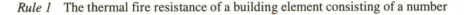

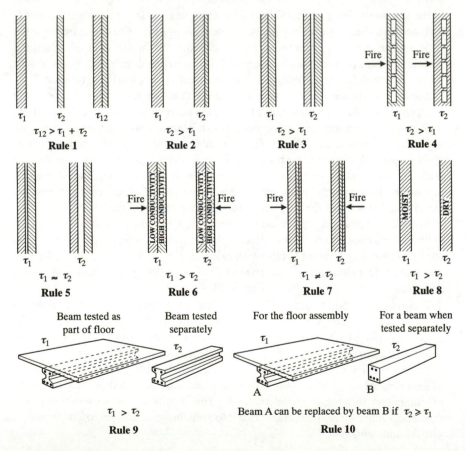

Fig. 12.4 Illustration of Harmathy's ten rules (Harmathy 1965b) (τ is fire resistance). Reprinted with permission from *Fire Technology* (Issue 1), copyright © 1965, National Fire Protection Association, Quincy, MA 02269

of parallel layers is greater than the sum of the thermal fire resistances of the individual layers.

Rule 2   The thermal fire resistance of a building element does not decrease with the addition of one or more layers of materials.[1]

Rule 3   The thermal fire resistance of building elements containing air gaps or cavities is greater than that of similar elements of the same mass but containing no air gaps or cavities.

Rule 4   The further the air gap or cavity is located from the exposed surface, the more beneficial is its effect on the thermal fire resistance.

Rule 5   The thermal fire resistance of a building element is practically independent of the thickness of a completely enclosed air layer.

Rule 6   From the point of view of thermal fire resistance, layers of materials of low thermal conductivity are better utilized on the side of the element on which fire is more likely to occur.

Rule 7   The thermal fire resistance of an asymmetrical building element depends on which side is exposed to fire.

Rule 8   The presence of moisture, if it does not result in explosive spalling, increases the fire resistance.

Rule 9   The load-supporting elements (girders, beams, trusses, joists; see Fig. 12.5) of floor, roof, or ceiling assemblies yield higher fire resistances when subjected to fire resistance tests as parts of the assemblies than they would if tested separately.

Rule 10   The load-supporting elements of a floor, roof, or ceiling assembly can be replaced by such other load-supporting elements which, when tested separately, yield fire resistances not less than that of the assembly.

Extension formulas provide means for estimating changes in fire resistance on account of moderate changes in dimensions or materials. Analogy between some electrical and heat transmission phenomena suggested the formulation of several useful formulas (McGuire 1958; Robertson and Gross 1958; McGuire *et al.* 1975). A few of them will be presented here, not necessarily in the forms in which they

---

1   Rule 2 is clearly a corollary of Rule 1. Certain restrictions apply to these two rules and to some others [see Harmathy (1965b)]. It is important to remember that the structural fire resistance of an element may decrease with the addition of layers of materials to the unexposed side.

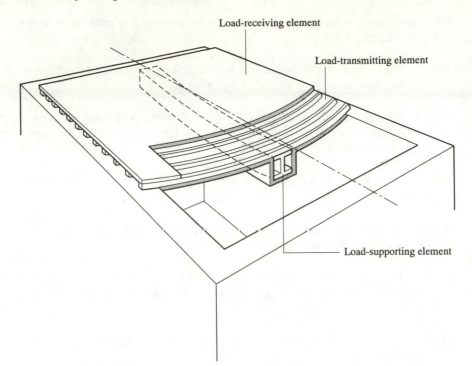

Fig. 12.5 Interpretation of the term 'load-supporting element'

originally appeared. The symbol * denotes information related to the reference building element, namely the one of which the fire resistance is known from a standard fire test. Attention is called to the illustrations in Fig. 12.6 (Harmathy 1979a).

For *slab-like* vertical or horizontal *separating elements* [walls, floors; Fig 12.6(a)], if their failure in a fire test is expected to take place by loss of insulating capability (see Chapter 3), the extension formula (Dunham *et al.* 1942; Jones 1962) is:

$$\frac{\tau}{\tau^*} = \left(\frac{L}{L^*} \sqrt{\frac{\kappa^*}{\kappa}} \right)^{1.7} \qquad\qquad [12.1]$$

where τ = thermal fire resistance of the building element, sought by calculation, h;

L = thickness of the building element, m;

κ = thermal diffusivity of the building element, $m^2\,s^{-1}$;

and τ^*, L^*, and κ^* relate to the reference building element.

For slab-like building elements [walls, floors; Fig. 12.6(b)], which are expected

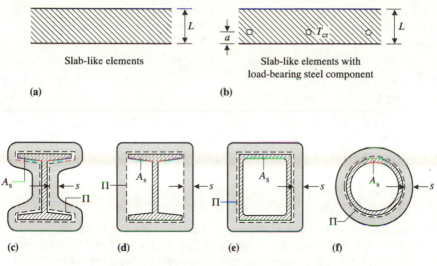

(a) Slab-like elements

(b) Slab-like elements with load-bearing steel component

Columns: contour protection and box protection

(c)   **(d)**   **(e)**   **(f)**

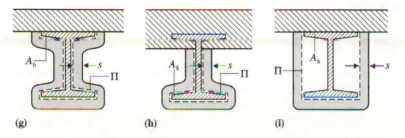

Beams: contour protection and box protection

(g)   **(h)**   **(i)**

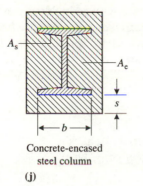

Concrete-encased
steel column

(j)

- - - - -   Heated perimeter, Π

A_s Cross-sectional area
of steel core

A_e Cross-sectional area
of encasement

Fig. 12.6 Various kinds of building elements (Harmathy 1979a)

to fail in fire test on the attainment of a critical temperature level, T_{cr} (K), by an important load-bearing component (e.g. reinforcing or prestressing steel) at a distance a (m) from the fire-exposed surface (Jones 1962), the formula is

$$\frac{\tau}{\tau^*} = \left(\frac{a}{a^*} \sqrt{\frac{\kappa^*}{\kappa}} \right)^n \qquad [12.2]$$

where a^* (m) is the distance of the important load-bearing (steel) component from the fire-exposed surface in the reference building element (whose fire resistance is τ^*), and the value of exponent n is a function of T_{cr}.

$$n = \begin{cases} 1.53 & \text{if } T_{cr} = 672\,\text{K } (399\,°\text{C}) \\ 1.36 & \text{if } T_{cr} = 811\,\text{K } (538\,°\text{C}) \\ 1.08 & \text{if } T_{cr} = 950\,\text{K } (677\,°\text{C}) \end{cases} \qquad [12.3]$$

For other values of T_{cr}, n can possibly be estimated by interpolation.

For *protected steel columns or beams* [Figs 12.6(c)–12.6(j)], which are expected to fail in fire tests when the steel component attains a temperature level of 811 K (538 °C) in the case of columns, or 866 K (593 °C), in the case of beams[2] (McGuire 1958; McGuire *et al.* 1975; Anonymous 1976), the following formula applies:

$$\frac{\tau}{\tau^*} = \left(\frac{s}{s^*} \frac{A}{A^*} \frac{\Pi^*}{\Pi} \right)^{0.8} \qquad [12.4]$$

where s = cover thickness over the load-bearing steel component (i.e. minimum thickness of material between the steel and the fire-exposed surface), m;

A = cross-sectional area of the load-bearing core of column or beam (see Eqs 12.5 and 12.6), m²;

Π = heated perimeter of the core [shown as a broken line in Figs 12.6(c)–12.6(i)], m;

and, again, τ^*, s^*, A^* and Π^* relate to the reference building element.

In the case of *contour-* and *box-protected columns* or *beams* [Figs 12.6(c)–12.6(i)], if the protection is provided by some sprayed lightweight material, A is equal to the cross-sectional area of the steel core, A_s (m²), and $A^* = A_s^*$. Otherwise

$$A = A_s + \frac{s\Pi}{2} \frac{\rho_p c_p}{\rho_s c_s} \qquad [12.5]$$

where ρ_p = density of the protecting material, kg m⁻³;

ρ_s = density of the load-bearing steel core, kg m⁻³;

---

2   These temperatures are specified in ASTM E 119 as critical temperatures for the load-bearing steel components of columns and beams, respectively.

c_p = specific heat of the protecting material, $J\,kg^{-1}\,K^{-1}$;
c_s = specific heat of the load-bearing steel core, $J\,kg^{-1}\,K^{-1}$;

and A^* is to be calculated from an equation similar to Eq. 12.5, in which A_s, s, Π, ρ_p, and c_p are replaced by A_s^*, s^*, Π^*, ρ_p^*, and c_p^*, respectively.

For *columns with massive protection* [(Fig. 12.6(j)], Π is taken to be equal to b [where b (m) is the width of flange], and A is interpreted as

$$A = A_s + \frac{A_e}{2}\frac{\rho_e c_e}{\rho_s c_s} \qquad [12.6]$$

where A_e = cross-sectional area of the encasement, m^2;
ρ_e = density of the encasing material (usually concrete), $kg\,m^{-3}$;
c_e = specific heat of encasing material (usually concrete), $J\,kg^{-1}\,K^{-1}$.

Again, $\Pi^* \simeq b^*$, and A^* is to be calculated from an equation similar to Eq. 12.6, in which A_s, A_e, ρ_e, and c_e are replaced by A_s^*, A_e^*, ρ_e^*, and c_e^*, respectively.

Owing to the approximate nature of these relations, it is permissible to use in Eqs 12.1, 12.2, 12.5, and 12.6 material property values listed in Table 10.2.

Naturally, the preceding rules and extension formulas are applicable only if both the reference building element and the element for which fire resistance information is sought remain dimensionally sound up to the point of failure.

A more accurate technique for calculating the fire resistance of box-protected steel columns, and steel columns with massive protection, is discussed in the next section [see Figs 12.10(a) and 12.10(b), Eq. 12.12 and the relevant discussion].

12.3 Semiempirical and Empirical Formulas, Graphs

Many semiempirical and empirical formulas have been generated over the years by correlating the results of a multitude of experimental or computer studies. Although there are no basic differences between the techniques employed in development of the two kinds of formulas, the ones generated by more comprehensive studies, involving wide ranges of geometric variables and material properties, will be referred to here as semiempirical formulas.

Nearly 1200 computer calculations were performed in a study concerning the

Table 12.2 The values of constants in Eqs 12.7 and 12.8 (Harmathy 1970b)

Constant	Normal-weight concrete	Lightweight concrete
E	2.46×10^{-3}	6.50×10^{-3}
p	0.67	0.4
q	1.3	1.3
F	17.1×10^{-3}	70.2×10^{-3}
r	0.85	0.5
s	1.2	1.2

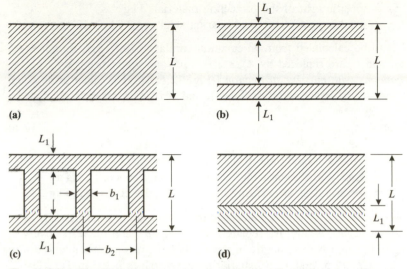

Fig. 12.7 Slab-like building elements: (a) monolithic slab; (b) double-layer configuration; (c) hollow-core slab; (d) composite slab (Harmathy 1979a)

thermal fire resistance of solid and hollow-core concrete slabs (walls, floors, roofs).[3] Covered in the study were four kinds of material (two normal-weight and two lightweight concretes, the volume specific heats and thermal conductivities of which are plotted in Figs 6.19 and 6.20), and four geometric variables. The following semiempirical formulas have been generated (Harmathy 1970b, 1979a).

● For *monolithic (solid) slabs* [Fig. 12.7(a)]:

$$\tau_{|} = E\left(\frac{k}{L}\right)^p\left(\frac{L^2}{\kappa}\right)^q \qquad [12.7]$$

● For *double-layer configurations* [assemblies of two slabs separated by an air layer of any thickness; Fig. 12.7(b)]:

$$\tau_{\|} = F\left(\frac{k}{L_1}\right)^r\left(\frac{L_1^2}{\kappa}\right)^s \qquad [12.8]$$

● For *hollow core slabs* [Fig. 12.7(c)]:

$$\tau_{\mathrm{H}} = \left(\frac{1}{\dfrac{b_1/b_2}{\tau_{|}^{1/2}} + \dfrac{1 - b_1/b_2}{\tau_{\|}^{1/2}}}\right)^2 \qquad [12.9]$$

---

3   Owing to somewhat higher convective heat losses from the unexposed surface of horizontal slabs, the thermal fire resistance is slightly greater for horizontal slabs than for vertical slabs. However, under practical conditions the differences are hardly noticeable.

In the above equations, $\tau_|$, $\tau_\|$, and τ_H (s)[4] are the thermal fire resistances for the three geometric configurations shown in the figures, and

L = thickness [Figs 12.7(a)−12.7(c)], m;
L_1 = thickness [Figs 12.7(b) and 12.7(c)], m;
k = thermal conductivity of concrete at room temperature, $W\,m^{-1}\,K^{-1}$;
κ = thermal diffusivity of concrete at room temperature, $m^2\,s^{-1}$;
b_1, b_2 = thickness, separation [Fig. 12.7(c)], m.

The coefficients E and F, and the exponents p, q, r, and s are empirical constants; their values are listed in Table 12.2.

Equations 12.7, 12.8, and 12.9 express the fire resistance of concrete slabs in *moistureless* condition.[5]

The thermal fire resistance of *composite slabs* [Fig. 12.7(d)] consisting of two layers, one normal-weight concrete and the other lightweight concrete, and containing *normal amounts of moisture*, can be estimated from the following two empirical equations (Lie 1978).

● If the layer of normal-weight concrete is on the side exposed to fire:

$$\tau = 0.318 \times 10^6 \left(2L^2 - L_1L + \frac{98.3 \times 10^{-6}}{L} \right) \qquad [12.10]$$

● If the layer of lightweight concrete is on the side exposed to fire:

$$\tau = 0.352 \times 10^6 \left(L^2 + 2L_1L - L_1^2 + \frac{65.6 \times 10^{-6}}{L} \right) \qquad [12.11]$$

where L (m) is the total thickness of the slab, and L_1 (m) is the thickness of the exposed layer; 0.318×10^6 and 0.352×10^6 $(s\,m^{-2})$, 98.3×10^{-6} and 65.6×10^{-6} (m^3) are empirical constants.

Abrams and Gustaferro (1968, 1969) studied experimentally the thermal fire resistance of monolithic and composite concrete slabs. Displayed in Fig. 12.8 is a plot of the fire resistance of *monolithic slabs* of various concretes (at *normal moisture content*) against slab thickness. The fire resistance of *composite (two-layer) slabs*, consisting of normal-weight and lightweight concretes, is plotted in Fig. 12.9. Abrams and Gustaferro's studies also covered concrete slabs used in combination with sprayed vermiculite, sprayed mineral fiber, sprayed

---

4   In engineering practice, the dimension of fire resistance is h or min. However, for the sake of dimensional consistency, using the dimension s is often necessary.
       It has been customary to express the fire resistance of hollow-core slabs in terms of *equivalent thickness*. The equivalent thickness of a slab (or a masonry unit) is interpreted as the volume of the slab (or masonry unit) divided by the area to be exposed to fire. A technique of measuring the equivalent thickness of masonry units was described by Harmathy and Oracheski (1970). Although correlating fire resistance with equivalent thickness is not strictly correct, it often provides an accuracy adequate for engineering practice.
5   The effect of the presence of moisture on the fire resistance may be taken into account by means of Eq. 4.13b.

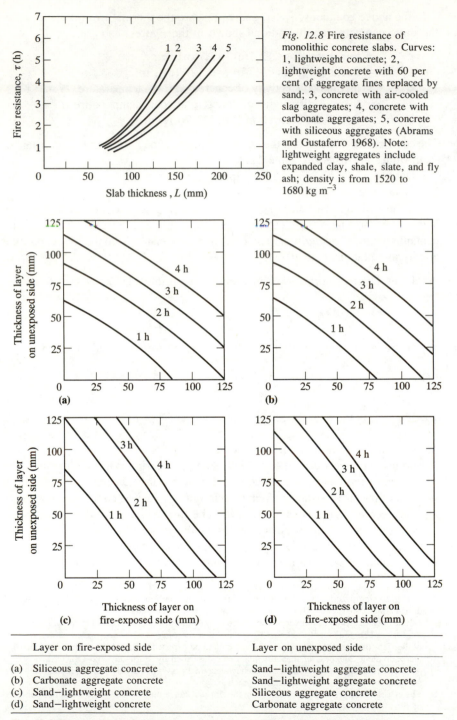

Fig. 12.8 Fire resistance of monolithic concrete slabs. Curves: 1, lightweight concrete; 2, lightweight concrete with 60 per cent of aggregate fines replaced by sand; 3, concrete with air-cooled slag aggregates; 4, concrete with carbonate aggregates; 5, concrete with siliceous aggregates (Abrams and Gustaferro 1968). Note: lightweight aggregates include expanded clay, shale, slate, and fly ash; density is from 1520 to 1680 kg m^{-3}

	Layer on fire-exposed side	Layer on unexposed side
(a)	Siliceous aggregate concrete	Sand−lightweight aggregate concrete
(b)	Carbonate aggregate concrete	Sand−lightweight aggregate concrete
(c)	Sand−lightweight concrete	Siliceous aggregate concrete
(d)	Sand−lightweight concrete	Carbonate aggregate concrete

Fig. 12.9 Fire resistance of two-layer concrete slabs (Abrams and Gustaferro 1969);

intumescent mastic, mineral board, glass fiber board, and terrazzo. The results of these studies were presented graphically in the cited publications and in the *ACI Guide* (ACI Committee 216 1989).

Dividing the normal-weight concretes into siliceous aggregate concretes and carbonate aggregate concretes, as in Figs 12.8 and 12.9, is common practice in reporting fire resistance test data. On account of the high thermal conductivity of silica, siliceous aggregate concretes can indeed be expected to yield somewhat lower thermal fire resistances than other concretes. However, the popular belief that carbonate aggregates are inherently superior to all other kinds of aggregate is of questionable validity. The heat absorbed in the decomposition of calcite, dolomite, or magnesite is likely to be beneficial for the results of long fire tests only, since, as Table 6.7 shows, the decomposition of these minerals takes place at fairly high temperatures. The thermal properties of the various minerals furnished in Chapter 6 seem to indicate that most rocks of nonsiliceous silicates, especially those in the basalt−gabbro group,[6] are suitable for use in fire-resistant concretes.

If the gain due to the presence of a normal amount of moisture is taken into account, the fire resistance values calculated from Eq. 12.7 are comparable with those read from Fig. 12.8. However, Eq. 12.7, by expressing the fire resistance in terms of thermal properties, offers the advantage of allowing the design of fire-resistant concrete slabs for specific applications.

The fire resistance of *concrete-protected steel columns* [Figs 12.10(a) and 12.10(b)] was the subject of a computer study conducted by Lie and Harmathy (1974). They assumed, in line with the ASTM E 119 test specification, that failure occurs when the temperature of the steel core reaches 811 K (538 °C). Their findings were condensed into the following semiempirical formula:

$$\tau = 19 \times 10^3 \left(\frac{A_s}{\Pi}\right)^{0.7} + 409 \times 10^3 \frac{s^{1.6}}{k_e^{0.2}} \left[1 + 1.5 \left(\frac{C}{\rho_e c_e s \Pi} \frac{b}{b+s}\right)^{0.8}\right]$$

$$[12.12]$$

The first term on the right-hand side of this equation is the fire resistance of the steel column without fire protection (Stanzak and Lie 1973). In the bracketed expression, C ($\text{J m}^{-1}\,\text{K}^{-1}$) is the heat capacity of the core per unit length. It is to be calculated

- in the case of box-protected steel columns [Fig. 12.10(a)] as

$$C = \rho_s c_s A_s \qquad [12.13a]$$

- in the case of steel columns with massive protection [Fig. 12.10(b)], as

$$C = \rho_s c_s A_s + \rho_e c_e (b_1 b_2 - A_s) \qquad [12.13b]$$

For columns with massive protection, s and b are interpreted as

---

6   Minerals in this group are, in fact, the most abundant in the Earth's crust.

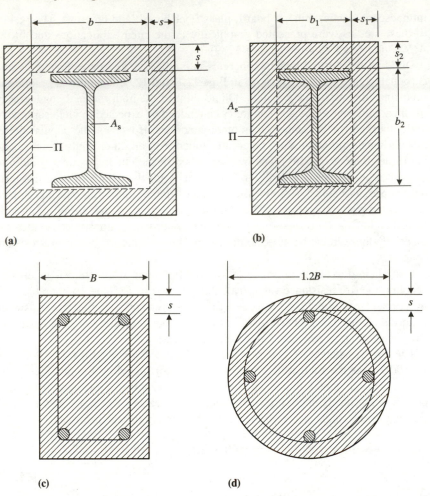

Fig. 12.10 Various concrete–steel columns: (a) box-protected steel column; (b) steel column with massive protection; (c) rectangular reinforced concrete column; (d) circular reinforced steel column

$$s = \frac{1}{2}(s_1 + s_2)$$ [12.14]

$$b = \frac{1}{2}(b_1 + b_1)$$ [12.15]

In Eqs 10.12 to 10.15, 19×10^3, 409×10^3 are dimensional constants, τ (s) is the fire resistance of the columns in *moistureless* condition,[7] A_s (m^2) is the cross-sectional area of the steel core, Π (m) is the heated perimeter of core [to

---

7   The effect of moisture content (within the usual limits) on the fire resistance of columns is insignificant (Chapter 14).

be interpreted as shown in Figs 12.10(a) and 12.10(b)], k (W m^{-1} K^{-1}) is thermal conductivity, ρ (kg m^{-3}) is density, and c (J kg^{-1} K^{-1}) is specific heat; the s subscripts relate to the steel core, and the e subscripts to the encasing material (concrete). The dimensions s, b, s_1, s_2, b_1, and b_2 (m) are explained in the figures.

Theoretical and experimental studies conducted by Lie and Allen (1972) provided the basis for a set of empirical formulas for *reinforced concrete columns*, which were presented in the form of dimensional requirements for satisfactory performance in fire.[8]

Two of the requirements are related to the minimum dimension of the column, B (m). With rectangular columns [Fig. 12.10(c)], the condition of attaining a fire resistance of τ (s) (at *normal moisture content*) is:

- if the column is made from normal-weight concrete:

$$B \geq 28.2 \times 10^{-6}(\tau + 3600)$$ [12.16a]

- if the column is made from lightweight concrete:

$$B \geq 21.2 \times 10^{-6}(\tau + 3600)$$ [12.16b]

Two other requirements concern the minimum concrete cover over the vertical reinforcements, s (m).

$$s \geq 7.1 \times 10^{-6}\tau \qquad \text{if } \tau \leq 7200 \text{ s (2 h)}$$ [12.17a]

$$s \geq 3.5 \times 10^{-6}\tau + 0.025 \quad \text{if } \tau > 7200 \text{ s (2 h)}$$ [12.17b]

In the case of circular columns [Fig. 12.10(d)], the diameter must not be less than $1.2B$, with B determined from Eq. 12.16a or 12.16b.

12.4   Tabulated Information

Tempted to make short work of fire safety design, building designers often turn to ready-made solutions, arranged in tables or lists. Building elements made from specified materials and with the observance of certain minimum dimensions (tabulated in authoritative documents) are guaranteed to yield specified fire resistances. Most of these dimensions have been derived from the results of a multitude of fire resistance tests. Unfortunately, fire testing laboratories are rarely equipped with facilities for studying material properties[9] (thermal properties in particular). In the test reports, the construction materials are usually described in generic terms, which allow little insight into the anatomy of the test results.

The tables usually contain two kinds of information:

(1) minimum thicknesses of (slab-like) vertical or horizontal separating elements

---

8   In newer versions of these formulas, the type of concrete, the length of the column, the end conditions, and the area of vertical reinforcement are also taken into account (Lie 1989, 1992). See Chapter 14.
9   In fact, few testing laboratories could afford the cost of conducting a comprehensive investigation into the properties of all the materials of which a test specimen is made.

(walls, floors, roofs) which ensure specified levels of thermal performance, and

(2) the minimum cover thicknesses over some important steel components of load-bearing building elements which ensure specified levels of structural performance.[10]

Several publications provide tabulated information (e.g. Cement and Concrete Association 1975; Bobrowski 1978; Harmathy 1979a; Read *et al.* 1980; Gustafson 1980; Comité Euro-International du Béton 1982; Associate Committee on the National Building Code 1990). The information published in the *Supplement to the National Building Code of Canada* (Associate Committee on the National Building Code 1990) is one of the most comprehensive. Excerpts from the *Supplement* are presented in three tables. Tables 12.3 and 12.4 list the minimum thicknesses of slab-like concrete elements (walls, floors, roofs) required for various levels of thermal performance. Table 12.5 summarizes information on the

Table 12.3 Minimum thickness (or minimum equivalent thickness[a]) (in mm) of (load-bearing and non-load-bearing) walls made from solid or hollow concrete masonry units, required for the given fire resistance (in h)[b] (Associate Committee on the National Building Code 1990)

Material	Fire resistance (h)						
	0.5	0.75	1.0	1.5	2.0	3.0	4.0
Type S or N concrete	44	59	73	95	113	142	167
Type L_1 concrete	42	54	64	82	97	122	143
Type L_2 concrete	42	54	63	79	91	111	127
Type $L_1$20S concrete	42	54	66	87	102	129	152
Type $L_2$20S concrete	42	54	64	81	94	116	134

[a] See Footnote 4 for the definition of equivalent thickness.

[b] See Table 6.12 for the types of concrete.

Table 12.4 Minimum thickness (in mm) of monolithic (load-bearing and non-load-bearing) concrete walls, and reinforced or prestressed concrete floor or roof slabs, required for the given fire resistance (in h)[a] (Associate Committee on the National Building Code 1990)

Material	Fire resistance (h)						
	0.5	0.75	1.0	1.5	2.0	3.0	4.0
Type S concrete	60	77	90	112	130	158	180
Type N concrete	59	74	87	108	124	150	171
Type L or type L40S concrete	49	62	72	89	103	124	140

[a] See Table 6.12 for the types of concrete.

10 The thermal and structural performance requirements are to be interpreted according to the fire resistance test standard. They were discussed in some detail in Chapter 3.

minimum cover thicknesses over important load-bearing steel components, required for various levels of structural performance.[11]

The wall and slab thicknesses given in Tables 12.3 and 12.4 are in fair agreement with those that one can read from Fig. 12.8 or calculate from Eqs 12.7–12.9. Up to a fire resistance of 2 h, these wall and slab thicknesses do not seem to be out of line with what a designer would choose with constructibility, structural

Table 12.5 Minimum thickness of cover (in mm) over steel in reinforced and prestressed concrete floor or roof slabs, reinforced and prestressed concrete beams, and concrete-protected steel beams and columns, required for the given fire resistance (in h)[a] (Associate Committee on the National Building Code 1990)

Construction and material	Fire resistance (h)						
	0.5	0.75	1.0	1.5	2.0	3.0	4.0
Reinforced concrete floor or roof slabs							
Type S, N, L, or L40S concrete	20	20	20	20	25	32	39
Prestressed concrete floor or roof slabs[b]							
Type S, N, L, or L40S concrete	20	25	25	32	39	50	64
Reinforced concrete beams							
Type S, N, or L concrete	20	20	20	25	25	39	50
Prestressed concrete beams[bc]							
Type S and N concrete							
$0.026\,\mathrm{m}^2 < A_b \leq 0.097\,\mathrm{m}^2$	25	39	50	64	—	—	—
$0.097\,\mathrm{m}^2 < A_b \leq 0.194\,\mathrm{m}^2$	25	26	39	45	64	—	—
$A_b > 0.194\,\mathrm{m}^2$	25	26	39	39	50	77	102
Type L concrete, $A_b > 0.097\,\mathrm{m}^2$	25	25	25	39	50	77	102
Concrete protected steel beams[d]							
Type S concrete	25	25	25	25	32	50	64
Type N or L concrete	25	25	25	25	25	39	50
Steel columns, monolithic concrete protection[d]							
Type S concrete	25	25	25	25	39	64	89
Type N or L concrete	25	25	25	25	32	50	77
Steel columns, concrete masonry protection[e]							
Type S concrete	50	50	50	50	64	89	115
Type N or L concrete	50	50	50	50	50	77	102

[a] See Table 6.12 for the types of concrete.

[b] For slabs (beams) with several tendons, the given thickness is to be interpreted as the average of those of the individual tendons, except that the cover over any tendon cannot be less than half of the value given, or less than 20 (25) mm.

[c] A_b (m^2) is the cross-sectional area of the beam.

[d] Applies to cast-in-place concrete reinforced by 5.21 mm diameter wire or 1.57 mm diameter wire mesh with 100 mm by 100 mm openings.

[e] Applies to concrete masonry reinforced by 5.21 mm diameter wire or 1.19 mm diameter wire mesh 10 mm by 10 mm openings, laid in every second course.

---

11 The minimum cover thickness requirements for reinforced concrete columns are given by Eqs 12.17a and 12.17b.

stability, and sound and heat insulation in mind. Similarly, the tabulated cover thicknesses also seem reasonable up to a 2 h fire resistance level.

As discussed earlier (see Section 10.5 in Chapter 10 and Section 11.4 in Chapter 11), the usual duration of fully developed fires is from 0.3 to 1 h, and the realistic range of fire resistance requirements is from 0.25 to 2.25 h. Since fire resistance requirements well over 2 h are rarely justifiable, it is not surprising that the tabulated wall and slab thicknesses, and cover thicknesses, for fire resistance levels above 2 h appear to be greater than what engineering judgment would suggest.

12.5   Performance Requirements for Key Elements

There is one area, however, where some extra margin of safety is fully justified. As emphasized earlier in this chapter, it is imperative that all the key elements of a building be designed to remain structurally sound even if, owing to fire spread, they become exposed to fire from more than one direction.

Figure 12.11 illustrates situations that may arise in spreading fires (Harmathy 1977/1978). At a time $t = 0$, a fairly severe fire (characterized by a normalized heat load of $5.6 \times 10^4 \, \mathrm{s}^{1/2} \, \mathrm{K}$) develops in Compartment I, lying on Side I of a reinforced concrete slab 152.4 mm thick. The fire, spreading by convection, may reach Compartment II, lying on Side II of the slab, right at the start, i.e.

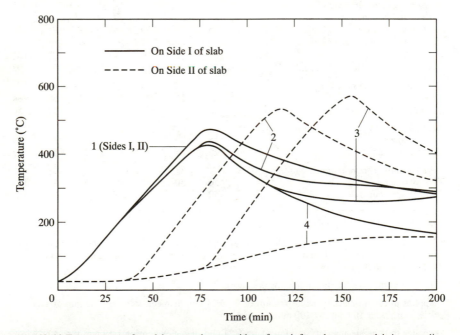

Fig. 12.11 Temperature of steel bars on the two sides of a reinforced concrete slab in spreading fires. Curves: 1, simultaneous exposure of two sides; 2, exposure of Side II delayed by 37.5 min; 3, exposure of Side II delayed by 75 min; 4, exposure of Side II delayed indefinitely (Harmathy 1977/1978)

at $t = 0$. It is more likely, however, that the spread will be delayed, and the fire in Compartment II will begin at $t = t_1, t_2, t_3$, etc. following the outbreak of fire in Compartment I.

The figure shows the temperature history of steel reinforcing bars located on both sides of the slab at 30.5 mm from the surface. Apparently, the maximum temperature reached by the reinforcement (after the cessation of fire in the two compartments) will be higher on Side II than on Side I, and the maximum on Side II will depend on the delay in the spread of fire to Compartment II.

No test facilities exist that would provide for the exposure of key building elements to fire from both the obverse and reverse sides, with the possibility of delayed exposure on the reverse side. If the steel reinforcing bars on both sides of a key concrete element are designed to carry substantial stresses, the designer is advised to specify larger-than-usual cover thicknesses over the principal steel bars. In critical cases, it may be necessary to determine the safe cover thicknesses from numerical analyses, similar to that described by Harmathy (1977/1978), making use of the concept of stress-modified critical temperature.

Nomenclature

Note: The symbols marked with asterisk, *, denote information related to a reference building element, i.e. an element whose fire resistance is known from a standard fire test.

a	distance of the load-bearing steel component from the fire-exposed surface, m
A	cross-sectional area; area of the load-bearing core, m^2
b, b_1, b_2	dimensions (Figs 12.6, 12.7 and 12.10), m
B	minimum dimension of column, m
c	specific heat, $J\,kg^{-1}\,K^{-1}$
C	heat capacity per unit length, $J\,m^{-1}\,K^{-1}$
E, F	coefficients, various dimensions
k	thermal conductivity, $W\,m^{-1}\,K^{-1}$
L	thickness of slab-like building element, m
L_1	thickness (Fig. 12.7), m
n, p, q, r, s	exponents, dimensionless
s	cover thickness, m
t	time, s (min, h)
T	temperature, K (°C)
κ	thermal diffusivity, $m^2\,s^{-1}$
Π	heated perimeter, m
ρ	density, $kg\,m^{-3}$
τ	fire resistance; thermal fire resistance, s (h)

Subscripts

b	of beam
cr	critical
e	of encasement; of encasing material
p	of protecting material

s	of steel
I	of monolithic (solid) slab
II	of double-layer configuration
H	of hollow-core slab

13 Temperature History of Building Elements in Fire

As discussed in the preceding chapter, the calculation of the fire resistance of building elements from first principles usually consists of two steps. In the first step, information is developed on the temperature history of the building element in the fire test. The second step is aimed at determining the element's structural response to heating, as manifested by the decline of its load-bearing capacity, up to the point of structural failure.

It was also mentioned in Chapter 12 that for many types of building elements fire resistance is identical with *thermal fire resistance*, and therefore there is no need for the second-step calculations.

In this chapter, the various techniques of calculating the temperature history of building elements during exposure to a standard test fire (or perhaps to real-world fires) are reviewed. The discussion is addressed to the theoretically minded but not highly mathematically trained practitioner. Sufficient details are presented to enable the practitioner to perform hand calculations (in a few simpler cases) or to prepare a program for computer execution. Those who would like to be fully conversant with the topics to be discussed may refer to the publications cited.

13.1 Some Basic Correlations

In rectangular coordinates (x, y, z), the law of conservation of energy is expressed by the following equation:

$$-\left(\frac{\partial q_x}{\partial x} + \frac{\partial q_y}{\partial y} + \frac{\partial q_z}{\partial z}\right) + Q = \rho c \frac{\partial T}{\partial t} \qquad [13.1]$$

where Q = internal heat generation per unit volume, $W\,m^{-3}$;
ρ = density of the solid, $kg\,m^{-3}$;
c = specific heat of the solid, $J\,kg^{-1}\,K^{-1}$;
T = temperature, K;
t = time, s;

and q_x, q_y and q_z are the components of the heat flux vector (local heat flow per unit area, $W\,m^{-2}$). For an anisotropic solid,

$$q_x = - \left(k_{11} \frac{\partial T}{\partial x} + k_{12} \frac{\partial T}{\partial y} + k_{13} \frac{\partial T}{\partial z} \right)$$

$$q_y = - \left(k_{21} \frac{\partial T}{\partial x} + k_{22} \frac{\partial T}{\partial y} + k_{23} \frac{\partial T}{\partial z} \right) \qquad [13.2]$$

$$q_z = - \left(k_{31} \frac{\partial T}{\partial x} + k_{32} \frac{\partial T}{\partial y} + k_{33} \frac{\partial T}{\partial z} \right)$$

and k_{11}, k_{12}, ... k_{33} ($\text{W m}^{-1}\text{K}^{-1}$) are the components of the (symmetric) thermal conductivity tensor.

In a macroscopic sense, building materials are, by and large, isotropic materials. Yet, since their properties are temperature-dependent, in a numerical formulation of heat flow problems their thermal conductivities appear to be quantities exhibiting different values in different directions.

As discussed in Chapter 6, it is practical to regard the ρc product as a single material property: ρc ($\text{J m}^{-3}\text{K}^{-1}$) is usually referred to as volume specific heat.

Equations 13.1 and 13.2 are the fundamental equations of heat conduction. The left-hand side of Eq. 13.1 represents the sum of the rates of energy gain by conduction and heat generation at a point in the solid; the right-hand side is the rate of energy storage. The solution of the equations of heat conduction for temperature response requires specifying the temperature distribution in the solid body at the start of the heat flow process (initial condition), and the behavior of the boundaries of the solid during the process (boundary conditions).

As to the initial condition, it is usually assumed that the temperature distribution in the body is uniform and equal to the temperature of the environment, T_e (K).

$$T(x, y, z) = T_e \qquad\qquad \text{at } t = 0 \qquad [13.3]$$

With respect to the boundary conditions, any of the following conditions may apply (Fig. 13.1):

- The temperature along a section S_1 of the boundary is constant and equal to the temperature of the environment.

$$T(x, y, z) = T_e \qquad\qquad \text{on } S_1 \qquad [13.4]$$

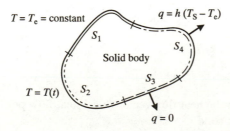

$T = T_e = \text{constant}$ $q = h(T_S - T_e)$

S_1

S_4

Solid body

S_3

$T = T(t)$ S_2

$q = 0$

Fig. 13.1 Various boundary conditions

- The temperature along a section S_2 of the boundary is a function of time.

$$T(x, y, z) = T(t) \qquad\qquad \text{on } S_2 \qquad\qquad [13.5]$$

- The solid body is insulated along a section S_3 (so that there is no heat flow across S_3).

$$q_x n_x + q_y n_y + q_z n_z = 0 \qquad\qquad \text{on } S_3 \qquad\qquad [13.6]$$

- Heat is exchanged by convection and/or radiation between a section S_4 of the boundary surface (at a temperature T_S (K), a function of time) and the environment.

$$q_x n_x + q_y n_y + q_z n_z = h(T_S - T_e) \qquad \text{on } S_4 \qquad\qquad [13.7]$$

Here n_x, n_y, and n_z are direction cosines of the outward normal to surface S_3 (in Eq. 13.6) or to S_4 (in Eq. 13.7). S_1, S_2, S_3, and S_4 add up to the total surface of the solid, S (m^2).

In Eq. 13.7, h (W m^{-2} K^{-1}) is the coefficient of heat transfer. In general, it consists of two terms,

$$h = h_c + h_r \qquad\qquad [13.8]$$

where h_c = coefficient of heat transfer by convection (generally natural convection), W m^{-2} K^{-1};

h_r = coefficient of heat transfer by radiation, W m^{-2} K^{-1}.

The coefficient of heat transfer by natural convection between a solid surface and the ambient air can be calculated from the following empirical equations (McAdams 1954).

- For a vertical surface:

$$h_c = 1.313|T_S - T_a|^{1/3} \qquad\qquad [13.9]$$

- For a horizontal surface (hot surface facing up or cool surface facing down):

$$h_c = 1.521|T_S - T_a|^{1/3} \qquad\qquad [13.10]$$

In Eqs 13.9 and 13.10, 1.313 and 1.521 are dimensional constants, and

T_S = temperature of the solid's surface, K;
T_a = temperature of the ambient air, K.

The coefficient of heat transfer by radiation between a gray surface[1] and a nonreflecting medium is

$$h_r = \sigma \epsilon_S \frac{T_S^4 - T_{nr}^4}{T_S - T_{nr}} \qquad\qquad [13.11]$$

and that between two parallel gray surfaces (Surfaces I and II) is

---

1   A gray surface is a radiator whose spectral emissivity is independent of wavelength.

$$h_r = \sigma \frac{1}{\dfrac{1}{\epsilon_{SI}} + \dfrac{1}{\epsilon_{SII}} - 1} \frac{T_{SI}^4 - T_{SII}^4}{T_{SI} - T_{SII}}$$ [13.12]

In Eqs 13.11 and 13.12,

σ = Stefan–Boltzmann constant = 5.67×10^{-8} W m^{-2} K^{-4};
ϵ_S = emissivity of the surface(s), dimensionless;
T_S = temperature of the surface(s), K.
T_{nr} = temperature of a nonreflecting medium, K.

The following expressions are recommended for use in fire safety design (see Eqs 13.7–13.12 and Fig. 13.2).

• For the coefficient of heat transfer from the fire gases to the fire-exposed surface (surface E in the figure) of a building element, h_E (W m^{-2} K^{-1}),

$$h_E = \sigma \epsilon_E \frac{T_F^4 - T_E^4}{T_F - T_E}$$ [13.13]

where ϵ_E = emissivity of the fire-exposed surface, dimensionless;
T_F = temperature of the fire gases (for test fires see Eq. 3.1), K;
T_E = temperature of the fire-exposed surface, K.

Explanation   On account of the very high temperatures arising in fire tests (and also in real-world fires), radiation is the dominant mode of heat transmission between the fire gases and the building element. Theoretical and experimental studies have indicated that the heat transmission can be

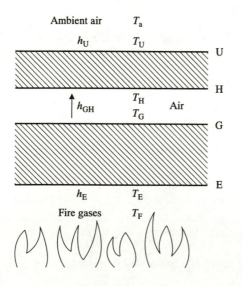

Fig. 13.2 Heat flow through a building element with enclosed air layer

modeled as radiation to the surface of the building element from a black body whose temperature follows the temperature history of the fire gases.[2]

- For the coefficient of heat transfer from the unexposed surface (surface U in the figure) of a building element to the air, h_U (W m^{-2} K^{-1}):

In the case of vertical building elements (walls, partitions):

$$h_U = 1.313 \, (T_U - T_a)^{1/3} + \sigma \epsilon_U \frac{T_U^4 - T_a^4}{T_U - T_a} \qquad [13.14]$$

and in the case of horizontal building elements (floors, roofs):

$$h_U = 1.521 \, (T_U - T_a)^{1/3} + \sigma \epsilon_U \frac{T_U^4 - T_a^4}{T_U - T_a} \qquad [13.15]$$

where T_a (K) is the temperature of the ambient air, and

ϵ_U = emissivity of the unexposed surface,[3] dimensionless;
T_U = temperature of the unexposed surface, K.

Explanation   At the unexposed surface of the building element, heat transfer takes place by a combination of natural convection and radiation into a nonreflecting air mass.

- For the coefficient of heat transfer across an air layer (bounded by surfaces G and H, see the figure) inside a building element, h_{GH} (W m^{-1} K^{-1}):

In the case of a vertical building element:

$$h_{GH} = 1.313 \left(\frac{T_G - T_H}{2} \right)^{1/3} + \sigma \frac{1}{\dfrac{1}{\epsilon_G} + \dfrac{1}{\epsilon_H} - 1} \frac{T_G^4 - T_H^4}{T_G - T_H} \qquad [13.16]$$

and in the case of a horizontal building element:

$$h_{GH} = 1.521 \left(\frac{T_G - T_H}{2} \right)^{1/3} + \sigma \frac{1}{\dfrac{1}{\epsilon_G} + \dfrac{1}{\epsilon_H} - 1} \frac{T_G^4 - T_H^4}{T_G - T_H} \qquad [13.17]$$

where ϵ_G and ϵ_H = surface emissivities (dimensionless)
T_G and T_H = temperatures of the bounding surfaces (K)

Explanation   The heat capacity of the air layer is negligible, and its temperature, if the layer is completely enclosed, is equal to the arithmetic mean of the temperatures of the two bounding surfaces.

If the thickness of the air layer is small (less than, say, 5 mm), natural

---

2   A black body, having an emissivity of unity, emits the maximum possible amount of radiation. Apparently, the neglect of heat transmission by (natural and forced) convection is roughly compensated for by assuming black body radiation by the fire gases.

3   For many nonmetallic building materials, the surface emissivity at elevated temperatures is about 0.9.

convection cannot develop. Hence, h_{GH} is to be calculated by omitting the first term on the right-hand side of Eq 13.16 or 13.17.[4]

A rather complex problem arises with cavities, e.g. with cavities present in walls made from hollow masonry units. The temperature will vary along the surface of the cavities, and the heat flow pattern will become two- or three-dimensional. A numerical technique described by Gebhart (1961) may be used for the calculation of radiant heat exchange between the surface elements. Gebhart's technique was employed by Harmathy (1970b) in studying the fire resistance of masonry walls.

Nearly all heat flow problems that occur in fire engineering practice are one- or two-dimensional. For this reason, and because three-dimensional heat flow problems are difficult to handle even in this age of computers, only one- and two-dimensional problems will be discussed here.

13.2   Analytical Solutions

Finding analytical solutions to the equations of heat conduction is, for the most part, possible only if

(1) the solid is of a relatively simple geometry,
(2) the material properties are constant, and
(3) the boundary conditions are uncomplicated.

The last two conditions are unlikely to apply to building elements exposed to fire tests. However, analytical solutions may still prove useful, by allowing the designer to obtain preliminary insight into product behavior, or to extend the validity of experimental information.

In standard test fires, the temperature of the furnace gases rises sharply to a high value at first, then increases relatively slowly. Since the coefficient of heat transfer to the fire-exposed side of the test specimen is, on account of intense radiation, very high, it is often permissible to assess the temperature distribution in the specimen by assuming that the temperature of the exposed surface, T_E, is constant and equal to a nominal temperature of the fire gases, T_F, during the test. That nominal temperature may be selected as the average of T_F over the duration of standard fire exposure: $(T_F)_{av}$ (see Eq. 3.2 and Fig. 3.6).[5]

At the start of a fire test, the building element is required to be in thermal equilibrium with the ambient atmosphere. T_a will, therefore, denote not only the (constant) temperature of the ambient atmosphere, but also the (uniform) initial temperature of the test specimen.

---

4   Clearly, there is no foundation to the belief that the fire resistance of a building element can be improved by increasing the thickness of an enclosed air layer. See Rule 5 in Section 12.2 of Chapter 12.

5   When dealing with analytical solutions, T_F is to be interpreted as identical with $(T_F)_{av}$.

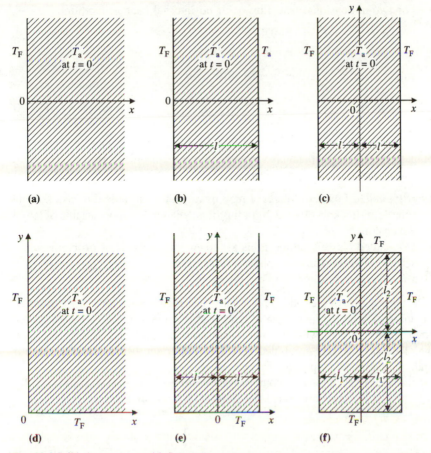

Fig. 13.3 Solid shapes: (a) semi-infinite solid; (b) slab heated on one side; (c) slab heated on both sides; (d) rectangular corner; (e) semi-infinite rectangle; (f) rectangle

Figure 13.3 shows six solid shapes. They are representative of the cross-sectional areas of the most common building elements. The flow of heat is one-dimensional in shapes (a), (b) and (c), and two-dimensional in shapes (d), (e) and (f). Solutions to nearly all problems that can be handled by mathematical analysis have been presented in Carslaw and Jaeger's book (1959). Those related to the shapes in Fig. 13.3 will be reviewed here.

For isotropic materials with constant properties, the equation of heat conduction (see Eqs 13.1 and 13.2) assumes the following forms:

- For one-dimensional heat flow with no internal heat generation:

$$\frac{\partial^2 T}{\partial x^2} = \frac{1}{\kappa} \frac{\partial T}{\partial t} \qquad\qquad [13.18]$$

- For two-dimensional heat flow with no internal heat generation:

$$\frac{\partial^2 T}{\partial x^2} + \frac{\partial^2 T}{\partial y^2} = \frac{1}{\kappa} \frac{\partial T}{\partial t}$$

[13.19]

where κ (m^2 s^{-1}) is the thermal diffusivity of the solid,

$$\kappa = \frac{k}{\rho c}$$

[13.20]

The analysis of the process of heat flow in solid bodies inevitably leads to the compaction of the independent variables into dimensionless groups. Three are relevant to the subjects to be discussed.

- $\kappa t/l^2$, called Fourier number: l may be any characteristic dimension of the solid (e.g. thickness of a slab, radius of a cylinder), or any variable of length dimension.
- hl/k, called Nusselt number: this group enters all heat flow problems where surface heat transmission plays a role.
- x/l, y/l, etc. are dimensionless distances.

The temperature of the solid, T, is usually the dependent variable. It is customary to combine T with the input data of temperature dimension into a dimensionless group. When dealing with fire resistance problems, T_F and T_a are such input data. The way they are combined makes an important difference in the ease of handling a number of problems.

It has been conventional to combine them into the group $(T - T_a)/(T_F - T_a)$. Harmathy (1963) recommended the following grouping:

$$\frac{T_\infty - T}{T_\infty - T_a}$$

For the sake of brevity, this expression will be referred to as the *T-group*. In it, T_∞ (K) is the steady-state temperature (i.e. the temperature as $t \to \infty$) at any point under consideration. As illustrated in Fig. 13.4, $(T_\infty - T)/(T_\infty - T_a)$ signifies the degree of remoteness of the solid from the steady-state condition. If all the boundaries of the solid are kept at T_F [Fig. 13.4(b)], $T_\infty = T_F$, and therefore the *T*-group becomes

$$\frac{T_F - T}{T_F - T_a}$$

As will be shown later, using the *T*-group offers the convenience that the solution of two-dimensional (or three-dimensional) problems with constant-temperature boundary conditions can be expressed as the product of solutions to one-dimensional problems. Furthermore, it simplifies the plotting and tabulating of solutions of complex heat conduction problems.

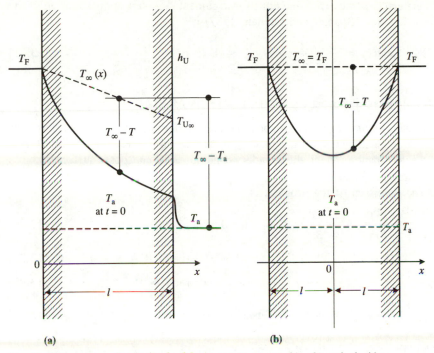

Fig. 13.4 Temperature distribution in slabs heated (a) on one side, (b) on both sides

13.2.1 Semi-infinite Solid [Fig. 13.3(a)]

When a building element is exposed to a test fire, the temperatures in a region near its exposed surface (where the important load-bearing components, e.g. reinforcing steel bars, are usually located) are, for a long time, not noticeably affected by the boundary condition at its unexposed surface. Consequently, many fire safety problems can be studied by modeling the building element as a semi-infinite solid.

The statement of the problem is

$$\frac{\partial^2 T}{\partial x^2} = \frac{1}{\kappa}\frac{\partial T}{\partial t} \qquad \text{for } x > 0 \qquad\qquad [13.21]$$

$$T = T_a \qquad \text{when } t = 0; \text{ for } x \geq 0 \qquad\qquad [13.22]$$

$$T = T_F \qquad \text{at } x = 0; \text{ for } t > 0 \qquad\qquad [13.23]$$

Since $T_\infty = T_F$, the solution, in terms of the T-group, is given by

$$\frac{T_F - T}{T_F - T_a} = \text{erf}\left[\frac{1}{2}\left(\frac{x^2}{\kappa t}\right)^{1/2}\right] \qquad\qquad [13.24]$$

where erf(ζ) is the error function[6] for argument ζ. A fair approximation of the error function (Spanier and Oldham 1987) is

$$\text{erf}(\zeta) \simeq \frac{3.372\zeta}{3 + \zeta^2} \qquad \zeta \le 1 \qquad\qquad\qquad [13.25]$$

13.2.2 Slab Heated on One Side [Fig. 13.3(b)]

The majority of vertical or horizontal separating elements can be modeled as slabs (solids bounded by two parallel planes). If the thermal fire resistance of the element is the information sought, the boundary condition at its unexposed surface becomes an important factor.

The statement of the problem is

$$\frac{\partial^2 T}{\partial x^2} = \frac{1}{\kappa}\frac{\partial T}{\partial t} \qquad\qquad \text{for } 0 < x < l \qquad\qquad [13.26]$$

$$T = T_a \qquad\qquad \text{when } t = 0;\ \text{for } 0 \le x \le l \qquad [13.27]$$

$$T = T_F \qquad\qquad \text{at } x = 0;\ \text{for } t > 0 \qquad\qquad [13.28]$$

$$-k\frac{\partial T}{\partial x} = h_U(T - T_a) \ \text{ at } x = l;\ \text{for } t > 0 \qquad [13.29]$$

The last equation is based on Eqs 13.2 and 13.7.

In terms of the T-group, the solution is

$$\frac{T_\infty - T}{T_\infty - T_a} = 2\frac{T_F - T_a}{T_\infty - T_a}\sum_{n=1}^{\infty}\frac{\left[\alpha_n^2 + \left(\dfrac{h_U l}{k}\right)^2\right]\sin\left(\alpha_n\dfrac{x}{l}\right)}{\alpha_n\left[\alpha_n^2 + \dfrac{h_U l}{k} + \left(\dfrac{h_U l}{k}\right)^2\right]}\exp\left(-\alpha_n^2\frac{\kappa t}{l^2}\right)$$

$$[13.30]$$

where [see Fig. 13.4(a)] the steady-state temperature, T_∞, is a function of x:

$$T_\infty = T_a + (T_F - T_a)\frac{1 + \dfrac{h_U l}{k}\left(1 - \dfrac{x}{l}\right)}{1 + \dfrac{h_U l}{k}} \qquad\qquad [13.31]$$

and the α_n are the positive roots of

$$\alpha\cot\alpha + \frac{h_U l}{k} = 0 \qquad\qquad\qquad [13.32]$$

With $h_U l/k$ varying from 0 to ∞, the first root of Eq. 13.32, α_1, increases from

---

6 Tabulated values of this function are available from numerous handbooks (e.g. Carslaw and Jaeger 1959; Abramowitz and Stegun 1970).

Table 13.1 The first root of the equation $\alpha \cot \alpha + h_{\mathrm{u}}l/k = 0$ (Carlslaw and Jaeger 1959)

$\dfrac{h_{\mathrm{u}}l}{k}$	α_1	$\dfrac{h_{\mathrm{u}}l}{k}$	α_1	$\dfrac{h_{\mathrm{u}}l}{k}$	α_1
0	1.5708				
0.1	1.6320	1.0	2.0288	10	2.8628
0.2	1.6887	2.0	2.2889	20	2.9930
0.3	1.7414	3.0	2.4557	30	3.0406
0.4	1.7906	4.0	2.5704	40	3.0651
0.5	1.8366	5.0	2.6537	50	3.0801
0.6	1.8798	6.0	2.7165	60	3.0901
0.7	1.9203	7.0	2.7654	80	3.1028
0.8	1.9586	8.0	2.8044	100	3.1105
0.9	1.9947	9.0	2.8363	∞	3.1416

$\pi/2$ to π, the second root, α_2, from $3\pi/2$ to 2π, the third root, α_3, from $5\pi/2$ to 3π, etc.

Thanks to the presence of an exponential expression in Eq. 13.30, only the first term under the summation sign is significant if $\kappa t/l^2 > 0.3$. That term can be calculated employing the values of α_1 listed in Table 13.1.

In the earliest stage of the fire test, characterized by small values of $\kappa t/l^2$ (smaller than, say, 0.05), the heat penetration is confined to the vicinity of the fire-exposed surface of the test specimen. Consequently, the specimen can be regarded as a semi-infinite solid, and its temperature can be calculated from Eq. 13.24.

Equation 13.30 is tabulated in Table 13.2, where values of the T-group are listed for seven values of the Fourier number ($\kappa t/l^2 = 0$, 0.05, 0.1, 0.2, 0.4, 0.8, and 1.6), six values of the Nusselt number ($h_{\mathrm{u}}l/k = 0$, 0.5, 1, 2, 5, and ∞), and six values of the dimensionless distance ($x/l = 0$, 0.2, 0.4, 0.6, 0.8, and 1).

Since the thermal fire resistance is defined in terms of the temperature of the specimen's unexposed surface, determining the variation of temperature at $x = l$ is of primary interest. That temperature be read from the graph in Fig. 13.5, in which $(T_\infty - T)/(T_\infty - T_a)$ is plotted against $\kappa t/l^2$ for $x/l = 1$ and 15 values of $h_{\mathrm{u}}l/k$, ranging from 0 (meaning perfect insulation) to ∞ (meaning contact with a perfect conductor).

The calculation of the thermal fire resistance, τ (s), of a (slab-like) building element, employing the analytical solution presented by Eqs 13.30 and 13.31, is illustrated though a worked example in Section 13.6.1.

13.2.3 Slab Heated on Both Sides [Fig. 13.3(c)]

A slab-like building element may be heated on both sides in spreading fires. The statement of the problem is

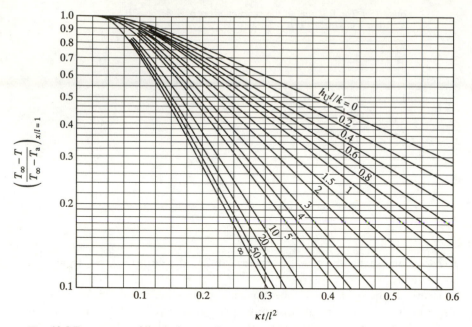

Fig. 13.5 Temperature of the surface at $x/l = 1$ of the slab heated on one side (Harmathy 1963)

$$\frac{\partial^2 T}{\partial x^2} = \frac{1}{\kappa} \frac{\partial T}{\partial t} \qquad \text{for } -l < x < l \qquad [13.33]$$

$$T = T_a \qquad \text{when } t = 0; \text{ for } -l \leq x \leq l \qquad [13.34]$$

$$T = T_F \qquad \text{at } x = -l, l; \text{ for } t > 0 \qquad [13.35]$$

Since the heating of the slab is symmetrical, there is no heat flow across the $x = 0$ plane. The $-l \leq x \leq 0$ and $0 \leq x \leq l$ regions may therefore be regarded as two slabs heated on one side and insulated on the other side. This problem is, in fact, a special case of that discussed previously, and the information in the uppermost segment of Table 13.2 (related to the $h_U l/k = 0$ condition) is directly applicable, provided the order of the x/l values is inverted (i.e. 0 replaced by 1, 0.2 by 0.8, etc.).

If the surface of the slab shown in Fig. 13.3(b) is insulated at $x = l$ (i.e. $h_U l/k = 0$), $T_\infty = T_F$ for any value of x/l (according to Eq. 13.31), and $\alpha_1 = \pi/2$, $\alpha_2 = 3\pi/2$, $\alpha_3 = 5\pi/2$, etc. Substituting these values into Eq. 13.30 and moving the zero of the x-axis to the right by a distance l, the solution of the problem is obtained, after some rearrangement, as

$$\frac{T_F - T}{T_F - T_a} = \frac{4}{\pi} \sum_{n=0}^{\infty} \frac{(-1)^n}{2n+1} \cos\left[(2n+1)\frac{\pi}{2}\frac{x}{l} \right] \exp\left[-(2n+1)^2 \left(\frac{\pi}{2}\right)^2 \frac{\kappa t}{l^2} \right]$$

$$[13.36]$$

or, in shorthand,

Table 13.2 Tabular presentation of Eq. 13.30 (Harmathy 1963)

		$(T_\infty - T)/(T_\infty - T_a)$						
$\dfrac{h_U l}{k}$	$\dfrac{x}{l}$	$\kappa t/l^2$						
		0	0.05	0.1	0.2	0.4	0.8	1.6
0	0	1.0	0	0	0	0	0	0
	0.2	1.0	0.4729	0.3452	0.2443	0.1467	0.0547	0.0076
	0.4	1.0	0.7941	0.6286	0.4616	0.2790	0.1040	0.0144
	0.6	1.0	0.9422	0.8185	0.6304	0.3839	0.1431	0.0199
	0.8	1.0	0.9884	0.9191	0.7363	0.4513	0.1682	0.0234
	1.0	1.0	0.9969	0.9493	0.7723	0.4745	0.1769	0.0246
0.5	0	1.0	0	0	0	0	0	
	0.2	1.0	0.4353	0.2985	0.1910	0.0956	0.0248	0.0017
	0.4	1.0	0.7624	0.5714	0.3812	0.1921	0.0498	0.0034
	0.6	1.0	0.9277	0.7734	0.5447	0.2768	0.0718	0.0048
	0.8	1.0	0.9842	0.8909	0.6577	0.3368	0.0874	0.0059
	1.0	1.0	0.9955	0.9295	0.6996	0.3593	0.0932	0.0063
1	0	1.0	0	0	0	0	0	
	0.2	1.0	0.4143	0.2724	0.1616	0.0697	0.0134	0.0003
	0.4	1.0	0.7425	0.5357	0.3318	0.1441	0.0278	0.0010
	0.6	1.0	0.9174	0.7412	0.4860	0.2130	0.0411	0.0015
	0.8	1.0	0.9807	0.8680	0.5988	0.2645	0.0510	0.0019
	1.0	1.0	0.9941	0.9123	0.6433	0.2851	0.0550	0.0020
2	0	1.0	0	0	0	0	0	
	0.2	1.0	0.3918	0.2445	0.1305	0.0451	0.0055	0.0001
	0.4	1.0	0.7193	0.4937	0.2749	0.0955	0.0117	0.0002
	0.6	1.0	0.9037	0.6988	0.4120	0.1444	0.0178	0.0003
	0.8	1.0	0.9754	0.8335	0.5181	0.1829	0.0225	0.0003
	1.0	1.0	0.9921	0.8846	0.5633	0.1996	0.0245	0.0004
5	0	1.0	0	0	0	0	0	
	0.2	1.0	0.3675	0.2143	0.0978	0.0237	0.0014	0.0000
	0.4	1.0	0.6911	0.4432	0.2098	0.0511	0.0031	0.0000
	0.6	1.0	0.8844	0.6398	0.3185	0.0779	0.0047	0.0000
	0.8	1.0	0.9655	0.7753	0.4047	0.0996	0.0060	0.0000
	1.0	1.0	0.9869	0.8319	0.4445	0.1096	0.0066	0.0000
∞	0	1.0	0	0	0	0	0	
	0.2	1.0	0.3411	0.1817	0.0651	0.0090	0.0002	0.0000
	0.4	1.0	0.6568	0.3821	0.1403	0.0195	0.0004	0.0000
	0.6	1.0	0.8556	0.5551	0.2101	0.0292	0.0006	0.0000
	0.8	1.0	0.9437	0.6683	0.2593	0.0361	0.0007	0.0000
	1.0	1.0	0.9660	0.7071	0.2771	0.0386	0.0007	0.0000

$$\frac{T_F - T}{T_F - T_a} = \Phi\left(\frac{x}{l}, \frac{\kappa t}{l^2}\right)$$

[13.37]

Again, on account of the exponential expression, only the first term under the summation sign needs to be considered if $\kappa t/l^2 > 0.3$.

13.2.4 Rectangular Corner [Fig. 13.3(d)]

This shape, as well as those shown in Figs 13.3(e) and 13.3(f), is a geometric model that may be used to find the maximum temperatures reached during a fire test by the steel components of reinforced or prestressed concrete beams or columns.

The statement of the problem is

$$\frac{\partial^2 T}{\partial x^2} + \frac{\partial^2 T}{\partial y^2} = \frac{1}{\kappa}\frac{\partial T}{\partial t} \quad \text{for } x > 0 \text{ and } y > 0 \qquad [13.38]$$

$$T = T_a \qquad\qquad \text{when } t = 0; \text{ for } x \geq 0 \text{ and } y \geq 0 \; [13.39]$$

$$T = T_F \qquad\qquad \text{at } x = 0 \text{ and } y = 0; \text{ for } t > 0 \qquad [13.40]$$

A rectangular corner is formed by the superposition of two semi-infinite solids. If the temperature is expressed in terms of the T-group, the solution to this problem can be obtained as the product of the solutions for the pertinent one-dimensional problems, presented by Eq. 13.24.

$$\frac{T_F - T}{T_F - T_a} = \mathrm{erf}\left[\frac{1}{2}\left(\frac{x^2}{\kappa t}\right)^{1/2}\right] \mathrm{erf}\left[\frac{1}{2}\left(\frac{y^2}{\kappa t}\right)^{1/2}\right] \qquad [13.41]$$

13.2.5 Semi-infinite rectangle [Fig. 13.3(e)]

This shape may be used to model concrete beams. The statement of the problem is

$$\frac{\partial^2 T}{\partial x^2} + \frac{\partial^2 T}{\partial y^2} = \frac{1}{\kappa}\frac{\partial T}{\partial t} \quad \text{for } -l < x < \text{ and } y > 0 \qquad [13.42]$$

$$T = T_a \quad \text{when } t = 0; \text{ for } -l \leq x \leq l \text{ and } y \geq 0 \qquad [13.43]$$

$$T = T_F \qquad\qquad \text{at } x = -l, l \text{ and } y = 0; \text{ for } t > 0 [13.44]$$

A semi-infinite rectangle is formed by superposing a semi-infinite solid on a slab (heated on both sides). Thus, provided the temperature is expressed in terms of the T-group, the solution to Eqs 13.42 to 13.44 is obtained, by means of Eqs 13.24, 13.36 and 13.37, as

$$\frac{T_F - T}{T_F - T_A} = \Phi\left(\frac{x}{l}, \frac{\kappa t}{l^2}\right) \mathrm{erf}\left[\frac{1}{2},\left(\frac{y^2}{\kappa t}\right)^{1/2}\right] \qquad [13.45]$$

13.2.6 Rectangle [Fig. 13.3(f)]

This shape is representative of the cross section of concrete columns. The statement of the problem is as follows:

$$\frac{\partial^2 T}{\partial x^2} + \frac{\partial^2 T}{\partial y^2} = \frac{1}{\kappa}\frac{\partial T}{\partial t} \quad \text{for } -l_1 < x < l_1 \text{ and } -l_2 < y < l_2 \quad [13.46]$$

$$T = T_a \quad \text{when } t = 0; \text{ for } -l_1 \le x \le l_1 \text{ and } -l_2 \le y \le l_2 \quad [13.47]$$

$$T = T_F \quad \text{at } x = -l_1, l_1 \text{ and } y = -l_2, l_2; \text{ for } t > 0 \quad [13.48]$$

A rectangle is formed by the superposition of two slabs (heated on both sides). Its temperature history is given by the following equation:

$$\frac{T_F - T}{T_F - T_A} = \Phi\left(\frac{x}{l_1}, \frac{\kappa t}{l_1^2}\right)\Phi\left(\frac{y}{l_2}, \frac{\kappa t}{l_2^2}\right) \quad [13.49]$$

Equations 13.36 and 13.37 have been used.

13.3 Finite Difference Method

13.3.1 One-dimensional Heat Flow

Two kinds of numerical techniques have been adapted to the calculation of the temperature history of building elements in fire tests or in real-world fires: the finite difference method and the finite element method. The latter is claimed to be especially suitable for studying the temperature history of solids of complex geometrical shapes, e.g. turbine blades. However, the use of the finite element method requires some familiarity with a number of fields of advanced mathematics. Since building elements are usually composed of rectangular shapes, and since the finite difference method is very easy to understand and use, only the finite difference method will be discussed here in detail.

An essential feature of all numerical techniques is *discretization*: subdivision of the solid into elementary volumes, and the time scale into small intervals (steps). With the finite difference technique, one point in each volume element, the *nodal point* (or *node*), is looked upon as representative (with respect to temperature and material properties) of the volume element as a whole. The temperature history of the solid is developed by following up step by step the temperature of the volume elements.

It goes without saying that the discretization of the space and time variables does introduce some degree of inaccuracy into the calculations. However, the question of calculation accuracy should be viewed in a broader context: as one component in the overall accuracy of problem solving.

To begin with, the accuracy of the fire resistance test, which the numerical technique is intended to simulate, is not adequately known. Sultan *et al.* (1986) pointed out that, on account of marked differences in the operating characteristics of existing fire test furnaces, the reproducibility of the test results is not satisfactory. An ASTM study (American Society for Testing and Materials 1982) revealed that the coefficient of variation for the results may be as high as 15 per cent.

Another major problem is how to select the properties of the materials from which the test specimen is made. The thermal conductivity is an especially elusive property. It is usually assumed to depend only on the temperature. In fact, being a structure-sensitive property, it depends on a number of factors that are not readily identifiable on the basis of customary material specifications.

The difficulties in acquiring reliable information on the apparent specific heat of physicochemically unstable materials are also numerous, and the presence of moisture in building materials, especially in concrete and wood, is an additional complicating factor. All in all, it can be confidently stated that, in fire performance calculations, the error due to the use of numerical techniques is insignificant in comparison with the errors associated with the inadequate definability of the problem to be simulated.

There are two ways of introducing the finite difference method. The formal way consists of replacing the space and time derivatives in the equations of heat conduction (Eqs 13.1 and 13.2) by finite difference quotients. This method is favored by the mathematician who is interested not only in problem solving, but also in such questions as *truncation error* (i.e. error due to the finite difference approximation) and *convergence* (i.e. approach to the exact solution as the volume elements and time steps become smaller and smaller).

Engineers usually prefer to formulate the finite difference procedure by writing energy conservation equations for each volume element (region) of the discretized solid, assuming that for a short period quasi-steady-state conditions prevail. In the final analysis, the two methods lead essentially to the same results. Because of its graphicness and intelligibility, the latter method will be employed in this chapter.

Figure 13.6 shows the subdivision of a slab-like solid (a building element) into N elementary volumes, and the location of the N nodal points. (The dimensions of each volume element perpendicular to the x-axis are assumed to be equal to

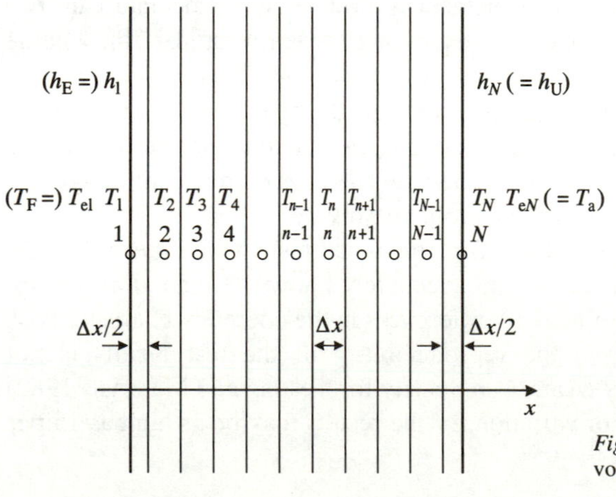

Fig. 13.6 Subdivision of a slab into volume elements

unity.) Since the temperature history of the solid is followed up by monitoring the temperatures at the nodes, and since the surface temperatures play an important part in the surface heat transmission, it is practical to place nodal points on the surfaces. Accordingly, $(N - 2)$ nodes (points 2 to $(N - 1)$) are located in the solid's interior, and two (points 1 and N)) on the bounding surfaces. The thickness of the inside volume elements is Δx (m), and that of the elements adjacent to the surfaces is $\Delta x/2$.

The 'nodes' along the temperature scale are: $t = 0, \Delta t, 2\Delta t, 3\Delta t, \ldots j\Delta t, (j + 1)\Delta t, \ldots$ where Δt (s) is the time step.

If quasi-steady-state conditions prevail over the time period $j\Delta t < t \leq (j + 1)\Delta t$, the conservation of energy for the nth volume element (in the slab's interior) can be written as

$$k^j_{(n-1):n} \frac{T^j_{n-1} - T^j_n}{\Delta x} \Delta t + k^j_{n:(n+1)} \frac{T^j_{n+1} - T^j_n}{\Delta x} \Delta t = (\rho c)^j_n \Delta x (T^{j+1}_n - T^j_n)$$

[13.50]

where $(n - 1), n, (n + 1)$ subscripts indicate that the value of the variable pertains to the $(n - 1)$th, nth, $(n + 1)$th nodal point (or the corresponding volume element), respectively;

$j, (j + 1)$ superscripts indicate that the value of the variable pertains to the $t = j\Delta t, (j + 1)\Delta t$ time level, respectively;

$k^j_{(n-1):n}$ and $n^j_{n:(n+1)}$ indicate the thermal conductivity of the solid along the paths from the $(n - 1)$th to the nth element, and from the nth to the $(n + 1)$th element, respectively.

The values of the solid's thermal conductivity along the paths from the $(n - 1)$th to the nth, and from the nth to the $(n + 1)$th elements may be regarded as arithmetic means for k for the two adjacent regions. Thus

$$k^j_{(n-1):n} = \frac{k^j_{n-1} + k^j_n}{2}; \quad k^j_{n:(n+1)} = \frac{k^j_n + k^j_{n+1}}{2}; \quad \text{etc.} \qquad [13.51]$$

Since the material properties are looked upon as functions of the temperature only,

$$k^j_{n-1} = k(T^j_{n-1}); \quad k^j_n = k(T^j_n); \quad k^j_{n+1} = k(T^j_{n+1}); \quad \text{etc.} \qquad [13.52]$$

and

$$(\rho c)^j_n = \rho c(T^j_n); \quad \text{etc.} \qquad [13.53]$$

The left-hand side of Eq. 13.50 describes the energy received by the nth volume element[7] by conduction from the two adjacent regions over a Δt period of time. The right-hand side represents the storage of sensible heat by the element during the same time. The evolution or absorption of heat due to physical or chemical

---

7   The volume element has the dimensions of Δx (m) along the x-axis and 1 m by 1 m perpendicular to the x-axis.

changes ('reactions', represented by Q in Eq. 13.1) may be taken into account by adding another term to Eq. 13.50 (Harmathy 1961) or, more conveniently, by including the latent heat effects in the apparent specific heat of the solid.

In formulating the conservation of energy equations for the first and last volume elements (represented by nodes 1 and N in Fig. 13.6), account has to be taken of the prevailing boundary conditions at the two surfaces (exchange of heat with the environment by convection and/or radiation). The following equations apply:

$$h_1^j(T_{e1}^j - T_1^j)\Delta t + k_{1:2}^j \frac{T_2^j - T_1^j}{\Delta x} \Delta t = (\rho c)_1^j \frac{\Delta x}{2}(T_1^{j+1} - T_1^j) \quad [13.54]$$

and

$$k_{(N-1):N}^j \frac{T_{N-1}^j - T_N^j}{\Delta x} \Delta t + h_N^j(T_{eN}^j - T_N^j)\Delta t = (\rho c)_N^j \frac{\Delta x}{2}(T_N^{j+1} - T_N^j)$$

$$[13.55]$$

where the subscripts and superscripts are to be interpreted as explained before, and

T_{e1}, T_{eN} = temperatures of the environment on the left and right sides of the solid, respectively, K;

h_1, h_N = coefficients of heat transfer at the left and right sides, respectively, $W\,m^{-2}\,K^{-1}$.

T_{e1} and T_{eN} may be either constant or functions of time. The coefficients of heat transfer depend on both the surface temperature and the temperature of the surroundings (see Eqs 13.13 to 13.15).

$$h_1^j = h_1(T_1^j, T_{e1}^j); \qquad h_N^j = h_N(T_N^j, T_{eN}^j) \quad [13.56]$$

Four other kinds of boundary conditions are of practical interest:

(a) air layer between two slabs;
(b) insulation (or symmetry);
(c) interface between two slabs; and
(d) prescribed (constant or time-dependent) surface temperature.

A building element with a completely enclosed air layer consists of two parallel slabs [slab I and slab II, Fig. 13.7(a)], which may or may not be of the same material and, in the mathematical treatment, may or may not be subdivided into volume elements of the same thickness. The two slabs are coupled by heat transmission that takes place through the air layer.

This problem can be handled using the already introduced equations with the following modifications:

(1) When formulating the conservation of energy equation for the Nth (surface) nodal point (see Eq. 13.55) for slab I, take the temperature of the 'environment' as equal to that of the first nodal point of slab II [i.e. make $(T_{eN}^j)_I = (T_1^j)_{II}$].
(2) When formulating the conservation of energy equation for the first (surface)

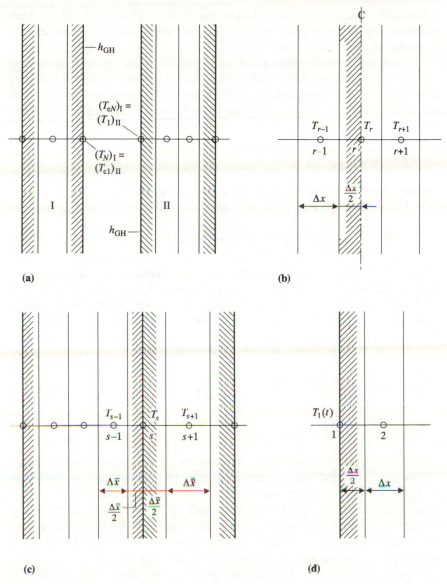

Fig. 13.7 Miscellaneous boundary conditions: (a) air layer between two slabs; (b) insulation (or symmetry); (c) interface between two slabs; (d) prescribed (constant or time-dependent) surface temperature

nodal point (see Eq. 13.54) for slab II, take the temperature of the 'environment' as equal to that of the Nth nodal point of slab I [i.e. make $(T^j_{e1})_{II} = (T^j_N)_I$].

(3) Replace h^j_N in Eq. 13.55 and h^j_1 in Eq. 13.54 by h^j_{GH}, to be calculated from Eq. 13.16 or 13.17, where $T_G \equiv (T^j_N)_I$ and $T_H \equiv (T^j_1)_{II}$.

Harmathy (1961) showed how the process of heat flow in fire resistance tests performed on slab-like specimens and on specimens with air layers can be simulated by desk calculations.

Although perfect insulation of a surface is not realizable in practice, a condition analogous to insulation does arise, namely along planes of symmetry. If, for example, a slab is heated from both sides [Fig. 13.7(b)], there is no heat flow across the central plane. It is justified, therefore, to assume that the slab is insulated along the central plane, and to study the temperature distribution only in the left or right half of the slab.[8]

Let r be a nodal point on an insulated surface (or in a plane of symmetry), as shown in Fig. 13.7(b). Since there is no heat flow across that plane, the surface volume element (of $\Delta x/2$ thickness) receives heat by conduction only from the $(r - 1)$th volume element. Thus, the conservation of energy equation becomes

$$k^j_{(r-1):r} \frac{T^j_{(r-1)} - T^j_r}{\Delta x} \Delta t = (\rho c)^j_r \frac{\Delta x}{2} (T^{j+1}_r - T^j_r) \qquad [13.57]$$

In Fig. 13.7(c), the node s represents an interface between two slabs made from different materials. Let the thickness of the volume elements be $\Delta \bar{x}$ for the slab to the left, and $\Delta \bar{\bar{x}}$ for the slab to the right; and let the material properties be $\bar{k}$ and $\overline{\rho c}$ for the material to the left, and $\bar{\bar{k}}$ and $\overline{\overline{\rho c}}$ for the material to the right. The conservation of energy equation for the interfacial volume element, which is of $(\Delta \bar{x} + \Delta \bar{\bar{x}})/2$ thickness and straddles the interface, now takes the form

$$\bar{k}^j_{(s-1):s} \frac{T^j_{s-1} - T^j_s}{\Delta \bar{x}} \Delta t + \bar{\bar{k}}_{s:(s+1)} \frac{T^j_{s+1} - T^j_s}{\Delta \bar{\bar{x}}} \Delta t$$

$$= \left((\overline{\rho c})^j_s \frac{\Delta \bar{x}}{2} + (\overline{\overline{\rho c}})^j_s \frac{\Delta \bar{\bar{x}}}{2} \right)(T^{j+1}_s - T^j_s) \qquad [13.58]$$

As to the prescribed (constant or time-dependent) temperature boundary condition: specifying the temperature at a surface point is equivalent to specifying the temperature of a surface volume element of $\Delta x/2$ thickness [Fig. 13.7(d)].

Equations 13.50, 13.54, 13.55, 13.57 and 13.58 are the finite difference formulations of the conservation of energy equation for all kinds of situation that may occur when dealing with one-dimensional heat flow problems. To make them suitable for numerical heat flow studies, one has to rearrange them to obtain expressions for T^{j+1}, i.e. for the temperatures at the end of the $j\Delta t \leq t \leq (j + 1)\Delta t$ period. The following equations result:

- For the interior nodes (or volume elements) of a slab ($n = 2, 3 \ldots (N - 2), (N - 1)$), from Eq. 13.50:

$$T^{j+1}_n = \frac{k^j_{(n-1):n}\Delta t}{(\rho c)^j_n \Delta x^2} T^j_{n-1} + \left[1 - \frac{k^j_{(n-1):n} + k^j_{n:(n+1)}}{(\rho c)^j_n \Delta x^2} \Delta t \right] T^j_n$$

---

8   The same assumption was used in the derivation of Eq. 13.36.

$$+ \frac{k^j_{n:(n+1)}\Delta t}{(\rho c)^j_n \Delta x^2} \; T^j_{n+1} \qquad\qquad\qquad\qquad [13.59]$$

- For a node on the left surface of the slab ($n = 1$), where heat is exchanged with the environment by convection and/or radiation, from Eq. 13.54

$$T^{j+1}_1 = \frac{2h^j_1 \Delta x \Delta t}{(\rho c)^j_1 \Delta x^2} \; T^j_{e1} + \left[1 - 2 \, \frac{h^j_1 \Delta x + k^j_{1:2}}{(\rho c)^j_1 \Delta x^2} \, \Delta t \right] T^j_1$$

$$+ \frac{2k^j_{1:2} \Delta t}{(\rho c)^j_1 \Delta x^2} \; T^j_2 \qquad\qquad\qquad\qquad [13.60]$$

- For a node on the right surface ($n = N$), where heat is also exchanged with the environment by convection and/or radiation, from Eq. 13.55

$$T^{j+1}_N = \frac{2k_{(N-1):N} \Delta t}{(\rho c)^j_N \Delta x^2} \; T^j_{N-1} + \left[1 - 2 \frac{k_{(N-1):N} + h^j_N \Delta x}{(\rho c)^j_N \Delta x^2} \, \Delta t \right] T^j_N$$

$$+ \frac{2h^j_N \Delta x \Delta t}{(\rho c)^j_N \Delta x^2} \; T^j_{eN} \qquad\qquad\qquad\qquad [13.61]$$

- For a node r on an insulated surface or in a plane of symmetry, from Eq. 13.57

$$T^{j+1}_r = \frac{2k_{(r-1):r} \Delta t}{(\rho c)^j_r \Delta x^2} \; T^j_{r-1} + \left[1 - \frac{2k_{(r-1):r} \Delta t}{(\rho c)^j_r \Delta x^2} \right] T^j_r \qquad [13.62]$$

- For a node s on an interface between two slabs, from Eq. 13.58

$$T^{j+1}_s = \frac{\overline{k}^j_{(s-1):s} \Delta t}{C^j_s \Delta \overline{x}} \; T^j_{s-1} + \left[1 - \left(\frac{\overline{k}^j_{(s-1):s}}{C^j_s \Delta \overline{x}} + \frac{\overline{\overline{k}}^j_{s:(s+1)}}{C^j_s \Delta \overline{\overline{x}}} \right) \Delta t \right] T^j_s$$

$$+ \frac{\overline{\overline{k}}^j_{s:(s+1)} \Delta t}{C^j_s \Delta \overline{\overline{x}}} \; T^j_{s+1} \qquad\qquad\qquad\qquad [13.63]$$

where

$$C^j_s = \tfrac{1}{2}[(\overline{\rho c})^j_s \Delta \overline{x} + (\overline{\overline{\rho c}})^j_s \Delta \overline{\overline{x}}] \qquad\qquad\qquad [13.64]$$

Comparing Eqs 13.59 and 13.62, one will see that the condition of insulation at, or symmetry about, a plane in which nodal point r is located may also be stated as

$$T^j_{r-1} = T^j_{r+1}; \quad T^{j+1}_{r-1} = T^{j+1}_{r+1} \qquad\qquad\qquad [13.65]$$

where, in the case of insulation, the nodal point ($r + 1$) is a virtual point that lies outside the boundaries of the solid.

The condition of prescribed surface temperature [at node 1 or N, see Fig. 13.7(d)] is

$$T^{j+1}_1 = T_1(t) \qquad \text{or} \qquad T^{j+1}_N = T(t) \qquad\qquad [13.66]$$

In simulations of fire tests performed on simple slab-like building elements,

$T_{e1}^j \equiv T_F$, variable temperature of the furnace gases (see Eq. 3.1);

$T_{eN}^j \equiv T_a$, constant temperature of the surrounding air;

$h_1^j \equiv h_E^j$, variable (temperature-dependent) coefficient of heat transfer on the fire-exposed side (see Eq. 13.13):

$h_N^j \equiv h_U^j$, variable (temperature-dependent) coefficient of heat transfer on the unexposed side (see Eqs 13.14 and 13.15).

Using an appropriate set of equations from among Eqs 13.59 to 13.63 and 13.66, one can calculate the temperature at all nodes in a simple or composite slab for the $t = (j + 1)\Delta t$ time level, provided one knows the temperatures at all nodes for the $t = j\Delta t$ level. The starting point is the $t = 0$ level, for which the temperature distribution is specified in the initial condition. For example, the slab illustrated in Fig. 13.6 is assumed to be in thermal equilibrium with the environment on the right-hand side, whose temperature is T_a. Hence, $T_1^0 = T_2^0 = T_3^0 = \ldots = T_{N-1}^0 = T_N^0 = T_a$.

The finite difference method consists of the repeated application of an appropriate set of equations to the $t = \Delta t, 2\Delta t, 3\Delta t, \ldots$ (i.e. $j = 1, 2, 3, \ldots$) time levels.

As Eqs 13.59 to 13.63 indicate, to learn about the temperature of a volume element at the time level $t = (j + 1)\Delta t$, one has to know the temperature of that element and the temperatures of the two adjacent elements at the $t = j\Delta t$ level. In fact, by suitably selecting the values of Δx and Δt, one can make the bracketed term in these equations equal to zero, thereby achieving that the temperature of the volume element at $t = (j + 1)\Delta t$ will not depend on its own temperature at $t = j\Delta t$. However, one cannot make the bracketed term less than zero. A negative value would mean that the higher the temperature of the volume element at the present time, the lower it will be at a future instant, in violation of thermodynamic principles.

Should the bracketed terms in Eqs 13.59 to 13.63 become even slightly negative, the computed values of the temperature will oscillate with an increasing magnitude. This phenomenon, called 'instability', has nothing to do with the question of accuracy.

The condition that the bracketed term in Eqs 13.59 to 13.63 be non-negative, leads to the following *stability criteria*:

- For the interior nodes of a slab (nodes 2 to $(N - 1)$ in Fig. 13.6):

$$\Delta t \le \frac{1}{2} \frac{(\rho c)_n^j}{k_n^j} \Delta x^2 \qquad [13.67]$$

- For the nodes on the two surfaces of the slab (nodes 1 and N in Fig. 13.6):

$$\Delta t \le \frac{1}{2} \frac{(\rho c)_1^j}{k_1^j + h_1^j \Delta x} \Delta x^2 \qquad [13.68]$$

$$\Delta t \le \frac{1}{2} \frac{(\rho c)_N^j}{k_N^j + h_N^j \Delta x} \Delta x^2 \qquad [13.69]$$

- For a node on an insulated surface or in a plane of symmetry [node r in Fig. 13.7(b)]:

$$\Delta t \leq \frac{1}{2} \frac{(\rho c)_r^j}{k_r^j} \Delta x^2 \qquad\qquad [13.70]$$

- For an interfacial node [node s in Fig. 13.7(c)]:

$$\Delta t \leq \frac{1}{2} \frac{(\overline{\rho c})_s^j \Delta \overline{x} + (\overline{\overline{\rho c}})_s^j \Delta \overline{\overline{x}}}{\dfrac{\overline{k}_s^j}{\Delta \overline{x}} + \dfrac{\overline{\overline{k}}_s^j}{\Delta \overline{\overline{x}}}} \qquad\qquad [13.71]$$

It has been assumed here that

$$\frac{k_{(n-1):n}^j + k_{n:(n+1)}^j}{2} \simeq k_n^j; \qquad k_{1:2}^j \simeq k_1^j; \qquad k_{(N-1):N}^j \simeq k_N^j$$

$$k_{(r-1):r}^j \simeq k_r^j; \qquad\qquad \overline{k}_{(s-1):s}^j \simeq \overline{k}_s^j; \qquad \overline{\overline{k}}_{s:(s+1)}^j \simeq \overline{\overline{k}}_s^j \quad [13.72]$$

To ensure stability, the most stringent of the applicable criteria among 13.67 to 13.71 must be satisfied. It should also be kept in mind that, as a rule, the material properties of the solid and the heat transfer coefficients vary during the heat flow process. To be on the safe side, one should use the lowest value of ρc and the highest values of k, h_1 and h_N that may occur during the entire process.

When dealing with fire test simulations, the high value of the coefficient of heat transfer on the fire-exposed side, h_E ($\equiv h_1$), is the principal factor that controls the stability of the calculation procedure. Consequently, according to Eq. 13.68, the criterion of stability is

$$\Delta t \leq \frac{1}{2} \frac{\Delta x^2}{\left(\dfrac{k}{\rho c} + \dfrac{h_E}{\rho c} \Delta x \right)_{\max}} \qquad\qquad [13.73]$$

Calculating h_E from Eq. 13.13, one will see that its value steadily increases as the fire test progresses, and by two hours into the test it may be as high as $300 \text{ W m}^{-2} \text{ K}^{-1}$. The maximum for the term in the denominator usually arises toward the end of the simulation of fire test. For example, if the building element is made from normal-weight concrete and the test is expected to last two hours, at the end of the test simulation the denominator will be approximately $(0.6 + 105\Delta x) \times 10^6$. Thus, if $\Delta x = 0.02 \text{ m}$ is selected, the stability criterion will require that Δt be less than 74 s.

The need for small values for Δt may be circumvented in two ways:

(1) by assuming that $h_1 \to \infty$, i.e. that $T_1 = T_{e1}$ (or $T_E = T_F$);
(2) by choosing relatively large time increments at the beginning of the simulation of the test, and reducing them as required in the course of the calculations.

If the stability criteria are observed, round-off errors have little effect on the solution (Richtmyer and Morton 1967). They accumulate roughly in proportion to the square root of the number of calculation steps. The truncation error (error due to finite difference approximation) is slightly more significant; it is proportional to both Δt and Δx^2.

Figure 13.8 has been prepared to illustrate the accuracy of the finite difference method. It shows the temperature history of a brick wall (slab) during exposure to an 'idealized' fire resistance test.[9] The solid curves display the analytical solution (see Eqs 13.30 and 13.31); the discrete points represent the results of a numerical study. The following input data have been used:

Thermal conductivity	$k = 0.952 \text{ W m}^{-1}\text{K}^{-1}$,
Volume specific heat	$\rho c = 1.594 \times 10^6 \text{ J m}^{-3}\text{K}^{-1}$
Thickness of wall	$l = 0.2032 \text{ m}$
Coefficient of heat transfer	$h_U = 16.02 \text{ W m}^{-2}\text{K}^{-1}$
Initial temperature of wall and temperature of ambient air	$T_a = 297 \text{ K}$
'Idealized' fire temperature and temperature of exposed surface	$T_F = T_E = 1261 \text{ K}$
Thickness of volume elements	$\Delta x = 0.0254, 0.0508, 0.1016 \text{ m}$
Time increment	$\Delta t = 360, 720, 1080 \text{ s}$

Apparently, even if the subdivision of the slab is rather crude ($\Delta x = 0.1016 \text{ m}$), and the time steps are large ($\Delta t = 1080 \text{ s}$), the results yielded by the finite difference technique are of acceptable accuracy.

In a more realistic simulation of a fire resistance test, T_F is regarded as a variable quantity (see Eq. 3.1), and h_E is also taken as variable, to be calculated from Eq. 13.13.

Often simple polynomials are employed to describe dependence of material properties (thermal conductivity and volume specific heat) on the temperature (see, for example, Lie and Lin 1985). Inputting information in tabular form is preferable, however, if the property versus temperature relation is complex or may vary from case to case. Specific values may be obtained by linear interpolation between the tabulated values, or by the application of Lagrange's interpolation formula (see, for example, Korn and Korn 1961).

Let ψ be a material property tabulated for n values of the temperature, T. Given $\psi_0 = \psi(T_0)$, $\psi_1 = \psi(T_1)$, $\psi_2 = \psi(T_2)$, $\ldots$, $\psi_k = \psi(T_k)$, $\ldots$, $\psi_n = \psi(T_n)$,

$$\psi(T) = \sum_{k=0}^{n} \psi_k L_k(T) \qquad [13.74]$$

where the Ls are coefficients defined as

---

9   An idealized test means two things: (1) the coefficient of heat transfer on the fire-exposed side, $h_E \rightarrow \infty$, and therefore $T_E = T_F$, and (2) the temperature of the furnace gases, T_F, is regarded as constant, representative of the expected duration of the fire test. (See Eq. 3.2 and Fig. 3.6.)

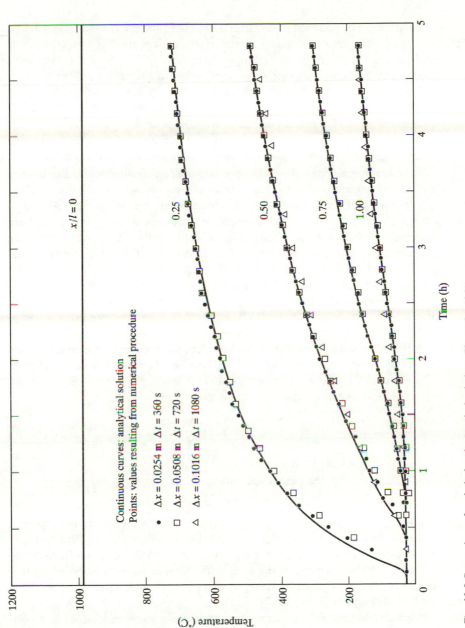

Fig. 13.8 Comparison of analytical and numerical solutions (Harmathy 1961)

$$L_k(T) = \prod_{\substack{m=0 \\ m \neq k}}^{n} \frac{T - T_m}{T_k - T_m}$$ [13.75]

The problem of presence of moisture in building materials cannot be handled in its full complexity. Although some mathematical models are capable of addressing the problem (see, for example, Lykov and Mykhaylov 1961; Harmathy 1969), they are too cumbersome for application in fire safety deliberations.

There are two relatively uncomplicated ways of taking account of the effect of moisture on fire resistance. One, an empirical approach (Harmathy 1965a), now part of the ASTM fire test standard, was discussed in Section 4.4 of Chapter 4. The other, a simplified theoretical approach, which treats moisture as an unstable component of the material, was described in Section 6.8 of Chapter 6.

The one-dimensional finite-difference procedure described so far can also be applied to the solution of quasi-one-dimensional problems. The calculation of the fire resistance of some protected steel columns and beams may be handled as a one-dimensional problem. For example, the columns whose cross-sections are depicted in Figs 12.6(e) and 12.6(f) may be looked upon as two-layer (protecting material plus steel) slabs, exposed to fire on the side of the protecting material and insulated along the inner surface of the steel form.

In fact, the one-dimensional procedure may be applied to all the geometric configurations shown in Figs 12.6(c) to 12.6(i). Since the thermal conductivity of steel is at least an order of magnitude higher than that of the protecting material, it is permissible to model the steel core as a heat-absorbing medium at essentially uniform temperature. Denoting, as in Fig. 13.6, the nodal point on the unexposed surface of the protection (i.e. the interface between the protection and steel core, or the interface between the protection and air space plus steel core) by N, the conservation of energy equation for this point can be written as

$$k^j_{(N-1):N} \frac{Tk^j_{N-1} - T^j_N}{\Delta x} \Delta t = \left((\rho c)^j_N \frac{\Delta x}{2} + \frac{M}{S}(c_{st})^j_N \right)(T^{j+1}_N - T^j_N)$$ [13.76]

where M = mass of the steel core, kg;
  S = area of the unexposed surface of protective layer, m²;
  c_{st} = specific heat of steel, $J\,kg^{-1}\,K^{-1}$.

Solved for T^{j+1}_N [temperature of the unexposed surface of protection, as well as that of the steel core, at $t = (j + 1)\Delta t$], Eq. 13.76, together with Eqs 13.59 and 13.60, can be used for developing information on the temperature histories of the protective layer and of the steel core. Similar techniques for predicting the fire resistance of protected steel members have been described by Pettersson *et al.* (1976) and Gamble (1989).

Before leaving this subject, it may be worth taking a second look at Eq. 13.50. Rearranged into the form

$$\frac{1}{\Delta x}\left(k^j_{(n-1):n} \frac{T^j_{n-1} - T^j_n}{\Delta x} - k^j_{n:(n+1)} \frac{T^j_n - T^j_{n+1}}{\Delta x} \right) = (\rho c)^j_n \frac{T^{j+1}_n - T^j_n}{\Delta t}$$

$$[13.77]$$

it can be recognized as the finite difference equivalent of Eq. 13.1, with q from Eq. 13.2.

In Eq. 13.77, the space derivatives (on the left-hand side) are approximated using temperatures prevailing at the $t = j\Delta t$ time level, whereas the time derivative (on the right-hand side) straddles the $j\Delta t \le t \le (j + 1)\Delta t$ interval. Thanks to such a formulation of the conservation of energy equation, there is only one unknown in Eq. 13.77: T^{j+1}_n, the temperature at the end of the interval. This formulation is, therefore, referred to as the explicit scheme.

There is nothing wrong, however, with approximating the space derivatives using temperature values prevailing at the $t = (j + 1)\Delta t$ time level (implicit scheme) or at the $t = (j + \frac{1}{2}\Delta t)$ level (Crank–Nicholson scheme). Such approximations do, in fact, offer some advantages. But there is one serious drawback. In order to develop information on the temperature at any point at the $t = (j + 1)\Delta t$ level, one has to solve at each time step a number of simultaneous algebraic equations (equal to the number of nodes). If the material properties are temperature-dependent and the boundary conditions are markedly variable, the implicit and Crank–Nicholson schemes offer no real advantages over the explicit scheme.

13.3.2 Two-dimensional Heat Flow

Concrete beams and columns, and walls made from hollow masonry units, are typical of building elements that present two-dimensional heat flow problems. Having acquired some skill in formulating the finite difference scheme for one-dimensional heat flow processes, the reader will have no difficulty in applying that skill to more complex processes.

The first step is again the discretization of space and time. The cross-section of the building element in question has to be mapped out in an $x-y$ plane, and subdivided into elementary areas. With a dimension of unity along the z-axis, these area elements will, in fact, represent volume elements. Again, one point of the volume element, the nodal point, will be looked upon as representative (with respect to temperature and material properties) of the volume element as a whole.

Figure 13.9 illustrates two ways of subdividing an object with a T-shaped cross-sectional area into volume elements. The object may be looked upon as a concrete beam–slab assembly. The axial heat flow in the assembly is assumed to be negligible.

The subdivision shown in Fig. 13.9(a) is probably the first that comes to mind.

However, there may be some difficulties with that arrangement. If, for reasons discussed earlier, nodal points are placed on the surfaces, four kinds of volume elements may arise: full rectangles (with a cross-sectional area $\Delta x\, \Delta y$; see element α), half rectangles (element β), quarter rectangles (element γ), and three-quarter rectangles (element δ). Clearly, with four kinds of volume elements to consider, the formulation of the conservation of energy equations becomes cumbersome.

The diagonal arrangement shown in Fig. 13.9(b) offers the convenience of having only two kinds of volume elements to deal with: inside elements with square cross-sectional areas, and surface elements with triangular cross-sectional areas. Since need for using networks other than diagonal rarely arises, only the diagonal arrangement will be discussed here.[10]

To be able to arrange a two-dimensional form on a diagonal grid, one has first to find the proper grid size: $\Delta\xi$ (m). This task amounts to finding common divisors for all the dimensions of the form.[11] If d (m) is a suitable common divisor, $\Delta\xi = d/\sqrt{2}$.

Having found a suitable grid size, one can now place the form on the grid and locate the nodal points. The form shown in Fig. 13.9(b) is symmetrical about the AB and EF planes. It is sufficient, therefore, to study the heat flow process in the ABCDEF section only.

It is recommended that the origin of the $x-y$ coordinate system be selected in such a way as to have the center of the nearest square element located at a $\Delta\xi/\sqrt{2}$ (i.e. $d/2$) distance from both coordinate axes. The coordinates of the nodal points will be multiples of $\Delta\xi/\sqrt{2}$. Each nodal point will be labeled with two numbers, m and n:

$$m = \frac{x}{\Delta\xi/\sqrt{2}} \quad \text{and} \quad n = \frac{y}{\Delta\xi/\sqrt{2}}$$

Thus, the point $P_{m,n}$ has the coordinates $x = m\Delta\xi/\sqrt{2}$ and $y = n\Delta\xi/\sqrt{2}$. Apparently, only those m,n pairs represent actual nodal points for which $(m + n)$ is an even number [see Fig. 13.9(b)].

Since two-dimensional forms come in an endless variety, it is impossible to discuss the finite difference formulation of two-dimensional problems in general terms. The calculation procedure can probably best be mastered by studying specific cases. The form in Fig. 13.9(b) has devised in such a way as to allow a good insight into a number of situations with which fire safety practitioners may have to deal.

Suppose the form in Fig. 13.9(b) represents a reinforced concrete beam−floor assembly with a wood deck, and the assembly is subjected to a standard fire

---

10 See Dusinberre's book (1949) for other kinds of network.

11 If the greatest common divisor appears to be too small to yield a reasonable grid size, one may think of arbitrarily changing some dimensions of the form by a few millimeters. Those familiar with the design of structures know that in practice the dimensions of concrete elements cannot be adhered to with an accuracy better than a few millimeters.

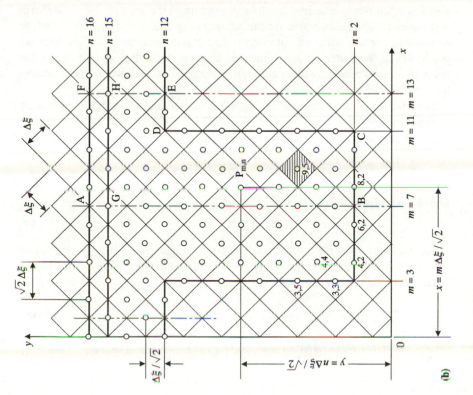

(b)

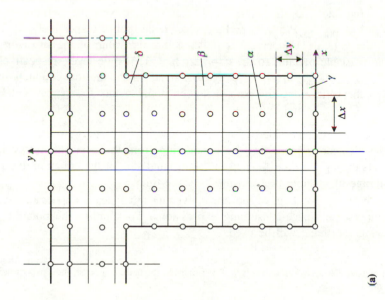

(a)

Fig. 13.9 Two ways of subdividing a T-shaped object (concrete beam) into volume elements

resistance test (from underneath). Its temperature at the start ($t = 0$) is uniform and equal to the temperature of the surrounding air, T_a. The upper surface remains in contact with the surrounding air throughout the test. The thermal properties of the materials are: concrete, k and ρc; wood, $\bar{k}$ and $\overline{\rho c}$. Since the mass of the steel bars is usually negligible in comparison with that of the concrete, the presence of steel may be disregarded.

The first step is to define the region to be studied and the initial temperature at the nodal points.

The region is the ABCDEF cross-sectional area:

$$7 \leq m \leq 11 \quad \text{and} \quad 2 \leq n \leq 11$$
$$7 \leq m \leq 13 \quad \text{and} \quad 12 \leq n \leq 16$$

and the initial condition is

$$T_{m,n}^0 = T_a \quad \text{at } t = 0 \tag{13.78}$$

The formulation of the conservation of energy equations will again depend on the location of the nodal point. Thus, for a volume element[12] represented by the nodal point $P_{m,n}$ inside the concrete,

$$k_{(m-1),(n-1):m,n}^j \frac{T_{(m-1),(n-1)}^j - T_{m,n}^j}{\Delta\xi} \Delta t\Delta\xi + k_{(m+1),(n-1):m,n}^j \frac{T_{(m+1),(n-1)}^j - T_{m,n}^j}{\Delta\xi} \Delta t\Delta\xi$$

$$+ k_{(m-1),(n+1):m,n}^j \frac{T_{(m-1),(n+1)}^j - T_{m,n}^j}{\Delta\xi} \Delta t\Delta\xi + k_{(m+1),(n+1):m,n}^j \frac{T_{(m+1),(n+1)}^j - T_{m,n}^j}{\Delta\xi} \Delta t\Delta\xi$$

$$= (\rho c)_{m,n}^j \Delta\xi^2 (T_{m,n}^{j+1} - T_{m,n}^j) \tag{13.79}$$

where the $(m - 1),(n - 1)$; $(m + 1),(n - 1)$; m,n; etc., pairs of subscripts indicate that the value of the subscripted variable pertains to the $P_{(m-1),(n-1)}$; $P_{(m+1),(n-1)}$; $P_{m,n}$; etc., nodal point, respectively;

the j, $(j + 1)$ superscripts indicate that the value of the superscripted variable pertains to the $t = j\Delta t$, $(j + 1)\Delta t$ time level, respectively;

$k_{(m-1),(n-1):m,n}$; $k_{(m+1),(n-1):m,n}$; etc., stand for the thermal conductivity of the solid along the paths between nodal points $P_{(m-1),(n-1)}$ and $P_{m,n}$; $P_{(m+1),(n-1)}$ and $P_{m,n}$; etc.

Again, the left-hand side of Eq. 13.79 describes the energy received by the volume element at point $P_{m,n}$ by conduction from the adjacent four volume elements during the time Δt, and the right-hand side of the equation describes the storage of sensible heat during the same time.

Clearly, the above usage of subscripts, even though logically impeccable, results in cumbersome, space-consuming expressions. To simplify the notation, the following shorthand will be used:

---

12   The volume element has the dimensions of $\Delta\xi$ (m) by $\Delta\xi$ (m) in the $x-y$ plane, and 1 m perpendicular to the $x-y$ plane.

p for $(m - 1)$; q for $(m + 1)$
r for $(n - 1)$; s for $(n + 1)$

and the commas will be omitted.[13] Thus, the temperature $T^j_{(m-1),(n-1)}$ will be denoted by T^j_{pr}, the thermal conductivity $k^j_{(m-1),(n-1):m,n}$ by $k^j_{pr:mn}$, the point $P_{(m-1),(n-1)}$ by P_{pr}, etc.

As before, the thermal conductivity along a path between two nodal points will be regarded as the arithmetic mean of the thermal conductivities at the two nodal points. Hence

$$k^j_{pr:mn} = \frac{k^j_{pr} + k^j_{mn}}{2}; \qquad k^j_{qr:mn} = \frac{k^j_{qr} + k^j_{mn}}{2}; \qquad \text{etc.} \qquad [13.80]$$

Since the material properties are functions of the temperature only,

$$k^j_{pr} = k(T^j_{pr}); \qquad k^j_{mn} = k(T^j_{mn}); \qquad \text{etc.} \qquad [13.81]$$

and

$$(\rho c)^j_{mn} = \rho c(T^j_{mn}) \qquad [13.82]$$

To save space, the conservation of energy equations will not be written out in their original forms. Only the equations to be used in the step-by-step calculation of the temperatures T^{j+1} (i.e. the temperatures at the successive $t = (j + 1)\Delta t$ levels) will be presented.

The following equations apply:

- For nodes to the left and to the right of the two planes of symmetry, AB and EF:[14]

$$m = 7 \qquad \text{and} \qquad 2 \leq n \leq 16$$
$$m = 13 \qquad \text{and} \qquad 12 \leq n \leq 16$$

$$T^j_{pn} = T^j_{qn} \qquad [13.83]$$

- For nodes in the interior of the concrete, i.e. within the regions:

$$7 \leq m \leq 10 \qquad \text{and} \qquad 3 \leq n \leq 12$$
$$7 \leq m \leq 13 \qquad \text{and} \qquad 13 \leq n \leq 14$$

$$T^{j+1}_{mn} = \frac{\Delta t}{(\rho c)^j_{mn} \Delta \xi^2} (k^j_{pr:mn} T^j_{pr} + k^j_{qr:mn} T^j_{qr} + k^j_{ps:mn} T^j_{ps} + k^j_{qs:mn} T^j_{qs})$$

$$+ \left[1 - \frac{\Delta t}{(\rho c)^j_{mn} \Delta \xi^2} (k^j_{pr:mn} + k^j_{qr:mn} + k^j_{ps:mn} + k^j_{qs:mn}) \right] T^j_{mn}$$

$$[13.84]$$

- For nodes along the BC and DE (fire-exposed) surfaces:

---

13 It is suggested that the reader prepare himself a sketch showing the location of points P_{pr}, P_{qr}, P_{ps}, and P_{qs} in relation to point P_{mn}.
14 The reader is reminded that nodal points are located only at those pairs of m,n for which $(m + n)$ is an even number.

$$7 \leq m \leq 11 \quad \text{and} \quad n = 2$$
$$11 \leq m \leq 13 \quad \text{and} \quad n = 12$$

$$T_{mn}^{j+1} = \frac{2\Delta t}{(\rho c)_{mn}^{j} \Delta \xi^2} (k_{ps:mn}^{j} T_{ps}^{j} + k_{qs:mn}^{j} T_{qs}^{j} + \sqrt{2}\Delta \xi (h_E)_{mn}^{j} T_F^{j})$$

$$+ \left[1 - \frac{2\Delta t}{(\rho c)_{mn}^{j} \Delta \xi^2} (k_{ps:mn}^{j} + k_{qs:mn}^{j} + \sqrt{2}\Delta \xi (h_E)_{mn}^{j}) \right] T_{mn}^{j}$$

$$[13.85]$$

- For nodes along the CD (fire-exposed) surface:

$$m = 11 \quad \text{and} \quad 2 \leq n \leq 12$$

$$T_{mn}^{j+1} = \frac{2\Delta t}{(\rho c)_{mn}^{j} \Delta \xi^2} (k_{pr:mn}^{j} T_{pr}^{j} + k_{ps:mn}^{j} T_{ps}^{j} + \sqrt{2}\Delta \xi (h_E)_{mn}^{j} T_F^{j})$$

$$+ \left[1 - \frac{2\Delta t}{(\rho c)_{mn}^{j} \Delta \xi^2} (k_{pr:mn}^{j} + k_{ps:mn}^{j} + \sqrt{2}\Delta \xi (h_E)_{mn}^{j}) \right] T_{mn}^{j}$$

$$[13.86]$$

- For nodes along the GH interface between concrete and wood:

$$7 \leq m \leq 13 \quad \text{and} \quad n = 15$$

$$T_{mn}^{j+1} = \frac{\Delta t}{C_{mn}^{j} \Delta \xi^2} (k_{pr:mn}^{j} T_{pr}^{j} + k_{qr:mn}^{j} T_{qr}^{j} + \bar{k}_{ps:mn}^{j} T_{ps}^{j} + \bar{k}_{qs:mn}^{j} T_{qs}^{j})$$

$$+ \left[1 - \frac{\Delta t}{C_{mn}^{j} \Delta \xi^2} (k_{pr:mn}^{j} + k_{qr:mn}^{j} + \bar{k}_{ps:mn}^{j} + \bar{k}_{qs:mn}^{j}) \right] T_{mn}^{j}$$

$$[13.87]$$

where

$$C_{mn}^{j} = \tfrac{1}{2}[(\rho c)_{mn}^{j} + (\bar{\rho c})_{mn}^{j}]$$

$$[13.88]$$

- For nodes along the AF (unexposed) surface:

$$7 \leq m \leq 13 \quad \text{and} \quad n = 16$$

$$T_{mn}^{j+1} = \frac{2\Delta t}{(\bar{\rho c})_{mn}^{j} \Delta \xi^2} (\bar{k}_{pr:mn}^{j} T_{pr}^{j} + \bar{k}_{qr:mn}^{j} T_{qr}^{j} + \sqrt{2}\Delta \xi (h_U)_{mn}^{j} T_a^{j})$$

$$+ \left[1 - \frac{2\Delta t}{(\bar{\rho c})_{mn}^{j} \Delta \xi^2} (\bar{k}_{pr:mn}^{j} + \bar{k}_{qr:mn}^{j} + \sqrt{2}\Delta \xi (h_U)_{mn}^{j}) \right] T_{mn}^{j}$$

$$[13.89]$$

In Eqs 13.85 and 13.86, T_F^j is the temperature of the furnace gases at $t = j\Delta t$, to be calculated from Eq. 3.1, and in Eqs 13.85, 13.86, and 13.89

$$(h_E)_{mn}^{j} = h_E(T_{mn}^{j}, T_F^{j}) \quad \text{and} \quad (h_U)_{mn}^{j} = h_U(T_{mn}^{j}, T_a) \qquad [13.90]$$

to be calculated from Eqs 13.13 and 13.15, respectively.

The criteria of stability have to be found again from the condition that the bracketed terms in Eqs 13.84 to 13.87 and 13.89 must be non-negative. As mentioned before, when dealing with fire test simulations the high value of the coefficient of heat transfer on the fire-exposed side is usually the controlling factor. With some simplifying assumptions, the following criterion is obtained from Eq. 13.85 or 13.86:

$$\Delta t \leq \frac{1}{4} \frac{\Delta \xi^2}{\left[\dfrac{k}{\rho c} + \dfrac{h_E}{\rho c} \dfrac{\Delta \xi}{\sqrt{2}} \right]_{max}} \qquad [13.91]$$

Again, the maximum for the term in the denominator is usually obtained with values of the material's thermal properties (at the temperature of the fire-exposed surface) and the coefficient of heat transfer arising toward the end of the test simulation.

In the preceding discussion, the problem of the presence of reinforcing steel bars was ignored. It is also possible, however, to take the presence of steel components into consideration. Suppose the elementary volume represented by nodal point $P_{9,5}$ [hatched in Fig. 13.9(b)] is a reinforcing bar. Since this point is in the interior of the concrete, Eq. 13.84 is applicable to the calculation of T^{j+1}. The simplest procedure is to have the computer alerted to use the thermal properties of steel [$(\rho c)_{st}$ and k_{st}] instead of those of concrete whenever 9 and 5 appear in any of the subscripts mn, pr, qr, ps or qs. This can be done by an instruction something like this:

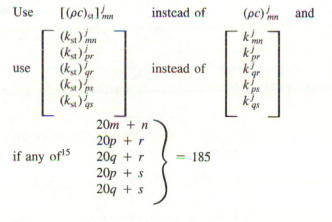

where

$$[(\rho c)_{st}]^j_{mn} = (\rho c)_{st}(T^j_{mn}) \qquad [13.92]$$

$$(k_{st})^j_{mn} = k_{st}(T^j_{mn}); \quad \text{etc.} \qquad [13.93]$$

---

15 Clearly, instructions such as '*if (m + n) = 14, (p + r) = 14*', etc. would not be satisfactory, since they would designate three nodal points, $P_{8,6}$, $P_{9,5}$, and $P_{10,4}$, for special treatment. With the above instruction, only one point, $P_{9,5}$ qualifies for special treatment within the ABCDEF region.

It is unlikely that the stability criterion imposed on the calculation procedure by the highly restrictive conditions at the fire-exposed boundaries has to be changed on account of the presence of steel bars in the concrete.

Sometimes definition of the region to be studied and the conditions of symmetry requires a bit of reflection. Figure 13.10 shows one-eighth of the cross-sectional area of a hollow-cored column, arranged on a diagonal grid. Clearly, on account of symmetry, it is sufficient to study the temperature history of the ABCD region only. This region (with the information given in the figure) is defined as

$$1 \leq m \leq 15 - n \quad \text{and} \quad 0 \leq n \leq 6$$

and the conditions of symmetry are:

- About the AB plane,

$$T^j_{0,n} = T^j_{2,n} \tag{13.94}$$

- About the CD plane,

$$T_{(14-n),n} = T_{(15-n),(n+1)} \tag{13.95}$$

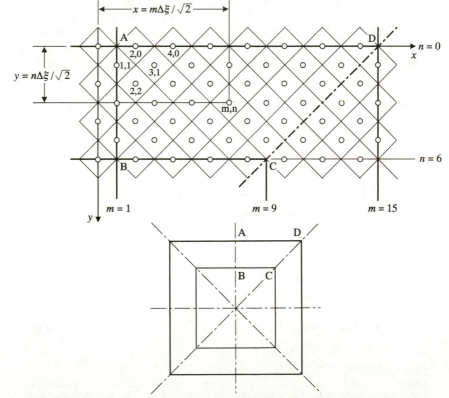

Fig. 13.10 The arrangement of one-eighth of the cross-section of a hollow-cored column on a diagonal grid

For a column with the same outside dimensions but without a hollow core, one-eighth of the cross-sectional area would be a right triangle, defined as

$$1 \leq m \leq 15 - n \quad \text{and} \quad 0 \leq n \leq 14$$

The conditions of symmetry remain the same as in Eqs 13.94 and 13.95.

The two-dimensional finite difference technique was used in some studies aimed at predicting the fire resistance of slab-like building elements and beams (Harmathy 1977/1978), concrete-protected steel columns (Lie and Harmathy 1972), concrete columns (Lie 1977, 1989; Lie *et al.* 1984; Lie and Lin 1985), concrete-filled steel columns (El-Shayeb 1986), and concrete slab—beam assemblies (Sultan *et al.* 1991).

Some difficulties may be encountered when applying the finite difference method to the assessment of the temperature history of building elements with enclosed cavities. The calculation of the radiant-heat interchange between the surface elements of the cavities involves solving for each time step a number of simultaneous algebraic equations (Gebhart 1961). Harmathy (1970b) used this procedure in studying the fire resistance of concrete masonry walls. The results were analyzed and correlated by semiempirical equations (Eqs 12.7 to 12.9).

Unfortunately, the finite difference method is not a felicitous choice for the assessment of temperature distribution in tapered beams, such as the one shown in Fig. 13.11(a). It may be necessary to perform computations first for a beam with a cross-section ABCDEF, and then for one with a cross-section AB′C′DEF, so that the temperature distributions for the desired time levels can be determined by interpolation.

Similar problems may arise with concrete I-beams. Again, the beam has to be modeled as composed of rectangular regions, e.g. in the way illustrated in Fig. 13.11(b).

Although building elements of circular cross-section (e.g. monolithic cylindrical columns) are two-dimensional forms, the problem they present is, in fact, quasi-one-dimensional.

Figure 13.12 shows the subdivision of a cylindrical column into $(N + 1)$ volume elements ($n = 0, 1, 2, \ldots, N$). The innermost element is of a circular cross-section with a diameter Δr (m). All the other elements are ring-shaped with Δr ring thickness, except the outside ring, whose thickness is $\Delta r/2$.

Based on conservation of energy equations, the following set of expressions can be developed for T^{j+1}.

- For the center of the cylinder ($n = 0$):[16]

$$T_0^{j+1} = \frac{4k_{0:1}\Delta t}{(\rho c)_0^j \Delta r^2} T_1^j + \left[1 - \frac{4k_{0:1}\Delta t}{(\rho c)_0^j \Delta r^2} \right] T_0^j \qquad [13.96]$$

---

16 Clearly, when dealing with one-dimensional and quasi-one-dimensional heat flow problems, using the previously introduced shorthand notation for the subscripts is not necessary.

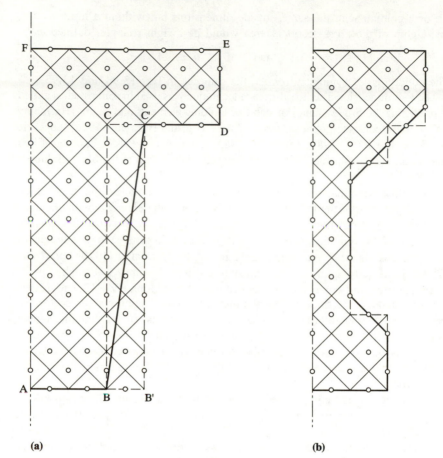

Fig. 13.11 Subdivision into volume elements of (a) a tapered concrete beam, (b) a concrete I-beam

- For the *n*th ring $[1 \leq n \leq (N - 1)]$:

$$T_n^{j+1} = \frac{2n-1}{2n} \frac{k_{(n-1):n}^j \Delta t}{(\rho c)_n^j \Delta r^2} T_{n-1}^j + \frac{2n+1}{2n} \frac{k_{n:(n+1)}^j \Delta t}{(\rho c)_n^j \Delta r^2} T_{n+1}^j$$

$$+ \left[1 - \frac{2n-1}{2n} \frac{k_{(n-1):n}^j \Delta t}{(\rho c)_n^j \Delta r^2} - \frac{2n+1}{2n} \frac{k_{n:(n+1)}^j \Delta t}{(\rho c)_n^j \Delta r^2} \right] T_n^j$$

[13.97]

- For the outside ring $(n = N)$:

$$T_N^{j+1} = \frac{2N-1}{N-\frac{1}{4}} \frac{k_{(N-1):N}^j \Delta t}{(\rho c)_N^j \Delta r^2} T_{N-1}^j + \frac{2N}{N-\frac{1}{4}} \frac{h_N^j \Delta r \Delta t}{(\rho c)_N^j \Delta r^2} T_F^j$$

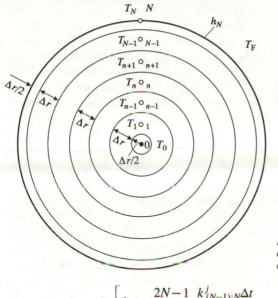

Fig. 13.12 Subdivision of the cross-section of a cylindrical column into volume elements

$$+ \left[1 - \frac{2N-1}{N-\frac{1}{4}} \frac{k_{(N-1):N}^j \Delta t}{(\rho c)_N^j \Delta r^2} - \frac{2N}{N-\frac{1}{4}} \frac{h_N^j \Delta r \Delta t}{(\rho c)_N^j \Delta r^2} \right] T_N^j \qquad [13.98]$$

Of course, when dealing with fire resistance problems, $h_N^j \equiv h_E^j$, to be calculated from Eq. 13.13.

Again, it is almost certain that the criterion of stability is controlled by the conditions at the fire-exposed surface. After some simplifications, the following criterion is obtained from Eq. 13.98:

$$\Delta t \leq \frac{1}{2} \frac{\Delta r^2}{\left[\frac{k}{\rho c} + \frac{h_E}{\rho c} \Delta r \right]_{max}} \qquad [13.99]$$

Lie (1984), El-Shayeb (1986), and Lie and Celikkol (1991) used the quasi-one-dimensional technique in studying the fire resistance of circular reinforced concrete columns and concrete-filled steel columns.

Those who wish to read more about the finite difference method may find valuable information in Emmons's paper (1943), in monographs by Douglas (1961) and Razelos (1973), and in a number of books, e.g. those by Dusinberre (1949), Hartree (1958), and Forsythe and Wasow (1960).

13.4   Finite Element Method

The finite element method is a more recent numerical technique. Although the underlying concepts have evolved over a period of at least 150 years, the method acquired practical utility only with the advent of computers.

The rising popularity of the finite element method is attributable to its great flexibility in handling problems involving complex geometrical shapes, several different materials, and mixed boundary conditions. As with the finite difference method, the discretization of the solid is the first step in the procedure. It consists of conveniently locating nodal points on the boundaries and in the interior of the region to be studied, not necessarily in a regular array. Connecting the nodal points with straight lines (in two dimensions), a multitude of triangular subregions (elements) are formed (Fig. 13.13).[17]

The unknown field variable (temperature) is expressed in terms of an assumed approximating function ('interpolation function') within each element. The nodal values of the field variable and the interpolation functions define the behavior of the field variable within the elements.

Once the nodal points have been located and the interpolation functions selected, matrix equations are formulated (making use of the variational principle; see Huebner and Thornton 1982) for the individual elements. These matrix equations are then combined for the overall system, taking account of the boundary conditions. The unknown nodal values of the field variable are finally obtained by solving a large set (sometimes in the hundreds or thousands) of simultaneous equations.

Those who may have to perform numerical heat flow studies more than just once or twice a year are advised to take the trouble of learning more about the finite element theory. A handbook edited by Kardestuncer and Norrie (1987) lists several user-oriented finite element software systems that have been developed for the solution of heat flow problems.

A number of handbooks are available on the subject, such as those by Zienkiewicz and Cheung (1967), Desai (1979), Huebner and Thornton (1982), Mori (1983), Segerlind (1984), and Burnett (1987). Papers by Wilson and Nickell

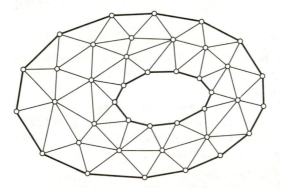

Fig. 13.13 Discretization of a complex shape for finite element studies

---

17 Three-node triangular elements are the most popular, because they can be easily assembled to form any complex geometry. Four-node rectangular elements are preferred, however, in the case of regions with straight-line boundaries. Quadrilateral elements (with or without an internal node) can be formed by combining two or four triangles.

(1966), and Emery and Carson (1969), address specifically the problem of heat conduction in solids. Several publications by the University of California, Berkeley (Becker *et al.* 1974; Becker and Bresler 1974; Iding *et al.* 1975; Bresler 1976b; Bresler and Iding 1983; Iding and Bresler 1984) deal with the application of the finite element method to fire safety problems.

13.5   Experimental Method

The Portland Cement Association (Skokie, IL) conducted an extensive research program to develop information on the temperature history of concrete slabs and beams during exposure to standard fire resistance tests. Temperatures at various distances from the fire-exposed surfaces of slabs made from six kinds of concrete (with normal moisture content) are shown in Figs 13.14 and 13.15 (Abrams and Gustaferro 1968; Gustaferro *et al.* 1971a). Except for very thin slabs, the thickness of the slab did not seem to affect noticeably the temperature distribution. Apparently, in fire test simulations slabs can usually be modeled as semi-infinite solids.

Figures 13.16(a), 13.16(b), and 13.16(c) display, at 1, 2, and 3 h into the fire test, the temperatures along the centerline of concrete beams (with widths up to 250 mm) at various distances from the bottom (ACI Committee 216 1989). The test specimens were made from normal-weight, carbonate-aggregate concrete, and contained normal amounts of moisture. As shown in the figures, for tapered beams the beam width, b, is to be interpreted as the width at a distance y from the bottom. Figures 13.16(d), 13.16(e), and 13.16(f) present similar information for lightweight concrete beams.[18]

Should it be necessary to know the temperatures at locations other than along the centerline, isotherms can be generated from the data given in Figs 13.14, 13.15, and 13.16, making use of the fact that solutions to certain two-dimensional problems can be written down in the form of products of T-groups representing solutions to one-dimensional problems (as discussed in Section 13.2). The calculation procedure is illustrated in worked examples in Section 13.6.2.

The temperature distribution in beams may also be assessed by means of an empirical technique described in a CRSI manual (Gustafson 1980).

Information on temperature history of concrete slabs and beams is available from a number of publications, e.g. from design manuals by The Institution of Structural Engineers (Bobrowski *et al.* 1978), by the CEB (Comité Euro-International du Béton 1982), and by the CRSI (Gustafson 1980), from a book by Kordina and Meyer-Ottens (1981), and from papers by Abrams (1979), Lin and Abrams (1983), and Bresler and Iding (1983).

---

18 Figures 13.16(a) to 13.16(f) were generated from the results of tests conducted at the Underwriters' Laboratories (Northbrook, IL) and at the Portland Cement Association (Skokie, IL). Temperature distributions obtained by other laboratories may be somewhat different, due to differences in the construction of the test furnace and the mode of heating.

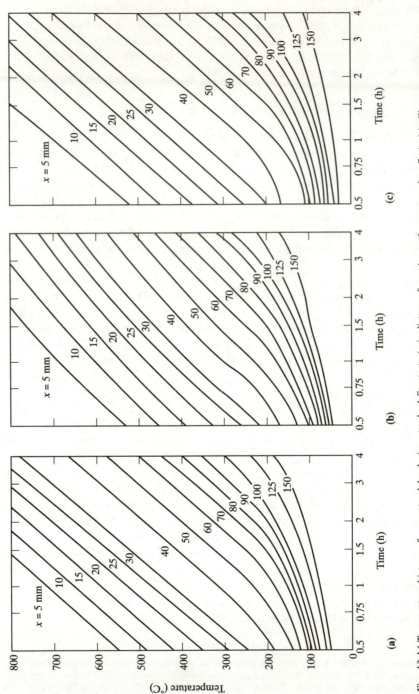

Fig. 13.14 Temperature history of concrete slabs during standard fire tests (x is distance from the surface exposed to fire): (a) siliceous-aggregate concrete; (b) carbonate-aggregate concrete; (c) lightweight concrete with 60 per cent of aggregate fines replaced by sand (Abrams and Gustaferro 1968)

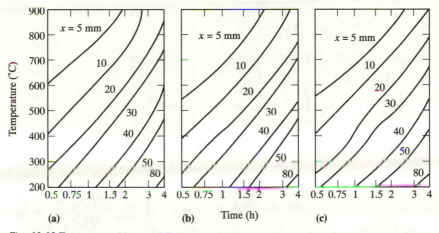

Fig. 13.15 Temperature history of lightweight insulating concrete slabs during standard fire tests (x is distance from the surface exposed to fire): (a) density 320–480 kg m^{-3}; (b) density 800–960 kg m^{-3}; (c) density 1120–1280 kg m^{-3} (Gustaferro *et al.* 1971a)

13.6   Worked Examples

13.6.1   Calculation of Fire Resistance

Calculate the thermal fire resistance of a brick wall specimen of 0.1 m thickness ($l = 0.1$ m).

Input information
Temperature of the ambient atmosphere and initial temperature of the specimen:

$$T_a = 20 °C$$

Temperature of the unexposed surface at the time of failure:

$$T_f = 20 + 140 = 160 °C$$

Thermal properties of brick (Table 10.2):

$$k = 1.0 \text{ W m}^{-1}\text{K}^{-1}; \quad \kappa = 0.529 \times 10^{-6} \text{ m}^2 \text{s}^{-1}$$

Coefficient of heat transfer at the unexposed surface:[19]

$$h_U = 14 \text{ W m}^{-1}\text{K}^{-1}$$

Nusselt number:

$$\frac{h_U l}{k} = \frac{14 \times 0.1}{1.0} = 1.4$$

---

19 Select $h_U = 14$ W m^{-2} K^{-1} for vertical surfaces, and $h_U = 15$ W m^{-2} K^{-1} for horizontal surfaces facing upward. These values have been obtained by means of Eqs 13.14 and 13.15, respectively, using $\epsilon_U \simeq 0.9$ and a nominal value of 425 K for T_U, which is a few degrees below the temperature of thermal failure.

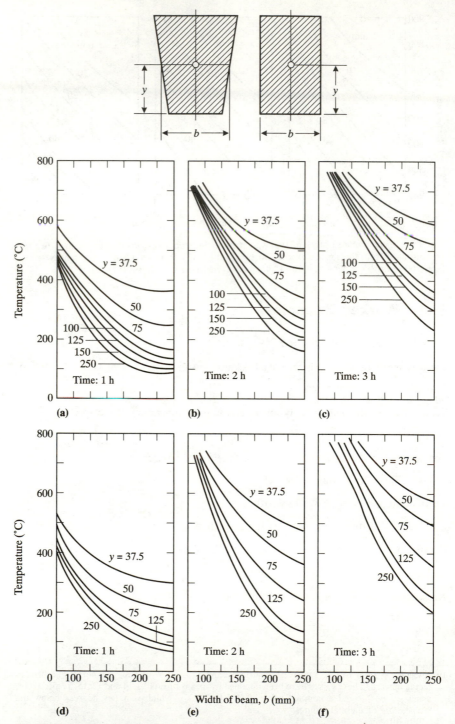

Fig. 13.16 Temperature distribution along the center line of rectangular and tapered concrete beams, at 1, 2, and 3 h into the fire test (for tapered beams the beam width, b, is to be interpreted as the width at a distance y from the bottom): (a), (b), and (c) normal-weight, carbonate-aggregate concrete; (d), (e), (f) lightweight concrete (ACI Committee 1989)

First approximation
Assume that the thermal fire resistance of the wall, τ (s), is 3600 s (1 h). The effective value for the temperature of the furnace gases, T_F [the value of $(T_F)_{av}$ from Fig. 3.6 for $\tau = 1$ h], is 785 °C.

Steady-state temperature of the unexposed surface at the assumed time of failure (Eq. 13.31 with $h_U l/k = 1.4$ and $x/l = 1$):

$$T_{U\infty} = 20 + (785 - 20)\frac{1}{1+1.4} = 339 \text{ °C}$$

The T-group for the unexposed surface at the assumed time of failure:

$$\frac{T_{U\infty} - T_f}{T_{U\infty} - T_a} = \frac{339 - 160}{330 - 20} = 0.561$$

Fourier number for the assumed time of failure (value of $\kappa t/l^2$ from Fig. 13.5 for $(T_{U\infty} - T_f)/(T_{U\infty} - T_a) = 0.561$ and $h_U l/k = 1.4$):

$$\frac{\kappa t}{l^2} = 0.218$$

Hence

$$t = \frac{0.218 \times 0.1^2}{0.529 \times 10^{-6}} = 4120 \text{ s} \ (1.14 \text{ h})$$

This is much larger than the assumed value: 3600 s.

Second approximation
Assume that the fire resistance of the wall is 4050 s (1.13 h). For this value of τ, $T_F = 800$ °C (Fig. 3.6), and the steady-state temperature at the unexposed surface is (Eq. 13.31):

$$T_{U\infty} = 20 + (800 - 20)\frac{1}{1+1.4} = 345 \text{ °C}$$

Second approximation for the T-group:

$$\frac{T_{U\infty} - T_f}{T_{U\infty} - T_a} = \frac{345 - 160}{345 - 20} = 0.569$$

Fourier number (Fig. 13.5):

$$\frac{\kappa t}{l^2} = 0.214$$

Hence

$$t = \frac{0.214 \times 0.1^2}{0.529 \times 10^{-6}} = 4045 \text{ s} \ (1.12 \text{ h})$$

The second approximation is acceptable. Consequently, the thermal fire resistance of the brick wall, τ, is 1.12 h.

Note that this method of calculating fire resistance is useful mainly in the case of monolithic specimens made from materials that are unable to hold substantial amounts of moisture under normal atmospheric conditions.

13.6.2 Calculation of Temperature Distribution in Concrete Beams

Two methods for developing information on the temperature distribution in concrete beams will be described here, making use of analytical solutions and experimental data presented in Figs 13.14 to 13.16.

Determine the temperature at Points 1, 2, 3, and 4 in the beam shown in Fig. 13.17, at 2 h into a standard fire test. The beam is made from normal-weight, carbonate-aggregate concrete.

Input Information

The (idealized) furnace temperature, T_F, for a 2 h test (value of $(T_F)_{av}$ at $\tau = 2$ h from Fig. 3.6) is 890 °C.

Initial temperature:

$$T_a = 20 \text{ °C}.$$

The coordinates of the points are:

Point 1: $x_1 = 0.04$, $y_1 = 0.04$;   Point 2: $x_2 = 0.08$, $y_2 = 0.04$;
Point 3: $x_3 = 0.12$, $y_3 = 0.04$;   Point 4: $x_4 = 0.12$, $y_4 = 0.16$.

Method 1

The temperatures at Points 1, 2, and 3 can be regarded as temperatures in a rectangular corner, whose surfaces are kept (under idealized test conditions) at a constant temperature (T_F). According to Eqs 13.24 and 13.41, the solution can be written down as the product of the appropriate T-groups.

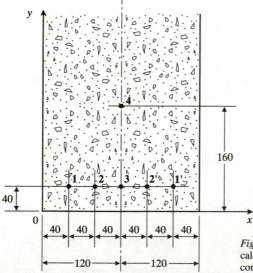

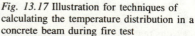

Fig. 13.17 Illustration for techniques of calculating the temperature distribution in a concrete beam during fire test

$$\frac{T_F - T}{T_F - T_a} = \frac{T_F - T_x}{T_F - T_a} \times \frac{T_F - T_y}{T_F - T_a}$$

where T is the temperature of the point in question, T_F is the (idealized) temperature of the test furnace, and T_x and T_y are temperatures that would arise at the given time in semi-infinite solids at distances of x and y, respectively, from the surfaces which are kept at T_F.

Information on the values of T_x and T_y can be obtained from Fig. 13.14(b) which, though developed for slabs, is approximately applicable to semi-infinite solids also. Thus, from Fig. 13.14(b), at 2 h

$$T_{x=0.04} = T_{y=0.04} = 450\ °C; \qquad T_{x=0.08} = 225\ °C; \qquad T_{x0.12} = 135\ °C;$$
$$T_{y=0.16} = 105\ °C$$

and therefore

$$\frac{T_F - T_{x=0.04}}{T_F - T_a} = \frac{T_F - T_{y=0.04}}{T_F - T_a} = \frac{890 - 450}{890 - 20} = 0.506$$

$$\frac{T_F - T_{x=0.08}}{T_F - T_a} = \frac{890 - 225}{890 - 20} = 0.764$$

$$\frac{T_F - T_{x=0.12}}{T_F - T_a} = \frac{890 - 135}{890 - 20} = 0.868$$

$$\frac{T_F - T_{y=0.16}}{T_F - T_a} = \frac{890 - 105}{890 - 20} = 0.902$$

Consequently

$$\frac{T_F - T_1}{T_F - T_a} = 0.506^2 = 0.256 \qquad \therefore\ T_1 = 667\ °C$$

$$\frac{T_F - T_2}{T_F - T_a} = 0.764 \times 0.506 = 0.387 \qquad \therefore\ T_2 = 553\ °C$$

$$\frac{T_F - T_3}{T_F - T_a} = 0.868 \times 0.506 = 0.439 \qquad \therefore\ T_3 = 508\ °C$$

$$\frac{T_F - T_4}{T_F - T_a} = 0.868 \times 0.902 = 0.783 \qquad \therefore\ T_4 = 209\ °C$$

Note that the values of T_3 (508 °C) and T_4 (209 °C) compare favorably with those read directly from Fig. 13.16(b) for points lying on the center line at 40 and 160 mm from the bottom of a beam 240 mm wide (490 °C and 225 °C, respectively).

Method 2
The temperature read from Fig. 13.16(b) for Point 3, T_3, is 490 °C. This temperature may be regarded as if it were developed using Eq. 13.45, according to which

$$\frac{T_F - T_3}{T_F - T_a} = \left(\frac{T_F - T_{x=0}}{T_F - T_a} \right)_{13.3(c)} \times \left(\frac{T_F - T_{y=0.04}}{T_F - T_a} \right)_{13.3(a)} \qquad [A]$$

where the 13.3(c) and 13.3(a) subscripts refer to the shapes shown in Figs 13.3(c) and 13.3(a) (i.e. slab heated on both sides and semi-infinite solid), respectively.

The value of $T_{y=0.04}$ is 450 °C [from Fig. 13.14(b); see previous example]. Thus, from Eq. A,

$$\left(\frac{T_F - T_{x=0}}{T_F - T_a} \right)_{13.3(c)} = \left(\frac{890 - 490}{890 - 20} \right) \div \left(\frac{890 - 450}{890 - 20} \right) = 0.909$$

As discussed in connection with Eqs 13.33 to 13.35, the case of a slab of thickness $2l$, heated on both sides, is equivalent to a slab of thickness l, heated on one side and insulated on the other side. Consequently, the T-group versus $\kappa t/l^2$ plot in Fig. 13.5 for the $h_U l/k = 0$ case is applicable to points on the center line of a ($2l$ thick) slab heated on both sides. Thus, from the figure,[20] for $(T_F - T_{x=0})/(T_F - T_a) = 0.909$,

$$\frac{\kappa t}{l^2} = 0.125$$

and therefore (remembering that $2l = 0.24$ m and $t = 7200$ s)

$$\kappa = 0.125 \times 0.12^2/7200 = 0.25 \times 10^{-6} \, \text{m}^2 \, \text{s}^{-1}$$

This is a 'notional value' of the thermal diffusivity;[21] it is valid only for the $t = 7200$ s time level.

Once the value of the Fourier number of the material is known, the temperature distribution in the beam (at $t = 7200$ s) can be developed by the application of Eqs 13.24, 13.36, 13.37, and 13.45.

Since the input information used has been obtained from two sources of experimental data, this method is probably less accurate than the previous one.

Nomenclature

b	width of concrete beam, m (mm)
$c, \bar{c}, \bar{\bar{c}}$	specific heat, J kg $^{-1}$K^{-1}
C	heat capacity, defined by Eq. 13.64 or 13.88, J m^{-2} K^{-1} or J m^{-3} K^{-1}
d	dimension, $= \sqrt{2}\Delta\xi$, m
h	coefficient of heat transfer, W m^{-2} K^{-1}
$k, \bar{k}, \bar{\bar{k}}$	thermal conductivity, W m^{-2} K^{-1}
l	characteristic dimension; thickness, m
L	coefficient in Lagrange's interpolation formula, dimensionless
M	mass of steel core, kg
n	direction cosine, dimensionless
q	heat flux, W m^{-2}
Q	internal heat generation, W m^{-3}
Δr	distance between nodal points, m

---

20 The $x/l = 1$ condition for a slab (of l thickness) heated on the side $x = 0$ and insulated on the other side corresponds to the $x/l = 0$ condition for a slab (of $2l$ thickness) heated on both sides ($x = -l$ and $x = l$). See Figs 13.3(b) and 13.3(c).

21 The relatively low value of κ is attributable to the high value of the apparent specific heat, resulting from the absorption of heat in the desorption of moisture. According to Table 10.2, the true value of thermal diffusivity is about 0.6×10^{-6} m^2 s^{-1} for normal-weight concretes.

S	boundary surface; surface area, m^2
t	time, s
Δt	time step, s
T	temperature, K
x, y, z	coordinates, m
$\Delta x, \Delta \bar{x}, \Delta \bar{\bar{x}}$	distance between the nodal points, m
Δy	distance between the nodal points, m
α	roots of Eq. 13.32, dimensionless
ϵ	emissivity, dimensionless
ζ	dummy variable
κ	thermal diffusivity, m^2 s^{-1}
$\Delta \xi$	distance between nodal points, m
$\rho, \bar{\rho}, \bar{\bar{\rho}}$	density, kg m^{-3}
σ	Stefan–Boltzmann constant $= 5.67 \times 10^{-8}$ W m^{-2} K^{-4}
τ	fire resistance time; time of fire exposure, h (min)
Φ	function
ψ	material property, various dimensions

Subscripts, superscripts

$1; 2; (n-1); n; (n+1); r; s; (N-1); N$; etc.	at the 1st; 2nd; $(n-1)$th; nth; $(n+1)$th; rth; sth; $(N-1)$th; Nth; etc. nodal point (in one dimension)
$1,2; (m-1),n; m,n; (m+1),n$; etc.	at the $P_{1,2}$; $P_{(m-1),n}$; $P_{m,n}$; $P_{(m+1),n}$; etc. nodal point (in two dimensions)
$(n-1){:}n; n{:}(n+1)$; etc.	along the path from the $(n-1)$th to the nth nodal point; along the path from the nth to the $(n+1)$th nodal point; etc. (in one dimension)
$(m-1),n{:}m,n; (m+1),(n-1){:}m,n$; etc.	along the path between nodal points $P_{(m-1),n}$ and $P_{m,n}$; along the path between nodal points $P_{(m+1),(n-1)}$ and $P_{m,n}$; etc. (in two dimensions)
m, n	$= x\sqrt{2}/\Delta \xi$; $= y\sqrt{2}/\Delta \xi$, respectively (in two dimensions)
p, q, r, s	shorthand for $(m-1)$, $(m+1)$, $(n-1)$, $(n+1)$, respectively
$0, j, (j+1)$	at the time levels $t = 0$, $t = j\Delta t$, $t = (j+1)\Delta t$, respectively
a	of the ambient air
av	average
c	by convection
e	of the environment
E	of/at the fire-exposed surface
f	of thermal failure
F	of the furnace gases
G, H, GH	of surface G; of surface H; between surfaces G and H; respectively
nr	of a nonreflecting medium
r	by radiation
st	of the steel component
S	of the surface
U	of/at the unexposed surface
x, y, z	along the coordinate axes x, y, z, respectively; at a distance of x or y from the fire-exposed surfaces
∞	steady-state value (as $t \to \infty$)

14 Structural Response of Concrete Elements to Fire

To be able to assess the performance of a reinforced or prestressed concrete element in fire, the designer must have at hand

(a) the preliminary design of the element for gravity and (possibly) lateral loads at normal temperatures,[1] and
(b) the temperature history of the element during fire exposure, at least in some critical regions.

The assessment consists of ascertaining if the rise of temperature in the critical regions is liable to cause the weakening of the element's resistance to deformation to such an extent as to result in structural failure in a time less than the desired fire resistance. If, with the preliminary design, the targeted fire resistance does not seem to be attainable, a number of options are available to the designer. Among them, increasing the cover thickness over the reinforcing or prestressing steel components will probably prove most expedient.

When exposed to fire, building elements are often subject to constraints imposed on them by the rest of the building. These constraints are usually beneficial for the elements' structural performance. The exact magnitude of their beneficial effect is, however, difficult to determine.

The mathematical models employed in the analyses of structural fire resistance may be quite detailed, encompassing techniques for determining the complete stress—deformation history of the element, or may be simplified, utilizing uncomplicated static models and concentrating mainly on the maximum stresses, strains, and deflections at the time of structural failure. Detailed or simplified, all models rely on a number of assumptions. The commonly used assumptions are as follows:

---

[1] It is usually required that the element be designed in accordance with an accepted code of practice, e.g. the ACI Building Code (ACI Committee 318 1989) in the United States, the CSA CAN3-A23.3-M84 standard (Canadian Standards Association 1984) in Canada, the Code of Practice BS 8110 (British Standards Institution 1985) in the United Kingdom, the CEB-FIP Model Code for Concrete Structures (Comité Euro-International du Béton 1978a, b), or the FIP Recommendations (Fédération International de la Récontrainte 1984).

- The Bernoulli—Navier hypothesis applies, according to which sections remain plane during deformation, and therefore the total strains are constrained to a linear or planar distribution.
- The concrete is not capable of carrying tensile stresses.
- The stresses are uniaxial.
- Slip between concrete and steel is not possible.

14.1   Flexural Elements

14.1.1   General

Flexural elements, such as slabs, beams, girders, and joists (often referred to simply as beams), represent the most important group of those building elements that have to be dealt with in fire safety design. In engineering practice, satisfaction of regulatory control often consists of selecting building elements from available listings, or using information on minimum dimensions tabulated in publications endorsed by the building code (see, for example, Tables 12.3, 12.4, and 12.5). Clearly, this approach does not promote optimum design.

The sophisticated mathematical models developed for studying the stress—deformation history of flexural elements in fire differ mainly in respect to

(a) the formulation of the mechanical properties of concrete and steel, and
(b) the formulation of strains.

All of them allow for the temperature-dependence and nonlinearity of the (quasi-instantaneous) stress—strain relations for concrete and steel. As to the formulation of strains (see Eq. 6.47), in some models the strains are assumed to develop instantaneously in response to stress or change of temperature; in some others time-dependent creep is also considered, without or with the inclusion of a strain component called 'transient strain' (Anderberg and Thelandersson 1976; Schneider 1976).

The stress—deformation history of the element is followed up by means of iterative procedures, performed in order to satisfy at each time step the equations of equilibrium between the external loads and internal stresses (Nizamuddin and Bresler 1979; Anderberg and Forsén 1982; Bresler and Iding 1983; Rudolph *et al.* 1986). In addition to providing information on the stress—deformation history of the element, these studies may also offer an insight into such aspects of the process as the redistribution of internal forces and moments, the progress of cracking or crushing of the concrete, the influence of varying degrees of restraint, and the influence of diverse fire environments.

If only the prediction of the time of structural failure is of interest, such detailed investigations are probably not justified. The fire resistance of flexural elements can be calculated using simple models. To some extent, these models depend on the practice followed in structural engineering design. However, since all design guides on structural fire safety rely on a fair amount of information developed by the researchers of the Portland Cement Association (Skokie, IL), the differences between them are not essential.

Flexural elements adequately reinforced for normal temperature service rarely, if ever, fail by shear in fire tests or real-world fires (Gustaferro 1986; Lin *et al.* 1988). Consequently, the rather remote likelihood of shear failure is commonly ignored, and only failure by flexure is taken into consideration.

At an advanced stage of fire exposure, the concrete mass in all flexural elements is badly cracked by thermal stresses (to be discussed). Consequently, though reinforced and prestressed concrete flexural elements are designed by different principles, there is no fundamental difference between them as far as assessing their performance in fire is concerned. In the following discussions, the word 'steel' (without adjective) may mean either reinforcing steel bars or prestressing steel tendons.

In essence, engineering design for fire resistance consists of calculating the maximum *acting moments* and *resisting moments* for the time of structural failure. The acting moments include

(1) the *service moments*, i.e. moments called forth by the *service load* (dead load plus live load), and possibly
(2) the *restraining moments*, i.e. moments generated by axial forces that arise on account of the element's thermal expansion.

The resisting moments are those associated with internal stresses. The ultimate resisting moments (those at incipient structural failure) are called *moment capacities*. Naturally, the moment capacities at normal temperatures differ from those under fire exposure conditions.

Since the *ultimate load* (the maximum load that can be placed on a flexural element before its failure) on which structural engineering design is based includes a substantial margin of safety, whereas in the assessment of fire resistance only the service load is considered, the service moment at the start of a fire resistance test rarely takes up more than 65 per cent of the element's moment capacity. The ratio of the service load to the ultimate load is referred to as the *load intensity*. As the design is based on the premise that at the ultimate the steel is stressed to its elastic limit, the load intensity can also be described as the ratio of the service stress (i.e. the maximum stress associated with the service load) to the yield strength. Thus

$$\mathcal{L} = \frac{w}{w_u} = \frac{\sigma_s^o}{(\sigma_s)_y^o} \qquad [14.1]$$

where $\mathcal{L}$ = load intensity, dimensionless;
  w = service load (dead load plus live load) per unit length of the flexural element, $\mathrm{N\,m^{-1}}$;
  w_u = ultimate load per unit length, $\mathrm{N\,m^{-1}}$;
  σ_s = service stress (maximum stress in the steel generated by the service load), Pa;
  $(\sigma_s)_y$ = yield strength of steel, Pa.

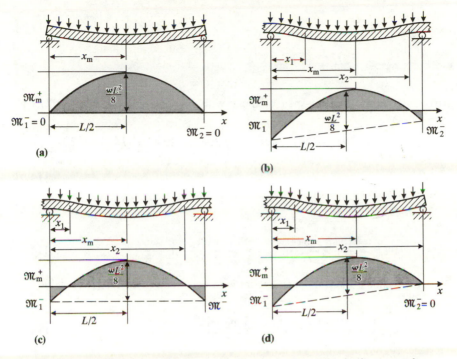

Fig. 14.1 Moments generated in flexural elements by gravity forces: (a) simply supported element; (b) continuous, interior-span element with unequal end moments; (c) continuous, interior-span element with equal end moments; (d) continuous, end-span element

and the superscripts o indicate normal temperature conditions.

The distribution of the service moment along a flexural element depends on the continuity of the element and on the nature of the supports. Four elements, subjected to uniformly distributed gravity loads, are depicted in Fig. 14.1. Each one represents a different set of conditions. The elements depicted are respectively,

(a) simply supported element,
(b) continuous, interior-span element with unequal end moments,
(c) continuous, interior-span element with equal end moments,
(d) continuous, end-span element.

Shown in the figure are, at normal temperatures, the deflected shape of the elements (on an enlarged scale), the distribution of the service moments, the location of the points of inflection, and the location of the maximum positive service moments.

The following equations apply.

(a) Simply supported element ($\mathfrak{M}_1^- = \mathfrak{M}_2^- = 0$):

$$x_1 = 0 \qquad x_2 = L \qquad\qquad\qquad [14.2]$$

$$x_m = \frac{L}{2} \tag{14.3}$$

$$\mathfrak{M}_m^+ = \frac{wL^2}{8} \tag{14.4}$$

(b) Continuous, interior-span element with unequal end moments ($\mathfrak{M}_1^- \neq \mathfrak{M}_2^-$):

$$x_{1,2} = \frac{2\mathfrak{M}_1^- - 2\mathfrak{M}_2^- + wL^2}{2wL} \mp \left[\left(\frac{2\mathfrak{M}_1^- - 2\mathfrak{M}_2^- + wL^2}{2wL}\right)^2 - \frac{2\mathfrak{M}_1^-}{w}\right]^{1/2} \tag{14.5}$$

$$x_m = \frac{L}{2} + \frac{\mathfrak{M}_1^- - \mathfrak{M}_2^-}{wL} \tag{14.6}$$

$$\mathfrak{M}_m^+ = \frac{(\mathfrak{M}_1^- - \mathfrak{M}_2^-)^2}{2wL^2} - \frac{\mathfrak{M}_1^- + \mathfrak{M}_2^-}{2} + \frac{wL^2}{8} \tag{14.7}$$

(c) Continuous, interior-span element with equal end moments ($\mathfrak{M}_1^- = \mathfrak{M}_2^- = \mathfrak{M}^-$):

$$x_{1,2} = \frac{L}{2} \mp \left[\left(\frac{L}{2}\right)^2 - \frac{2\mathfrak{M}^-}{w}\right]^{1/2} \tag{14.8}$$

$$x_m = \frac{L}{2} \tag{14.9}$$

$$\mathfrak{M}_m^+ = -\mathfrak{M}^- + \frac{wL^2}{8} \tag{14.10}$$

(d) Continuous, end-span element ($\mathfrak{M}_1^- \neq 0$, $\mathfrak{M}_2^- = 0$):

$$x_1 = \frac{2\mathfrak{M}_1^-}{wL} \qquad x_2 = L \tag{14.11}$$

$$x_m = \frac{L}{2} + \frac{\mathfrak{M}_1^-}{wL} \tag{14.12}$$

$$\mathfrak{M}_m^+ = \frac{(\mathfrak{M}_1^-)^2}{2wL^2} - \frac{\mathfrak{M}_1^-}{2} + \frac{wL^2}{8} \tag{14.13}$$

In Eqs 14.2 to 14.13 and in the figure,

x	= abscissa along the flexural element, m;
$x_1, x_2, x_{1,2}$	= abscissas of the points of inflection, m;
x_m	= abscissa of the maximum positive moment, m;
w	= uniformly distributed load (dead and live loads) on the flexural element, $N\,m^{-1}$;
L	= span length (distance between the supports), m;

$\mathfrak{M}_1^-$, $\mathfrak{M}_2^-$, $\mathfrak{M}^-$ = negative service moments at the left end, right end, both ends of the element,[2] respectively, N m;

$\mathfrak{M}_m^+$ = maximum positive service moment, N m.

For any of the four conditions depicted in Fig. 14.1, at the midspan, i.e. at $x = L/2$,

$$\mathfrak{M}_{ms}^+ + \frac{\mathfrak{M}_1^- + \mathfrak{M}_2^-}{2} = \frac{wL^2}{8} \qquad [14.14]$$

where $\mathfrak{M}_{ms}^+$ (N m) is the positive service moment at the midspan.

Figures 14.2(a) and 14.2(b) show a commonly accepted model for depicting the internal forces in some critical sections of reinforced concrete flexural elements

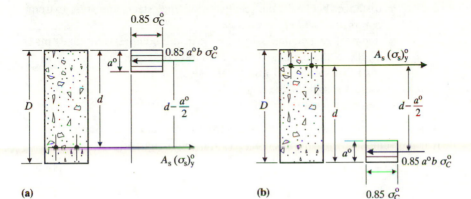

(a)   (b)

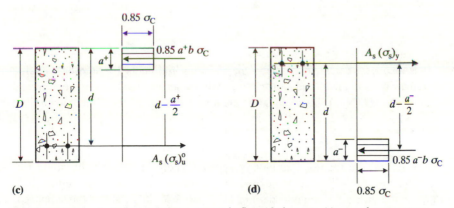

(c)   (d)

Fig. 14.2 Models of ultimate resisting moments in flexural elements: (a) normal temperatures, acting moment positive; (b) normal temperatures, acting moment negative; (c) fire exposure conditions, acting moment positive; (d) fire exposure conditions, acting moment negative

2   The negative service moments at the ends of continuous flexural elements bridging two, three, four, and five spans are listed in various handbooks, e.g. in the *Handbook of Engineering Fundamentals* (Eshbach 1952).

at the time of structural failure. The tensile force in the steel is opposed by an equal compressive force arising in a zone adjacent to the extreme top fiber. According to the model (at normal temperatures) the ultimate resisting moments (moment capacities, pertaining to some extreme values in the moment distribution) under ultimate load, M_u^o (N m), are given by[3]

$$M_u^o = A_s (\sigma_s)_y^o \left(d - \frac{a^o}{2} \right) \qquad [14.15]$$

where, on account of the equality of tensile and compressive forces,

$$a^o = \frac{A_s (\sigma_s)_y^o}{0.85 \sigma_C^o b} \qquad [14.16]$$

and d = distance between the centroid of the steel and the extreme compression fiber, m;

a = depth of rectangular stress block in the compression zone, m;

b = width of the element in the compression zone,[4] m;

A_s = total cross-sectional area of the steel, m^2;

$(\sigma_s)_y$ = yield strength of the steel, Pa;

σ_C = compressive strength of the concrete, Pa;

and, again, the superscripts o indicate normal temperature conditions.

It is convenient to rewrite Eq. 14.15 in the following form:

$$M_u^o = A_s (\sigma_s)_y^o \xi^o d \qquad [14.17]$$

where ξ^o (dimensionless) is defined as

$$\xi^o = \frac{d - a^o/2}{d} \qquad [14.18]$$

In general, $0.9 < \xi^o < 1$.

It is also convenient to describe the resisting moments (pertaining to some extreme values in the moment distribution) under service load (i.e. in nonultimate condition), M^o (N m), by an equation analogous to Eq. 14.17. Since, according to Eq. 14.1, $\sigma_s^o = \mathcal{L}(\sigma_s)_y^o$,

$$M^o = \mathcal{L} A_s (\sigma_s)_y^o \xi^o d \qquad [14.19]$$

where ξ^o may be regarded as having approximately the same value as in Eq. 14.17. Thus, $M^o/M_u^o = \mathcal{L}$ at normal temperatures. M^o may mean either $\mathfrak{M}_m^+$ (at or in the vicinity of the midspan) or $\mathfrak{M}^-$ (or $\mathfrak{M}_1^-$, $\mathfrak{M}_2^-$, above the supports).

---

3   Note that the symbols used by structural engineers for stresses and strengths are different from those used in this book and, in general, by mechanical engineers. In structural engineering, the following symbols are used: f_c = stress in concrete, f_c' = specified compressive strength of concrete, f_s = stress in steel, f_y = yield strength of steel, f_u = ultimate strength of steel.

4   In the case of one-way slabs, b means a 1 m wide section of the slab.

It is commonly assumed that, provided $(\sigma_s)_y$ and σ_C are taken to be representative of temperatures prevailing in fires, the model shown in Fig. 14.2(a) or 14.2(b) can also be used for assessing the moment capacities of flexural elements under fire exposure conditions. Unfortunately, that assumption is not necessarily valid.

Equations 14.15 and 14.16 have been based on the observation that structural failure occurs on account of the plastic deformation of reinforcing steel which, owing to a yield plateau in the stress–strain relation, proceeds for an extended period at an approximately constant stress, equal to the yield strength. However, at temperatures above 300 °C the yield plateau does not exist (see Fig. 7.5), and the plastic deformation of any steel (low-carbon or medium-carbon) takes place at steadily increasing stresses. There is some doubt, therefore, about the propriety in fire resistance studies of regarding the yield strength as the descriptor of the ultimate behavior of steel. It will be shown through a worked example (Section 14.4) that, if structural failure occurs when the steel temperature is well over 300 °C, more consistent results are obtained if the yield strength is replaced by the ultimate strength.

Since the principal steel components (reinforcing or prestressing), which are used to balance the positive service moment, are placed close to the fire-exposed surfaces of the flexural element, they do attain high temperatures. Hence, the following equations are recommended for the calculation of the positive moment capacities under fire exposure conditions [Fig. 14.2(c)]:

$$M_u^+ = A_s(\sigma_s)_u\left(d - \frac{a^+}{2}\right) \qquad\qquad [14.20]$$

and

$$a^+ = \frac{A_s(\sigma_s)_u}{0.85\sigma_C b} \qquad\qquad [14.21]$$

where $(\sigma_s)_u$ (Pa) stands for the ultimate strength of steel. The + superscripts indicate that the marked variables relate to the positive moment capacity, and the absence of o superscripts indicate that $(\sigma_s)_u$ and σ_C are functions of the temperatures that prevail in the steel and in the compression zone of concrete, respectively.

The reinforcement used for balancing negative service moments is embedded near the top of the element, where the temperature is likely to remain below 200 °C during a fire test or fire. Consequently, the yield strength is the appropriate property for characterizing the plastic behavior of steel near the point of structural failure. The following equations are applicable [Fig. 14.2(d)]:

$$M_u^- = A_s(\sigma_s)_y\left(d - \frac{a^-}{2}\right) \qquad\qquad [14.22]$$

and

$$a^- = \frac{A_s(\sigma_s)_y}{0.85\sigma_C b} \qquad\qquad [14.23]$$

where the $-$ superscripts indicate that the marked variables relate to the negative moment capacity and, again, the absence of o superscripts indicates that $(\sigma_s)_y$ and σ_C are functions of the prevailing temperatures.

Equations 14.20 and 14.22 may also be written in forms similar to Eq. 14.17.

$$M_u^+ = A_s(\sigma_s)_u \xi^+ d \qquad\qquad\qquad [14.24]$$

$$M_u^- = A_s(\sigma_s)_y \xi^- d \qquad\qquad\qquad [14.25]$$

where ξ^+ and ξ^- are defined analogously to ξ^o (Eq. 14.18), and their values also lie between 0.9 and 1.

14.1.2 Simply Supported, Unrestrained Flexural Elements

Since the fire resistance of simply supported flexural elements [the kinds shown in Fig. 14.1(a)] can be calculated with a fair degree of accuracy, it is convenient to start discussions on the design procedure with such elements.

The possible effect of thermal stresses on the load-induced stresses is a question that merits consideration. The magnitude and distribution of thermal stresses will be examined in the next section. Here it may suffice to say that, thanks to the ability of cement paste to absorb compressive stresses by plastic deformation, shrinkage, and crushing, and to react to tensile load by cracking, thermal stresses high enough to affect significantly the load-induced stresses are not expected to prevail at advanced stages of fire exposure.

Figure 14.3(a) shows the service moment (equal to the resisting moment) and the moment capacity at the beginning of a fire test conducted on a flexural element. The service moment is positive; its maximum is $\omega L^2/8$ at midspan. Since the bottom steel bars or tendons run full length, the moment capacity is constant along the length of the flexural element; it is considerably larger than the maximum positive service moment.

Contemplating the temperature-dependent material properties in Eqs 14.20 and

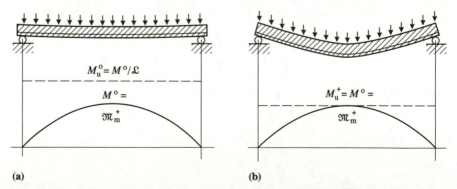

(a)   (b)

Fig. 14.3 Simply supported, unrestrained flexural element: service moment (——) and moment capacity (— — —); (a) at the beginning of a fire test; (b) at the point of structural failure

14.21, one will find that the strength of concrete in the compression zone (adjacent to the unexposed surface) will be barely affected by exposure to the fire until after the temperature of the unexposed surface has exceeded the level that normally represents thermal failure [see Fig. 6.15(d) and Eqs 6.45 and 6.46].[5] On the other hand, the strength of the principal steel bars or tendons, which are located close to the underside of the flexural element, will decline steadily as the temperature rises. Although the ratio $(\sigma_s)_u/\sigma_C$, which, according to Eq. 14.21, influences the depth of the compression zone, decreases as the test progresses, the variation of $(d - a^+/2)$ is usually insignificant, so that it is permissible to regard the lever arm as constant and approximately equal to $\xi^o d$.

The maximum positive service moment remains constant throughout the fire exposure. In contrast, the moment capacity will decrease as the ultimate strength of the steel declines with rising temperatures. The element will collapse when the temperature of the steel attains a level at which its moment capacity (M_u^+ from Eq. 14.24) is reduced to the level of the maximum service moment [see Fig. 14.3(b)]. Thus, on account of the equality of acting moment and resisting moment, the condition of collapse is $\mathfrak{M}_m^+ = M^o = M_u^+$ or, according to Eqs 14.19 and 14.24 (with $\xi^+ \simeq \xi^o$), $(\sigma_s)_u \simeq \mathcal{L}(\sigma_s)_y^o$.

Apparently, the time of collapse depends largely on the load intensity, yet it also depends on the rate of rise of steel temperature and, in turn, on the thickness of concrete cover.

Ensuring that the concrete cover stays in place is an important aspect of the design for fire resistance. Malhotra (1969), drawing on a series of tests conducted on reinforced and prestressed beams, observed that covers thicker than about 40 mm were liable to spall if supplementary reinforcement, such as welded-wire fabric, was not applied. On the other hand, Gustaferro et al. (1971b) found that the use of wire fabric within the cover was unnecessary, particularly in beams fabricated with carbonate-aggregate concretes. Since the beam specimens were conditioned differently in the two series of tests, it is possible that differences in the permeabilities of the concretes used by the two laboratories were responsible for the conflicting findings.[6]

The steel temperature is usually interpreted as the average temperature of the bars or tendons. That temperature can be calculated as described in Chapter 13,

---

5   As discussed in Chapter 3, the insulating capability of a building element is deemed to have been lost when the average temperature of the unexposed surface of the test specimen reaches a level 140 °C above the initial temperature (i.e. about 160 °C). Although, according to the ISO 834 standard, the loss of insulating capability does not always count as failure, some other standards, e.g. ASTM E 119, still require that the limiting temperature level not be exceeded.

6   Malhotra's (1969) beams were stored under cover for up to three years before being subjected to fire tests. The beams tested by Gustaferro et al. (1971b) were cured in the forms under damp burlap for seven days, then removed from the forms and stored in air maintained at 30 to 40 per cent relative humidity for two years or longer. It seems probable that the concretes used in Malhotra's experiments were more mature (and therefore had lower permeabilities) than those used in the studies of Gustaferro et al. See Chapter 4 on the effect of permeability on the susceptibility of concrete to spalling.

or assessed by means of various shortcut methods presented in several design guides (e.g. Bobrowski 1978; Gustafson 1980; Gustaferro and Martin 1989; ACI Committee 216 1989).

The practice of relating the point of structural failure to the load intensity relies on the validity of the concept of stress-modified critical temperature. That concept, as discussed in Chapter 12, is based on the assumption that the strength of steel is a unique function of the temperature. However, since the rapidly increasing deflection of the element near the point of collapse is caused by creep, using the creep concept would seem more appropriate.

The creep concept may be applied in two different ways. It may be assumed that structural failure will occur when the creep strain, ϵ_t (m m^{-1}), at the midspan becomes 'unacceptably large', larger than, say, 0.1 m m^{-1}. For large deformations, the creep strain can be written (using Eqs 5.25 and 5.21) as follows:[7]

$$\epsilon_t \simeq Z \int_0^t e^{-\Delta H_c/RT} dt \qquad [14.26]$$

where Z = Zener−Hollomon parameter, m m^{-1} h^{-1};
ΔH_c = activation energy of creep, J kmol^{-1};
R = gas constant = 8315 J kmol^{-1} K^{-1};
T = temperature of steel, K;
t = time, h.

The effort of numerically integrating Eq. 14.26 can be saved by specifying the point of failure in terms of a limiting creep rate at the midspan, $(d\epsilon_t/dt)$ (m m^{-1} h^{-1}), rather than in terms of a limiting creep strain. From the differentiated form of Eq. 14.26, the following expression is obtained for the critical temperature, T_{cr} (K), at which structural failure is imminent:

$$T_{cr} \simeq - \frac{\Delta H_c/R}{\ln\left[\dfrac{1}{Z}\left(\dfrac{d\epsilon_t}{dt}\right)_1\right]} \qquad [14.27]$$

A relatively large value, say 1.0 (= 1.7 per cent deformation per minute[8]) may be selected for $(d\epsilon_t/dt)_1$.

The concept of stress-modified critical temperature is compared with the creep concept through a worked example in Section 14.4.

---

7   The time-dependent plastic deformation of steel cannot really be called creep at temperatures lower than about 400 °C. However, one cannot object on logical grounds to extending the validity of the creep formulas to lower temperature domains, since the values they yield below 400 °C are hardly measurable. Clearly, there is no practical difference in saying that the steel does not creep at room temperature, or that its creep is exceedingly small.

8   By substituting various values of $(d\epsilon_t/dt)_1$ into Eq. 14.27, the reader will find that, within reasonable limits, the selection is not critical.

14.1.3 Simply Supported, Restrained Flexural Elements

Restraint against thermal expansion is generally beneficial for the performance of flexural elements in fire. Unfortunately, making reliable predictions of its beneficial effect is not possible. It is possible, however, to identify the conditions under which the effect of restraint is negligible or not beneficial.

A restrained flexural element is depicted schematically in Fig. 14.4. At the stage of imminent collapse, the equilibrium between the acting and resisting moments at the midspan can be written as

$$(d - n)\mathcal{S} + \left(n - \frac{a^+}{2}\right)\mathcal{C} = \frac{wL^2}{8} - r\mathcal{R} \qquad [14.28]$$

and the equilibrium of forces as

$$\mathcal{S} = \mathcal{R} + \mathcal{C} \qquad [14.29]$$

where

$$r = j - (Y + n) \qquad [14.30]$$

and $\mathcal{R}$ = restraining force (thrust), N;
$\mathcal{S}$ = tensile force in the steel, N;
$\mathcal{C}$ = compressive force in the concrete, N;
n = distance from the top of element to the neutral axis, m;
r = distance at the midspan between the neutral axis and the line of action of the restraining thrust, m;
Y = midspan deflection, m;
j = distance (at the supports) between the line of action of thrust and the top of beam, m;

and d and a^+ have the same meanings as before. Clearly, except $wL^2/8$, all terms in Eq. 14.28 are directly or indirectly dependent on the prevailing temperatures.

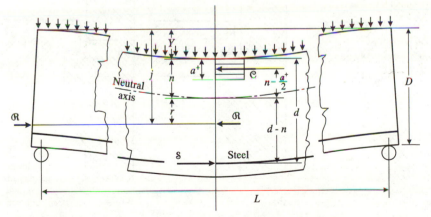

Fig. 14.4 Forces and moments acting on the midspan section of a simply supported, restrained flexural element

At the ultimate, $S = A_s(\sigma_s)_u$. If $\Re$, j, and Y were known near the point of structural failure, Eqs 14.28 to 14.30 could be used for calculating the location of the neutral axis, n, and the ultimate stress in the steel at failure, $(\sigma_s)_u$. With $(\sigma_s)_u$ known, the fire resistance time could be determined in the way discussed in the preceding section.

Although information for assessing $\Re$, j, and Y is available in the form of nomograms, formulas, graphs, and tabulated data (Carlson *et al.* 1966; Issen *et al.* 1970; Abrams *et al.* 1971; ACI Committee 216 1989; Gustafson 1980), the general applicability of that information is questionable. The value of Eqs 14.28−14.30 lies mainly in allowing an insight into the effects of various factors on the structural performance of restrained flexural elements.

The effectiveness of restraint is quantified by the $r\Re$ term on the right-hand side of Eq. 14.28. It depends on both the restraining thrust and the location of the thrust line. As to the latter, it is obvious from Eq. 14.30 that restraint can only be beneficial if $j > (Y + n)$. If $j < (Y + n)$, the thrust will increase the moment acting at the midspan section, and therefore it will have detrimental effect on the element's performance.

Whether the effect is beneficial or detrimental depends above all on the nature of the supports. If the element is supported as shown in Figs 14.5(a) and 14.5(b), it will, while expanding in fire, contact the supporting structure near its base first; hence the line of thrust at the supports will probably lie in the lower two-tenths of the element's depth. Although, according to experimental findings (Carlson *et al.* 1966; Issen *et al.* 1970; Abrams *et al.* 1971), the thrust line will rise as the test progresses, there is a good chance that the restraining moment at the midspan will remain negative long enough to augment the test result. On the other hand, if the flexural element is supported as shown in Fig. 14.5(c), the line of

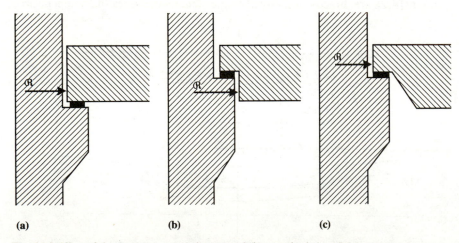

(a)   (b)   (c)

Fig. 14.5 Effect of the supports on restraint: (a) and (b) restraint is beneficial; (c) restraint is detrimental

thrust may lie above the neutral axis right from the beginning of the test and, therefore, the restraint may prove detrimental.

The restraining force depends on the thermal expansion and the deflection of the flexural element. Since, however, both the thermal expansion and the deflection may be influenced by the action of thermal stresses, the possible role of these stresses in the deformation process has to be examined first.

Figure 14.6 illustrates the distribution of strains (thermal strain, ϵ_{th}, m m^{-1}; total strain, ϵ, m m^{-1}) and stresses (σ, Pa) in a nonloaded, nonrestrained reinforced concrete flexural element at two stages of a fire test. The steel is treated as an equivalent concrete section. According to the Bernoulli–Navier hypothesis, the distribution of the total strains is linear.

Soon after the beginning of fire exposure, compressive stresses will arise in the AB and CD zones of the element, and tensile stresses in the BC zone [Fig. 14.6(a)]. The initially very high compressive stresses in the AB zone will be partially relieved by plastic deformation, shrinkage, and crushing, which take place primarily in the cement paste. With the progress of the test, the tensile stresses in the BC zone will grow to the limit allowed by the tensile strength of concrete, then subside on account of cracking. As the cracks widen, the compressive stresses in the AB zone will further be relieved. As Fig. 14.6(b) shows, at an advanced stage of the fire exposure the thermal stresses are low or moderate; it seems unlikely, therefore, that they can play an appreciable role in the deformation process.

The average thermal expansion (referred to original length) of a nonloaded, nonrestrained flexural element is given by

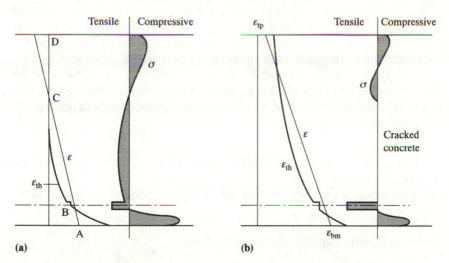

Fig. 14.6 Distribution of strains and stresses in a nonloaded, nonrestrained flexural element: (ϵ_{th} = thermal strain, ϵ = total strain, σ = stress): (a) at an early stage of fire test; (b) at an advanced stage of fire test

$$\frac{\Delta l}{l_o} = \frac{\epsilon_{bm} + \epsilon_{tp}}{2} \qquad\qquad [14.31]$$

where $\Delta l = l - l_o$, (m), and

l = expanded length of the element, m;

l_o = original length (equal to the span length, L), m;

ϵ_{bm} = strain at the bottom of the flexural element, $m\,m^{-1}$;

ϵ_{tp} = strain at the top of the flexural element, $m\,m^{-1}$.

The value of ϵ_{bm} is governed largely by the thermal expansion of steel, whereas ϵ_{tp} may be taken to be approximately equal to the thermal expansion of concrete in the upper compression zone. Their values at the appropriate temperatures can be read from the dilatometric curves of steel (Fig. 7.12) and of the particular concrete from which the element was made (see, for example, the curves in Fig. 6.18[9]).

Perfect restraint, i.e. one that would prevent the rotation of the ends of the element and force the element to remain within the original span, rarely occurs in real-world fires. There is usually a small gap (5–20 mm) at the supports into which the element can expand before making contact with the surrounding structures. In fire resistance studies performed on 25 double-tee reinforced normal-weight and lightweight concrete specimens, 5.45 m long, 1.27 m wide, and 0.38 m deep, Issen *et al.* (1970) found that the maximum thrust exerted by the restraining structure on the flexural element was (roughly) inversely proportional to the allowed expansion, and that, not unexpectedly, the thrust was greater for the specimens made from normal-weight concretes than for those made from lightweight concretes (Fig. 14.7).

With the (somewhat arbitrary) assumption that a flexural element reacts to restraint by yielding at a constant stress amounting to, say, 80 per cent of the compressive strength of concrete, one may obtain a rough estimate of the restraining force. However, such an estimate will probably prove several times higher than the values reported by Issen *et al.* (1970). In addition to the approximate nature of the calculation, the great discrepancy is partly attributable to the shrinkage and crushing of the cement paste. Another source of discrepancy is the deflection of the element.

Deflection lowers the effectiveness of restraint in two ways:

(1) it absorbs part of the element's thermal expansion, and thus softens the restraining thrust, and

(2) it reduces the lever arm of the thrust at the critical midspan section (Fig. 14.4).

The deflection of a flexural element consists of three components:

(1) thermal deflection due to uneven heating,

---

9   As discussed in Chapter 6 (in connection with Fig. 6.18), the thermal expansion of concrete is more or less a reflection of the dilatometric behavior of the principal aggregate.

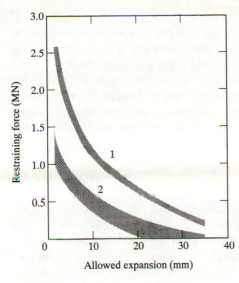

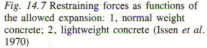

Fig. *14.7* Restraining forces as functions of the allowed expansion: 1, normal weight concrete; 2, lightweight concrete (Issen *et al.* 1970)

(2) deflection due to the restraint itself, and
(3) load-induced deflection due to elastic and plastic strains generated by the service load.

Uneven heating causes the beam to assume the shape of a circular arc. The radius of the arc diminishes fast at the beginning (usually in the first 45 min of the test), then at a decreasing rate. The midspan thermal deflection of a nonloaded, nonrestrained element, Y_{th} (m), can be estimated from the following equation:

$$\frac{Y_{th}}{L} \simeq \frac{1}{8}\frac{L}{D}(\epsilon_{bm} - \epsilon_{tp}) \qquad [14.32]$$

where D (m) is the depth of the element, and the span length, L, is taken to be equal to the original length of the element.

Restraint will cause the flexural element to assume a shape similar to a cosine curve. The midspan deflection due to restraint, Y_{rt} (m), may be written as

$$\frac{Y_{rt}}{L} \simeq 0.4 \sqrt{\alpha \frac{\epsilon_{bm}+\epsilon_{tp}}{2} - \frac{\delta}{L}} \qquad [14.33]$$

where α = empirical factor that takes account of the effect of restraining thrust on the thermal expansion, dimensionless;
δ = allowed expansion before the element makes contact with the surrounding structures, m.

The $(\epsilon_{bm} + \epsilon_{tp})/2$ term under the root sign is the average thermal expansion referred to the original length (Eq. 14.31). The factor α is probably somewhere between 0.8 and 1.

The load-induced deflection is parabolic at first, and small until the steel temperature reaches the 300−350 °C level. At higher temperatures, because of the plastic flow of steel, the rate of deflection accelerates at the highly stressed midspan section, and the deflected shape will approach a shallow V.

The load-induced deflection is very difficult to assess, because the acting moment depends on the instantaneous value of the restraining moment (the $r\mathcal{R}$ product in Eq. 14.28). For lack of experimental information, one may assume that the load-induced deflection at the midspan, Y_{li} (m), at a time reasonably close to the structural failure may be expressed as[10]

$$\frac{Y_{li}}{L} \simeq \frac{1}{4} \frac{L}{D} \epsilon_s^* \qquad\qquad [14.34]$$

where ϵ_s^* (m m^{-1}) is a notional strain for the steel, probably of the order of 0.03.

To take full advantage of the thermal restraint, the designer should aim at keeping the moment produced by restraint at the midspan negative for a period longer than the fire resistance would be in the absence of restraint. As pointed out earlier, the restraining moment is negative if $j > (Y + n)$. Y may be taken roughly to be equal to the sum of[11] Y_{th}, Y_{rt}, and Y_{li}, and n to be equal to about $2a^+$ (a^+ from Eq. 14.21). It does occur occasionally in fire tests that j remains larger than $(Y + n)$ long after the steel temperature has exceeded 800 °C (Gustaferro 1970). If that happens, the element will be likely to fail on account of the loss of its insulating capability, rather than by collapse.

Most fire tests of floor and roof assemblies are performed on rectangular specimens placed in heavy restraining frames. These specimens are subjected to biaxial restraint. From the work of Issen *et al.* (1970), it appears that effect of biaxial restraint is, in general, similar to that of uniaxial restraint.

Using concretes of high compressive strength to generate high restraining forces is not always expedient. Very high restraint may cause spalling and early failure of the element in fire test (Issen *et al.* 1970). According to Lin and Abrams's studies (1983), the performance of flexural elements is not greatly affected by the degree of restraint, except perhaps near the 0 and 100 per cent restraint conditions.

If strong restraint is anticipated, it is imperative to use negative moment reinforcement (to be discussed) at the supports, to counteract the negative moment associated with the restraint. It may also be necessary to examine whether the surrounding structures are capable of carrying the increased load resulting from the restraining forces.

---

10 Clearly, for lack of experimental support, Eq. 14.34 is of no real value to the designer.
11 Near the point of structural failure, the temperature of the steel is typically 600 °C, and that of concrete in the upper regions is typically 200 °C. For these temperatures, $\epsilon_{bm} \simeq 0.0093$ (from Fig. 7.12), and $\epsilon_{tp} \simeq 0.0015$ (from Fig. 6.18). Thus, near the point of failure, $Y_{th}/L \simeq 0.001 L/D$ (Eq. 14.32), and (for the case of perfect restraint, i.e. for $\delta = 0$), $Y_{rt}/L \simeq 0.03$ (Eq. 14.33).

14.1.4 Continuous Flexural Elements

Continuous flexural elements have considerably greater fire resistances than simply supported elements. Their superior performance is due to beneficial changes in the moment distribution that take place in response to fire exposure.

Figure 14.8(a) shows the distribution of service moment over a continuous, interior-span element with equal end moments. The reinforcing bars, embedded along the central sections near the bottom of the element, carry the stresses induced by the positive moment, and those over the supports near the top carry the stresses induced by the negative moments. To be effective, the bars must be cut to lengths that extend beyond the points of inflection [see Fig. 14.1(c) and Eq. 14.8].

The positive or negative moment capacities (at normal temperatures) are given by Eq. 14.15 or 14.17. They are constant along lengths of the respective bars, as shown in Fig. 14.8(a).

When exposed to fire from below, the element will deflect downward, and thereby additional stress will be generated in the top bars over the supports. With the deflection increasing, the stress will soon (usually within a half-hour into the test) reach the elastic limit. Hinges form, and the plastic deformation of the steel will proceed first at constant stress, then at a slowly decreasing stress, determined by the temperature dependence of the yield strength (see Figs 7.7 and 7.9). Since, however, the temperature of the top bars is expected to stay well below 300 °C, the decline of the yield stress rarely amounts to more than 20 per cent.

With the steel yielding, the capacity of the element to withstand the negative moments above the supports is exhausted. On account of the equality of resisting and acting moments, the negative moments, $\mathfrak{M}^-$, will rise to the level of the negative moment capacities, M_u^-, defined by Eq. 14.22 or 14.25. M_u^- is high at first, because $(\sigma_s)_y \simeq (\sigma_s)_y^0$, but later decreases slightly, reflecting the declining yield strength of steel as the temperature rises.

In step with changes in the negative moments above the supports, the maximum positive moment at the midspan, $\mathfrak{M}_m^+$, will first decrease and then slightly increase, in compliance with Eq. 14.10. Structural failure will occur when the temperature of the bottom bars reaches the level at which, owing to the diminished strength of steel, the positive moment capacity of the element becomes unable to match the prevailing positive midspan moment which, all in all, will still be much lower than it was before the test.

Figure 14.8(b) illustrates the moments and moment capacities at the point of structural failure. According to the figure, the positive moment, $\mathfrak{M}_m^+$, at the ultimate can be calculated by replacing in Eq. 14.10 the negative service moment, $\mathfrak{M}^-$, with the negative moment capacity, M_u^-, existing at the point of failure.

The effect of continuity on the fire resistance will be illustrated through a numerical example in Section 14.4.2.

Because of the redistribution of moments during the fire test or in a fire, the detailing should be done with circumspection. Figure 14.8(b) shows that early failure may occur at sections A if the top bars are cut to lengths that satisfy only

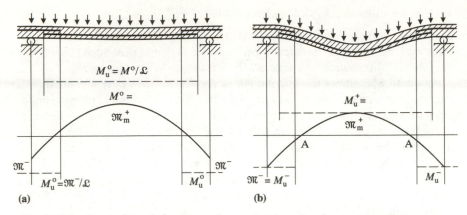

Fig. 14.8 Continuous flexural element: service moment (———) and moment capacity (— — —): (a) at the beginning of a fire test; (b) at the point of structural failure

the requirements of normal service. The designer should see to it that the negative moment reinforcement is extended a sufficient distance beyond the points of inflection (see Eqs 14.5, 14.8, and 14.11) even after the redistribution of the moments. In fact, it is recommended (ACI Committee 216 1989) that at least one-fifth of the top bars extend throughout the span.[12]

Continuous flexural elements are, as a rule, subjected to restraint during a fire exposure. The negative moment produced by the restraining thrust invariably adds to the fire resistance time. The factors that play some role in the beneficial effect of restraint have been discussed in the preceding section. However, the magnitude of that effect cannot as yet be reliably predicted.

Those who wish to learn more about the design and detailing of flexural elements for structural fire resistance are advised to consult some of the publications listed in the references, such as Bobrowski (1975, 1978), Canadian Prestressed Concrete Institute (1978), Gustafson (1980), DTU (1980), Comité Euro-International du Béton (1982), Boutin (1983), ACI Committee 216 (1989), and Gustaferro and Martin (1989).

14.2   Compression Elements

The fire resistance of reinforced concrete compression elements, columns in particular, has long been regarded as essentially a function of two variables: the minimum dimension of the element and the thickness of concrete cover over the vertical reinforcements. Until the 1950s, the results of column tests published by Hull and Ingberg (1926) provided the only guidance on which the designer could lean. By the 1980s, a wealth of information became available on the fire

---

12 The CIB–FIP document (Comité Euro-International du Béton 1982) describes a procedure for the calculation of the effect of top bars on the positive moment capacity of flexural members. It is common practice, however, to ignore that effect.

performance of columns from tests conducted in the United Kingdom (Thomas and Webster 1953), Germany (Seekamp 1959; Seekamp *et al.* 1964; Becker and Stanke 1970; Stanke 1970), and Canada (Lie and Woollerton 1988; Lie and Irwin 1990). However, it appears from a survey conducted by Bennetts (1982) that there is still a substantial lack of agreement concerning the minimum dimensions and cover thicknesses required by building codes or recommended in professional codes of practice.

Clearly, there are more than two essential variables on which the fire performance of compression elements depends. In order to explore the role of all identifiable variables, theoretical studies were undertaken in North America (Lie and Allen 1972; Allen and Lie 1974; Becker *et al.* 1974; Lie *et al.* 1984; Lie and Lin 1985) and in Europe (Klingsch 1976; Haksever 1982; Haksever and Anderberg 1982; Anderberg and Forsén 1982; Rudolph *et al.* 1986).

The mathematical models developed for studying the structural performance of compression elements differed mainly in respect to the formulation of the mechanical properties of concrete and, to a lesser extent, the properties of steel. To reflect the fact that crushed concrete still has some capability to carry compressive load, the stress−strain curve was extended in some models beyond the points of maximum stress. Figure 14.9 shows a family of idealized stress−strain curves at elevated temperatures for a concrete with a strength $\sigma_C = 35\ \text{MPa}$ at room temperature.

Although there were also some differences in the various models concerning the numerical technique of following up the process of deterioration of the compression element (to be referred to as a column, for simplicity) in fire, the method to be described here [used by Allen and Lie (1974)] may be regarded as typical.

Allen and Lie suggested that the process of deterioration of the column is governed by the strains in its most critical section: the midheight section. To be able to save the arduous work of studying the stress−strain history of the entire column, they assumed a relation between the deflected shape of the column and

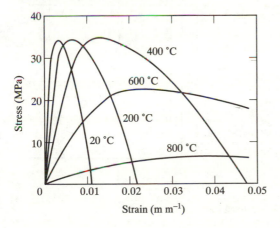

Fig. 14.9 Idealized stress−strain curves for a concrete ($\sigma_C^o = 35\ \text{MPa}$) at elevated temperatures (Lie *et al.* 1984)

the maximum deflection at midheight. They defined the maximum lateral deflection, Y (m), in terms of the radius of curvature at midheight, ρ_m (m), as

$$Y = \frac{1}{\rho_m} \frac{(kL)^2}{12} \qquad\qquad [14.35]$$

where L = unsupported length of the column, m;
$\quad k$ = length factor, dimensionless.

The factor k is defined as the fraction of L between the points of inflection; kL may be looked upon as the length of an equivalent pin-ended column. Thus, k = 1 for a hypothetical pin-ended column [Fig. 14.10(a)], and k = 0.5 for a column with fixed ends [Fig. 14.10(b)]. Depending on the end conditions, k may be larger than 1.

The curvature $(1/\rho)$ along the column was assumed to vary linearly between the ends and the midheight. As Fig. 14.10(b) shows, for columns with fixed ends curvature values were equal but opposite at the ends and at the midheight. This assumption was considered erroneous at the beginning of fire exposure, but reasonable at advanced stages of fire.

The procedure of following up the process of structural deterioration is based on the temperature history of the column, which has to be determined beforehand by any of the numerical techniques discussed in the preceding chapter, or by experiment. The first step is to assume that at the start of fire exposure ($t = 0$) the column is already deflected by a very small amount, Y^0, about the x-axis, which runs along the larger of the cross-sectional dimensions of the column. The procedure then consists of applying an iterative process to each and every time level, $t = \Delta t$, $t = 2\Delta t$, $t = 3\Delta t$, etc., to find ϵ_o and ρ_m from the following equation (see Fig. 14.11):

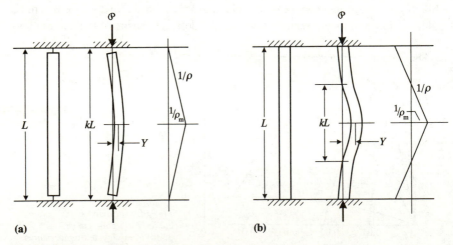

(a)   (b)

Fig 14.10 Column models: (a) pin ends, (b) fixed ends

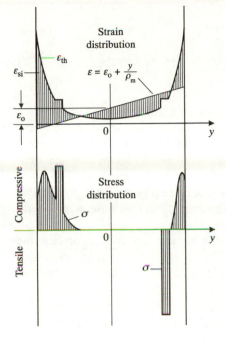

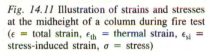

Fig. 14.11 Illustration of strains and stresses at the midheight of a column during fire test (ϵ = total strain, ϵ_{th} = thermal strain, ϵ_{si} = stress-induced strain, σ = stress)

$$\epsilon_o + \frac{y}{\rho_m} = \epsilon_{th} + \epsilon_{si} \qquad [14.36]$$

where ϵ_o = total strain at $y = 0$, $m\,m^{-1}$;
ϵ_{th} = thermal strain, $m\,m^{-1}$;
ϵ_{si} = stress-induced strain, $m\,m^{-1}$;
y = coordinate along the smaller cross-sectional dimension of the column, m.

The left-hand side of Eq. 14.36 describes the linear distribution of the total strain, ϵ, along the y-axis, as required by the Bernoulli–Navier hypothesis (Fig. 14.11). The solution is subject to the following conditions:

$$\mathcal{P} = \int_A \sigma dA + \int_{A_s} \sigma_s dA_s \qquad [14.37]$$

$$\mathcal{M} = \int_A y\sigma dA + \int_{A_s} y\sigma_s dA_s \qquad [14.38]$$

where σ = compressive stress in the concrete, Pa;
σ_s = stress (compressive or tensile) in the reinforcing steel, Pa;
A = area of concrete, m^2;
A_s = area of steel, m^2;
$\mathcal{P}$ = axial force applied to the column, N;
$\mathcal{M}$ = moment acting on the midheight section of the column, N m.

In Eqs 14.37 and 14.38

$$\sigma = \sigma(\epsilon_{si}, T) \qquad\qquad\qquad\qquad [14.39]$$

$$\sigma_s = \sigma_s(\epsilon_{si}, T) \qquad\qquad\qquad\qquad [14.40]$$

where,

$$T = T(x, y, t) \qquad\qquad\qquad\qquad [14.41]$$

The solution of Eq. 14.36 yields the values of ϵ_o and ρ_m at the end of each Δt time interval; ϵ_o is a measure of the overall expansion or shrinkage of the column. Radius of curvature ρ_m is used in Eq. 14.35 for calculating the lateral deflection, Y, which is needed for defining the moment, $\mathfrak{M}$, for the next time step.

When simulating standard fire tests, the axial load, $\mathcal{P}$, is constant and equal to the service load. However, $\mathcal{P}$ may be adjusted at every time step, to simulate conditions (e.g. restraint against thermal expansion) that may arise in real-world fires.

If the column can be treated as a pin-ended element, and if $\mathcal{P}$ is constant, the moment acting at the midheight section, $\mathfrak{M}$, is equal to $\mathcal{P}Y$ [Fig. 14.10(a)]. However, $\mathfrak{M}$ may also be adjusted from time to time to simulate more realistic real-world conditions.

Allen and Lie interpreted structural failure as the condition at which the radius of curvature at the midheight, ρ_m, decreased so fast that satisfying Eq. 14.36 became increasingly difficult.

Drawing on the results of 41 fire resistance tests and numerous theoretical investigations, Lie (1989) studied the fire resistance versus load intensity relation[13] for columns, and analyzed the influence of a number of factors on that relation. In general, the effect of load intensity on fire resistance proved less significant than it is in the case of flexural elements. Decreasing the load intensity from 1.0 to 0.5 usually resulted in a 25 per cent increase in fire resistance.

As to the factors studied, Lie offered the following conclusions.[14]

- With the increase of the percentage area of longitudinal reinforcement, the fire resistance increases somewhat faster than the assigned load capacity.
- With the increase of the strength of concrete, the fire resistance increases in proportion to the assigned load capacity.
- With the increase of the effective length, the fire resistance decreases more slowly than the assigned load capacity.

---

13 The load intensity is to be interpreted as the ratio of the load imposed on the column to the load capacity of (i.e. allowable maximum load on) the column. The load capacity consists of contributions from both the concrete and the steel. In addition to the axial load, the column is also expected to carry a moment corresponding to a 25 mm eccentricity in the application of the load. The effect of column slenderness is taken into account by increasing the eccentricity of load by an amount related to the Euler buckling load which, in turn, is inversely proportional to $(kL)^2$.

14 Unfortunately, Lie's studies did not include the effect of concrete cover over the longitudinal reinforcements. All test specimens were made with a cover thickness of 48 mm. It is known, however, from tests conducted by other researchers that the importance of the cover thickness increases with the percentage area of longitudinal reinforcement.

- With the increase of the cross-sectional area, the fire resistance increases faster than the assigned load capacity.
- If their minimum dimensions are identical, the fire resistance of rectangular columns is substantially higher than that of square columns.
- If their cross-sectional areas are identical, the fire resistance of columns with circular cross-sectional areas is about the same as that of square columns.
- The fire resistance of columns made with carbonate aggregates is substantially higher than that of columns made with siliceous aggregates.[15]
- The fire resistance of columns made with expanded shale aggregates is about the same as that of columns made with siliceous aggregates.
- The effect of load eccentricity on the fire resistance varies from highly adverse (in the case of carbonate-aggregate concretes) to negligible.
- The effect of axial restraint on the fire resistance is negligible.
- The effect of moisture content of concrete (within the usual limits) on the fire resistance is negligible.

A few simple empirical formulas recommended by Lie and Allen (1972) for designing reinforced concrete columns have been given in Chapter 12 (Eqs 12.16a to 12.17b). As a result of further Canadian studies (Allen and Lie 1974; Lie *et al.* 1984; Lie and Irwin 1990) the original design formulas have been amended as follows:

- For Type S concrete (see Table 6.12), provided the conditions stated in column 2 or column 4 of Table 14.1 apply:

$$B \geq 0.080f(\tau + 1) \qquad [14.42]$$

- For Type N concrete, provided the conditions stated in column 2 or column 4 of Table 14.1 apply:

$$B \geq 0.080f(\tau + 0.75) \qquad [14.43]$$

- For Type S and N concretes, provided the conditions stated in column 3 of Table 14.1 apply:

$$B \geq 0.100f(\tau + 1) \qquad [14.44]$$

- For all Type L and L40S concretes:

$$B \geq 0.075f(\tau + 1) \qquad [14.45]$$

In the case of columns with circular cross-sectional areas, the required minimum diameter is $1.2B$.

In Eqs 14.42 to 14.45, the numerals are dimensional constants, and

B = minimum dimension of rectangular columns, m;
τ = required fire resistance, h;
f = empirical factor, to be selected from Table 14.1, m h^{-1}.

---

15 See Section 6.9 in Chapter 6 on 'Types' of concrete.

Table 14.1 Values of the empirical factor f to be used in Eqs 14.42–14.45 (Associate Committee on the National Building Code 1990)

Column 1	Column 2	Column 3	Column 4
		f (m h^{-1})	
Overdesign factor[a]	If $kL \leq 3.7$ m	If $3.7 < kL \leq 7.3$ m — If $B \leq 0.3$ m $p \leq 0.03$[b]	All other cases
1.00	1.0	1.2	1.0
1.25	0.9	1.1	0.9
1.50	0.83	1.0	0.83

[a] Overdesign factor is the ratio of the load-carrying capacity of the column to the service load. It may be regarded as the reciprocal of load intensity.

[b] p is the ratio of the area of vertical reinforcement to the total area of the column.

These amended formulas have been adopted in the *Supplement to the National Building Code of Canada* (Associate Committee on the National Building Code 1990), together with the following requirements concerning the minimum thickness of concrete cover over the vertical reinforcement:

- If $\tau \leq 3$ h,

$$s = \begin{cases} 0.025\tau \\ 0.050 \end{cases} \quad \text{whichever is less} \qquad [14.46]$$

- If $\tau > 3$ h,

$$s = 0.050 + 0.0125(\tau - 3) \qquad [14.47]$$

where s (m) is cover thickness (least distance between the surface of reinforcement and the outer surface of concrete), and the numerals are dimensional constants.

14.3   Complex Structures and Frames

Ascertaining the fire resistance of buildings from information on the fire resistance of its elements is a long-established practice in fire safety engineering. There are many who believe that this practice is full of pitfalls, and claim that the building as a whole, or at least some major units of it, should be the subjects of fire safety considerations. A few instances can indeed be quoted where the fire performance of a structural system may not be deduced from information on the performance of the component elements. However, the merits of looking directly at the whole rather than looking at it through its constituent parts are often exaggerated.

It has been shown in the preceding sections that, in relation to the performance of the whole, continuity and interaction between the parts are either beneficial (as in the case of flexural elements) or not harmful (as in the case of compression elements). Thus, the assessment of the building's performance based on the performance of its components usually errs on the safe side.

The advantages of studying the fire performance of a building part by part are two-fold. First, the work involved is not overly onerous, and the tool for getting it done usually consists of no more than a personal computer. Second, if a novel solution is considered, it is relatively easy and inexpensive to prove or disprove the raw idea. Many fire research laboratories the world over are equipped for conducting validation tests.

Studying the behavior of structural systems and frames is, however, fully justified if the goal is to gain a deeper insight into the boundary conditions to which simple building elements may be subjected in fire. The forces and moments acting on flexural and compression elements are better understood from computer studies that reveal the deformations of the frame of a building. The result of such a computer study is illustrated in Fig. 14.12 (Bresler *et al.* 1976). Shown in the figure are the deformations on the lower floors in the frame of a 13-story building, due to a one-hour 'standard' fire. Papers and reports published by Becker *et al.* (1974), Bresler (1976a, 1977), Bresler *et al.* (1976), Haksever (1977a, 1977b), Iding and Bresler (1984), and Schleich (1987) are among the most significant contributions to the knowledge concerning this subject area.

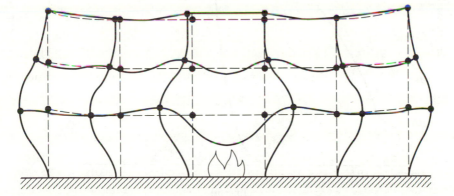

Fig. 14.12 Deformations (on a scale magnified 100 times) of the lower floors in the frame of a 13-story building, due to a 'standard' fire on the ground floor (Bresler *et al.* 1976)

14.4   Worked Examples

14.4.1   Fire Resistance of Simply Supported Concrete Beam

Calculate the structural fire resistance of a simply supported, reinforced concrete beam fabricated with carbonate-aggregate concrete, using both the concept of stress-modified critical temperature and the creep concept. The cross-section of the beam is similar to that illustrated in Fig. 13.17; it is shown in the inset to Fig. 14.13. The steel is assumed to be similar to 'Steel A', as described in Table 7.2. The temperature dependence of the yield and ultimate strengths of this steel are shown in Fig. 7.7, and its creep characteristics are described in Table 7.3. The history of the average temperature of the reinforcing steel bars has been calculated in a way described in Section 13.6.2 (worked example) of Chapter 13, and is shown in Fig. 14.13. The load intensity, $\mathcal{L}$, is 0.61.

Input Information
Yield strength of the steel at room temperature (Fig. 7.7):

$$(\sigma_s)_y^o = 320 \, \text{MPa}$$

Service stress at the midspan (Eq. 14.1):

$$\sigma_s^o = 0.61 \times 320 = 195.2 \, \text{MPa}$$

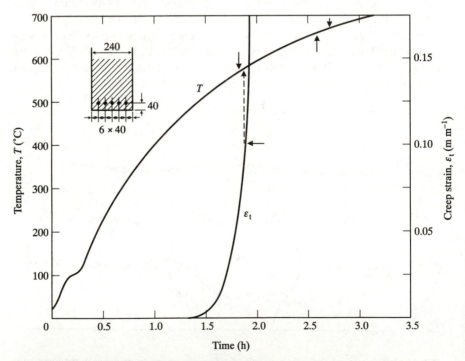

Fig. 14.13 Reinforced concrete beam used in the examples: curve T, average temperature of the reinforcing bars; curve ϵ_t, creep strain in the reinforcement at the midspan section (in the case of the simply supported beam)

Using the concept of stress-modified critical temperature
The beam is expected to collapse when the temperature of the steel reaches the level at which its ultimate strength is equal to 195.2 MPa. That level is 570 °C (Fig. 7.7). According to Fig. 14.13, 570 °C is reached at 1.83 h into the test. Hence, the predicted structural fire resistance of the beam is 1.83 h.

It may be worth examining what the result would be if, as suggested in several design guides, the yield strength was regarded as the property representative of the ultimate behavior of steel on the fire-exposed side. Figure 7.7 shows that the temperature at which the yield strength of Steel A diminishes to 195.2 MPa is about 275 °C. According to Fig. 14.13, the fire resistance time corresponding to this temperature would be unacceptably low; only 0.6 h.

Using the creep concept:
The creep properties of Steel A are as follows.
Activation energy of creep (Table 7.3):

$$\Delta H_c /R = 38890 \text{ K}$$

Zener–Hollomon parameter for $\sigma_s = 195.2$ MPa (Table 7.3):

$$Z = 1.23 \times 10^{16} \times \exp(4.35 \times 10^{-8} \times 195.2 \times 10^6)$$
$$= 6.0 \times 10^{19} \text{ m m}^{-1} \text{h}^{-1}$$

Method 1
Calculate the creep–strain (ϵ_t) versus time (t) relation, using Eq. 14.26.

$$\epsilon_t = 6.0 \times 10^{19} \int_0^t \exp\left(-\frac{38890}{T}\right) dt$$

The integration has been performed numerically. The ϵ_t versus time curve is also shown in Fig. 14.13. The creep strain is hardly noticeable up to about 1.25 h into the test. At 1.5 h, it begins to grow at an increasing rate, and reaches the specified limiting value, 0.1 m m^{-1} at 1.87 h (see section 14.1.2).

Method 2
Calculate the critical temperature (T_{cr}) on the assumption that failure occurs when the creep rate, $d\epsilon_t /dt$, reaches a limiting value of 1.0 m m^{-1} h^{-1} (see Footnote 8). From Eq. 14.27:

$$T_{cr} = -\frac{38890}{\ln\left[\dfrac{1}{6.0 \times 10^{19}} \times 1.0\right]} = 854 \text{ K } (581 \text{ °C})$$

As Fig. 14.13 shows, that temperature is reached at 1.92 h into the test.

Apparently, under conditions characteristic of fire exposures of 2–3 h duration, there are no appreciable differences in the fire resistance values calculated by means of the concept of stress-modified critical temperature and the creep concept.

14.4.2 Fire Resistance of Continuous Concrete Beam

Calculate the structural fire resistance of a continuous, five-span, reinforced concrete beam between the third and fourth supports (Fig. 14.8). The beam, with respect to dimensions, materials, and loading conditions, is similar to that described in the previous section. Thus: $(\sigma_s)_y^o = 320\,\text{MPa}$, $\mathcal{L} = 0.61$, and the service stress at the midspan at the start of test,

$$\sigma_s^o = 0.61 \times 320 = 195.2\,\text{MPa}.$$

Using the concept of stress-modified temperature
The negative moments over the supports before the fire test are (Eshbach 1952):

$$\mathfrak{M}^- = 0.079_w L^2$$

On account of Eq. 14.10, the maximum positive service moment is:

$$\mathfrak{M}_m^+ = 0.125_w L^2 - 0.079_w L^2 = 0.046_w L^2 \qquad \text{(before fire test)}$$

The negative moment capacity of the beam (above the supports) at normal temperatures [Fig. 14.8(a)] is:

$$M_u^o = 0.079_w L^2 / 0.61 = 0.130_w L^2$$

At the time of structural failure, the temperature of the top bars is about[16] 200 °C. At that temperature, the yield strength of the steel (from Fig. 7.7) is 245 MPa. With the steel yielding, the acting negative moment becomes equal to the prevailing negative moment capacity [Fig. 14.8(b)], which at 200 °C is

$$\mathfrak{M}^- = M_u^- = \frac{245}{320} \times 0.130_w L^2 = 0.099_w L^2 \qquad \text{(at failure)}$$

On account of Eq. 14.10,

$$\mathfrak{M}_m^+ = M_u^+ = 0.125_w L^2 - 0.099_w L^2 = 0.026_w L^2 \quad \text{(at failure)}$$

The ratio of $\mathfrak{M}_m^+$ at the time of imminent failure to $\mathfrak{M}_m^+$ before the test is $0.026/0.046 = 0.565$. Thus, when failure occurs, the stress in the bottom steel bars at the midspan is $0.565 \times 195.2 = 110.3\,\text{MPa}$. According to Fig. 7.7, the temperature at which the ultimate strength of the steel drops to the 110.3 MPa level is 665 °C. Figure 14.13 shows that this temperature is reached at 2.70 h into the test. This value compares with 1.83 h for the simply supported beam.

Using the creep concept
It is assumed again that collapse is imminent when the creep rate at the midspan reaches $1.0\,\text{m m}^{-1}\,\text{h}^{-1}$.
Zener–Hollomon parameter for $\sigma_s = 110.3\,\text{MPa}$ (Table 7.3):

$$Z = 1.23 \times 10^{16} \times \exp(4.35 \times 10^{-8} \times 110.3 \times 10^6)$$
$$= 1.49 \times 10^{18}\,\text{m m}^{-1}\,\text{h}^{-1}$$

---

16 It is assumed that the temperature history of the beam is available from previous calculations. However, approximate values for the temperature of the top bars usually suffice.

Critical temperature (Eq. 14.27, with $(d\epsilon_t/dt)_1 = 1.0\,\mathrm{m\,m^{-1}\,h^{-1}}$)

$$T_{cr} = -\frac{38890}{\ln\left[\dfrac{1}{1.47\times10^{18}}\times 1.0\right]} = 929\,\mathrm{K}\;(656\,°C)$$

According to the temperature versus time plot in Fig. 14.13, the average temperature of the reinforcing steel reaches the 656 °C level at 2.58 h into the test.

Nomenclature

a	depth of rectangular stress block in the compression zone, m
A	area; (without subscript) area of concrete, m^2
b	width of the element in the compression zone, m
B	minimum dimension of rectangular column, m
$\mathcal{C}$	compressive force in the concrete, N
d	distance between the centroid of the steel and the extreme compression fiber, m
D	depth of flexural element, m
f	empirical factor, dimensionless
ΔH_c	activation energy of creep, $\mathrm{J\,kmol^{-1}}$
j	distance (at the supports) between the line of action of the restraining thrust and the top of beam, m
k	length factor, dimensionless
l	expanded length of flexural element, m
Δl	$= l - l_o$, m
L	span length; unsupported length of column, m
$\mathcal{L}$	load intensity, dimensionless
M	resisting moment, N m
$\mathfrak{M}$	acting moment, N m
n	distance from the top of element to the neutral axis, m
p	ratio of the area of vertical reinforcement to the total cross-sectional area of the column, dimensionless
$\mathcal{P}$	axial force applied to the column, N
r	distance at the midspan between the neutral axis and the line of action of the restraining thrust, m
R	gas constant $= 8315\,\mathrm{J\,kmol^{-1}\,K^{-1}}$
$\mathcal{R}$	restraining thrust, N
s	cover thickness, m
$\mathcal{S}$	tensile force in the steel, N
t	time, h (s)
T	temperature, K (°C)
w	load per unit length; (without subscript) service load per unit length, $\mathrm{N\,m^{-1}}$
x, y	coordinates, m
$x_1, x_2, x_{1,2}$	abscissas of the points of inflection, m
Y	midspan/midheight deflection, m
Z	Zener–Hollomon parameter, $\mathrm{m\,m^{-1}\,h^{-1}}$
α	empirical factor, dimensionless

δ	allowed expansion before restraint takes effect, m
ϵ	strain; (without subscript) total strain, $m\,m^{-1}$
ϵ^*	notional strain, $m\,m^{-1}$
ξ	factor defined by Eq. 14.18, dimensionless
ρ	radius of curvature, m
σ	stress/strength; (without subscript) stress in concrete, Pa
τ	required fire resistance, h

Subscripts

bm	bottom
cr	critical
C	of concrete (strength)
l	limiting
li	load-induced
m	maximum; at midheight
ms	at the midspan
o	original; at $y = 0$
rt	due to restraint
s	of/in the steel
si	stress-induced
t	time-dependent (creep)
th	thermal
tp	top
u	ultimate
y	yield

Superscripts

o	related to conditions at normal temperatures
0	at $t = 0$
+	related to positive moment
−	related to negative moment

15 General Aspects of Fire Safety Design

A fire-safe building may be defined as one for which there is a high probability that all occupants will survive a fire without injury, and in which property damage will be confined to the immediate vicinity of the fire area (Harmathy 1977). The assessment of fire risk is, consciously or subconsciously, always an important part of fire safety design, and the design usually involves trading between various fire safety measures.

15.1 Fire Risk and Fire Hazard

Risk and *hazard* are poorly defined terms in fire science. Watts (1988) reviewed a number of ways of assessing fire risk, and Hall and Sekizawa (1991) developed a general conceptual framework to aid discussions on alternative approaches to fire risk analysis.

According to Kaplan and Garrick (1981), a risk analysis consists of answering the following three questions:

- What can happen?
- How likely is it that that can happen?
- If it does happen, how serious are the consequences?

The first question is concerned with the perilous scenario, the second with the probability of that scenario, and the third with the peril associated with it. Naturally, there can be several or many perilous scenarios. The risk assessment is a statement of all answers to these three questions.

The ASTM definition of fire risk (Roux 1982; American Society for Testing and Materials 1989b) is as follows:

$$\text{Fire risk} = \begin{pmatrix} \text{probability of} \\ \text{occurrence} \\ \text{of event} \end{pmatrix} \times \begin{pmatrix} \text{probability} \\ \text{of} \\ \text{exposure} \end{pmatrix} \times \begin{pmatrix} \text{potential} \\ \text{for} \\ \text{harm} \end{pmatrix} \qquad [15.1]$$

Hazard is usually regarded as an unacceptable level of risk,[1] as judged by the designer or defined by the authorities having jurisdiction.

---

1   According to Hall and Sekizawa, hazard is a measure of fire severity for a specified situation, characterized by such factors as fire deaths, injuries, direct monetary damages to property, etc.

15.2   Probabilistic and Deterministic Models

Mathematical models are called *probabilistic* or *deterministic*, depending on whether they are based mainly on the manipulation of statistical information or on the laws of nature. Although probabilistic models have been devised mainly for the benefit of the insurance companies, there are certain areas where information of considerable value to the designer can be derived by means of probabilistic considerations.

A technique developed by Yung and Beck (1989, p. 15), which employs a series of state-transition (see Section 9.8 in Chapter 9) and interrelated deterministic models, is claimed to be suitable for identifying cost-effective fire safety systems which are equivalent to or better than those prescribed in the building codes. Table 15.1 shows the 'expected risk-to-life' (ERL) and 'fire cost expectation' (FCE) figures for a six-story apartment building with various fire protection measures. (The values shown in the table have been normalized by dividing them by the ERL or FCE values corresponding to the building code options, i.e. nominal fire resistance, no sprinklers, central alarm.)

A fire safety design is usually the outcome of a stream of thoughts in which proficiency in fire science and practical experience play prominent parts. The National Fire Protection Association (NFPA) has devised a decision tree (National Fire Protection Association 1986) to guide the designer in choosing between various alternative strategies toward achieving the overall objective: fire safety. The decision tree consists of two main branches: Prevent Fire Ignition, and Manage Fire Impact. The first is concerned mainly with subjects outside the area of design decisions. A major part of the second branch is a kind of design code. It again consists of two branches: Manage Fire, and Manage Exposed. The Manage Fire branch is reproduced in Fig. 15.1.

The objectives shown at the various levels of the tree are connected with a gate. It is an *and-gate* (represented by the symbol $\odot$) if, to achieve an objective, all the items below the gate must be addressed. It is an *or-gate* (represented by $\oplus$) if considering one of the items below the gate is sufficient. As Fig. 15.1 shows, managing the fire can be achieved by controlling the combustion process, *or* by suppressing the fire, *or* by controlling the fire by passive or active safety measures.[2]

Although the decision tree was conceived primarily as a design guide, it can also be used as a basis for quantitative design decisions, provided the probabilities of success with the use of various alternatives are known. Thor and Sedin (1979/1980) and Ling and Williamson (1982) illustrated the use of decision trees in quantitative decision making.

---

2 Passive fire safety measures consist of properly laying out the building spaces, dimensioning the building's constituent elements, and avoiding certain construction materials. Thus, employing fire-resistant compartmentation, avoiding burnable lining in the corridors, etc., are passive measures. In contrast, employing smoke detectors, self-closing doors, and sprinkler systems are active measures.

Table 15.1 Expected risk-to-life (ERL) and fire cost expectation (FCE) values for a six-story apartment building (relative to the ERL and FCE values corresponding to the building code options), with various fire safety measures[a][b] (Yung and Beck 1989)

	0.5 × FRR		1.0 × FRR		1.5 × FRR	
	NS	WS	NS	WS	NS	WS
(a) Building without alarm system						
ERL	1.80	0.84	1.62	0.83	1.62	0.83
FCE	0.78	1.31	0.84	1.42	1.06	1.63
(b) Building with central alarm system						
ERL	1.08	0.60	1.00	0.60	1.00	0.60
FCE	0.94	1.47	1.00	1.57	1.21	1.79
(c) Building with higher reliability alarm system						
ERL	0.82	0.54	0.77	0.54	0.77	0.54
FCE	0.94	1.47	1.00	1.57	1.21	1.79

[a] Abbreviations: FRR, fire resistance required by building code; NS, no sprinkler system; WS, with sprinkler system.

[b] The building code options consist of 1.0 × FRR, central alarm system, no sprinkler system.

The foremost difficulty in applying the probabilistic models to fire safety design is that statistical data are not likely to come broken down in a manner suitable for specific design decisions. For lack of supporting data, the designer is tempted to substitute his own estimates of probabilities for factual information. If, on the other hand, an abundance of data happens to exist, the problem is how to pick the values appropriate for the given situation. For example, should the designer choose the values representing national averages, or rather those representative of the city, or perhaps of the income level of the expected occupants of the building? He will probably use the values that lead to results which he has had in mind all the time.

Further difficulties arise from the fact that statistical data reflect conditions characteristic of past construction practices and past patterns of human behavior, and may not be applicable to present conditions, let alone to conditions in the years to follow.[3] It comes as no surprise, therefore, that fire safety design is based mainly on experience and deterministic mathematical models.

15.3   Random Variables

Unfortunately, it is rarely possible to complete a design by relying solely on deterministic models. To illustrate the nature of the problems that may arise, the procedure for the calculation of fire resistance requirements (see Chapters 10 and 11) will be briefly reviewed.

---

3   For example, the decline in the number of smokers among educated people will, no doubt, lead to a substantial decrease in the human and property losses in the so-called 'better neighborhoods'. On the other hand, the increasing drug use in the poorer neighborhoods may create serious new design problems.

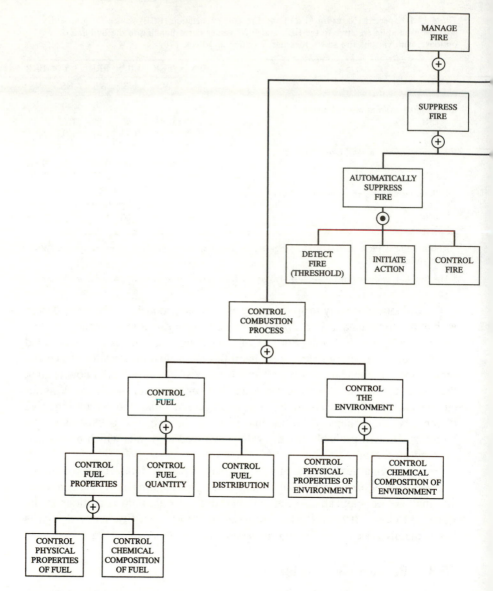

The potential of a fire to spread by the destruction of compartment boundaries (as quantified by the normalized heat load) was pointed out to depend, among others, on two random variables: the specific fire load, L (kg m^{-2}), and the rate of entry of air into the compartment, U_a (kg s^{-1}). As in other fields of engineering, the design should be based on the assumption of *reasonably adverse* conditions. Thus, the first step in developing a design procedure is to examine

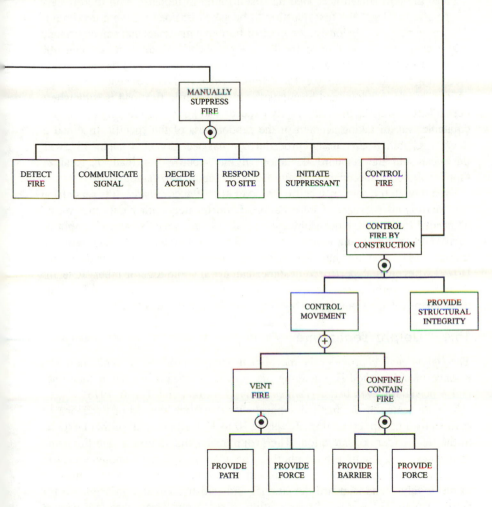

Fig. 15.1 The *Manage Fire* branch of the decision tree of the National Fire Protection Association (US) (National Fire Protection Association 1986). Reprinted with permission from *Fire Safety Concepts Tree*, copyright ©, National Fire Protection Association, Quincy, MA 02269

the ranges of the two random variables. If they have extreme values that represent reasonably adverse conditions, then these values can be used in the design as input information.

As discussed in Chapter 10 (see Eq. 10.4), the flow rate of air does have an extreme value, namely a minimum (quantified by the ventilation parameter, Φ, $kg\,s^{-1}$), which, as further discussed in connection with Eqs 11.3 and 11.4, indeed represents a reasonably adverse condition. Consequently, Φ qualifies as input information into the design (see Eqs 11.3 to 11.8).

Since the normalized heat load has been found to increase with the increase of specific fire load, the first question to be asked is: does L have a maximum? The answer is yes. Obviously, the mass of burnable materials that can be brought into a compartment is limited by the compartment's volume; it is about 900 $kg\,m^{-2}$. However, adopting this value as input information would be unrealistic.[4] According to statistical data (see, for example, Table 10.1), in residential buildings, offices, hotels, schools, and hospitals the mean specific fire load is somewhere between 10 and 50 $kg\,m^{-2}$, and rarely exceeds 100 $kg\,m^{-2}$. In this light, the only equitable way of taking account of the randomness of the specific fire load is to use a reliability-based design procedure (as has been done in Chapter 11), with the mean and the standard deviation of the specific fire load, L_m and σ_L (applicable to the occupancy in question), as input information.

Even though the data on specific fire load are often confusing, it is reassuring that there is no shortage of information. Unfortunately, other data that would be needed for objective decisions are, for all intents, nonexistent and cannot be derived by manipulating existing information. Consequently, experience (i.e. lessons learned from past mistakes) is the most important factor in design decisions. However, even an experienced designer can err at some time or other. It seems desirable, therefore, that, at least in the most important areas, there be some mechanism for distributing the burden of making decisions.

15.4   Delphi Technique

The *Delphi method* provides the mechanism for collective decision making. The essence of the method is practicing the adage: two heads are better than one, and n heads are even better (Dalkey 1969; Linstone and Turoff 1975).

The advocates of the method suggest that the input into major decisions be defined by a group, consisting of at least 10 to 15 members, all of them experts in the field under consideration. The group leader draws up the questions and distributes them among the members.[5] The answers are given anonymously in a sequence of rounds. Between each round, the group leader reduces the answers to mean values and standard deviations, and redistributes this information for further study. The statistical presentation of the group's response is a way of reducing the pressure for conformity. At the end of the exercise, there may still be significant spread in the opinions of the members.

---

4   If the room is completely filled with burnable materials, burning becomes impossible for lack of air. It is estimated that even if the compartment is used for storing burnable items, the maximum amount of fuel that can burn in a fire is about 150 $kg\,m^{-2}$.

5   It is often a prerequisite to a successful Delphi exercise that the members of the group be paid for their services. If some kind of an incentive does not exist, the enthusiasm of the members may wane with the passing of time, and the exercise may end in fiasco.

The Delphi technique is a viable way of bridging over some gray areas in fire statistics. It also offers the convenience of providing up-to-date estimates, perhaps even with some anticipation of future developments.

15.5   Trading between Fire Safety Measures

Based on a decision logic suggested by Harmathy (1986), a Delphi group (Harmathy *et al.* 1989)[6] examined the availability of factual data needed for making major decisions in fire safety design, and supplemented much of the missing data with Delphi estimates. The logic was built on the premise that, as long as the expected annual fire losses (human loss — deaths and injuries — and property loss, expressed in monetary terms) are the same, any set of fire safety measures is equivalent to any other set.[7]

According to the decision logic, the fire loss expectation is calculated from the following equation:

$$\mathcal{L} = N[P_{NF}\ell_{NF} + (1 - P_{NF})(P_{NS}\ell_{NS} + P_{SD}\ell_{SD} + P_{SC}\ell_{SC})] \quad [15.2]$$

where $\mathcal{L}$ = fire loss (human + property) expectation, dollars[8] per year and per square meter floor area;

N = expected number of fire incidents, per year and per square meter floor area;

P_{NF} = probability that, given ignition, no flashover will occur, dimensionless;

P_{NS} = probability that, given flashover, no spread to other compartment(s)[9] will occur, dimensionless;

P_{SD} = probability that, given flashover, spread by destruction of compartment boundaries will occur, dimensionless;

P_{SC} = probability that, given flashover, spread by convection (i.e. by the advance of flames and hot gases) will occur, dimensionless;

---

6   The Delphi group operated within the framework of ASTM Committee E5 but, to bypass the usually lengthy procedures required for ASTM approval, the results of the Delphi study were published without endorsement by the ASTM.

7   Note that the equality of fire loss expectations is taken to indicate only that one set of measures is equivalent to another set *from a fire safety point of view*. Of course, one or the other set may prove impractical on the basis of a cost–benefit study in which, in addition to the expected fire losses, the 'present worth' of purchase, installation, inspection, repair, etc., is also considered. (See, for example, Harmathy 1988a.)

   As discussed by Ramachandran (1988), the value of human life and the cost of injury are very controversial subjects. There are many who question the morality of putting a price tag on human life and suffering. Those charged with making design decisions, however, cannot be concerned with moral issues. The values reported by Ruegg and Fuller (1984) have been used by Harmathy *et al.* (1989), according to which a human life is worth US$0.558 × 10⁶, and the cost of injury is 0.022 × 10⁶. (These are figures adjusted for 1985, using the consumer price index.)

8   1985 US dollars. The currency of the appropriate country may be substituted for US dollar.

9   A compartment is defined as a building space which is most of the time separated from other building spaces by closed door(s). Consequently, a single-family dwelling or apartment consisting of two, three, or perhaps more rooms may often be regarded as a single compartment.

ℓ_{NF} = average loss (human + property) in fires where no flashover occurs, dollars per incident;

ℓ_{NS} = average loss (human + property) in fully developed fires where no spread occurs, dollars per incident;

ℓ_{SD} = average loss (human + property) in fully developed fires where spread by destruction occurs, dollars per incident;

ℓ_{SC} = average loss (human + property) in fully developed fires where spread by convection occurs, dollars per incident.

Among the listed four probabilities, P_{NS}, P_{SD}, and P_{SC} are conditional on the same event: flashover. Since no spread (NS), spread by destruction (SD), and spread by convection (SC) may be looked upon as mutually exclusive, and the only, alternatives,

$$P_{NS} + P_{SD} + P_{SC} = 1 \qquad [15.3]$$

Hence, Eq. 15.2 can be rewritten as follows:

$$\mathcal{L} = N[P_{NF}\ell_{NF} + (1 - P_{NF})(P_{NS}\ell_{NS} + P_{SD}\ell_{SD} + (1 - P_{NS} - P_{SD})\ell_{SC})] \qquad [15.4]$$

Furthermore, since the probability of spread by destruction, P_{SD}, can be assigned in the design for fire resistance (see Eqs 11.5 and 11.6, and Fig. 11.3; $P_{SD} \equiv P_f$), only two of the probabilities are truly independent variables: P_{NF} (probability of no flashover) and P_{NS} (probability of no spread).

P_{NF} and P_{NS} are conditional probabilities; the first is conditional on ignition, the second on flashover. In principle, both P_{NF} and P_{NS} may vary within the range of zero to one. However, to take account of unavoidable inaccuracies in the input information, it is advisable to restrict somewhat their ranges of variation, possibly as follows:

$$0 \leq P_{NF} \leq 0.9 \qquad [15.5]$$

$$0 \leq P_{NS} \leq 0.9 \qquad [15.6]$$

Since building code requirements for dwellings and low-rise apartment buildings are minimal, and yet these buildings have most of the features truly characteristic of community living, the statistical data related to these buildings can be regarded as representative of the 'basic' nonindustrial buildings. Consequently, the factors that characterize these basic buildings and their ties with communal facilities can be looked upon as constituting a set of 'reference conditions'.

According to the mentioned decision logic, the probabilities of no-flashover and no-spread for a specific building (characterized by one or more factors not present in basic buildings) are defined as

$$P_{NF} = (P_{NF})_r\left(1 + \sum_i \frac{(\Delta P_{NF})_i}{(P_{NF})_r}\right) \qquad [15.7]$$

$$P_{NS} = (P_{NS})_r\left(1 + \sum_i \frac{(\Delta P_{NS})_i}{(P_{NS})_r}\right) \qquad [15.8]$$

where $(P_{NF})_r$ $(P_{NS})_r$ (dimensionless) are the reference probabilities (i.e.

Table 15.2 Summary of information on the 'basic' versions of six types of buildings, derived from fire statistics for the USA (1985) (Harmathy *et al.* 1989)

Expected number of fire incidents (per year and per m^2 floor area)	
N	
Dwellings	40.1×10^{-6}
Apartments	49.4×10^{-6}
Hotels, motels	79.0×10^{-6}
Health care buildings	78.1×10^{-6}
Office buildings	24.5×10^{-6}
Educational buildings	23.6×10^{-6}
Reference probabilities	
$(P_{NF})_r$	0.787
$(P_{NS})_r$	0.350
$(P_{SD})_r$	0.100
Losses (human + property) (US\$ per fire incident)	
l_{NF}	2170
l_{NS}	13780
l_{SD}	63700
l_{SC}	58700

probabilities related to basic buildings), and $(\Delta P_{NF})_i$ and $(\Delta P_{NS})_i$ (dimensionless) are positive or negative probability increments, characterizing the various (nonbasic) features of the specific building. Naturally, the restrictions imposed on the values of P_{NF} and P_{NS} by Eqs 15.5 and 15.6 still apply.

The Delphi study covered six types of buildings, namely dwellings, apartments, hotels and motels, health-care buildings, office buildings, and educational buildings. The values of $(P_{NF})_r$, $(P_{NS})_r$, $(P_{SC})_r$, l_{NF}, l_{NS}, l_{SD}, and l_{SC} were developed from data on fire incidents in dwellings and low-rise apartments (i.e. in 'basic' buildings) supplied for the year 1985 by the TriData Agency (US),[10] and from other data scattered through a number of publications (see Harmathy *et al.* 1989). These values were assumed to be applicable to all six types of building covered in the study, as long as they complied (except for occupancy) with the reference conditions. In contrast, there seemed to be no justification for using a common value for N. According to the fire loss surveys published yearly by the National Fire Protection Association (US), there are substantial differences in the number of fire incidents in the different building types.

Table 15.2 summarizes the information derived for the 'basic' versions of the six types of buildings studied by the Delphi group.[11]

10 The information supplied by the TriData Agency was abstracted from the records of the National Fire Incident Reporting System (NFIRS). They covered some 274 000 fire incidents which, according to the statistics of the National Fire Protection Association (Karter 1986), represented 45 per cent of all fires that occurred in 1985 in dwellings and apartments.

11 Harmathy *et al.* (1989) noted that the available data gave no clues as to the probability of fire spread by the destruction of compartment boundaries. Fire investigation reports often make mention of structural damages sufficient to allow spread. There are good reasons to believe, however, that a great deal of those damages are consequences rather than causes of fire spread. Based on observations and theoretical studies (see Chapter 11), it seemed reasonable to assume that $P_{SD} \approx 0.1$, i.e. that under reference conditions about 10 per cent of postflashover fires spread by the destruction of compartment boundaries. Establishing a reference value for P_{SD} is of marginal interest only since, as mentioned earlier, the actual value of P_{SD} can be assigned by design.

Table 15.3 Incremental effect of various factors on the probabilities P_{NF} and P_{NS}
(Harmathy *et al.* 1989)

Factor i	$(\Delta P_{NF})_i/(P_{NS})_r$	$(\Delta P_{NS})_i\,(P_{NS})_r$
(1) Occupancy		
Dwellings, apartments	0	0
Hotels, motels	−0.027	−0.128
Heath-care buildings	0.090	0.246
Office buildings	0.064	0.149
Educational buildings	0.078	0.204
(2) Building height		
1 or 2 stories	0	0
3 to 8 stories	−0.016	−0.109
More than 8 stories	−0.057	−0.230
(3) Average floor area of compartments		
Less than 20 m^2	0	0
More than 20 m^2	0.064	−0.124
(4) Furniture arrangement		
Sparse	0.107	0.123
Normal (as in a dwelling)	0	0
Dense	−0.148	−0.147
(5) Combustible lining in compartments		
Not used	0	0
Used	−0.213	−0.238
(6) Combustible lining in corridors		
Not used	0	0
Used	0	−0.435
(7) Combustible lining on building exterior		
Not used	0	0
Used	0	−0.211
(8) Smoke detector(s)		
Not used	0	0
Used	0.118	0.268
(9) Sprinkler system		
Not used	0	0
Used	0.196	0.479
(10) Building pressurization		
Not used	0	0
Used	0	0.199
(11) Self-closing doors		
Not used	0	0
Used	0.030	0.530
(12) Municipality served by		
Full-time FD	0	0
Composite FD	−0.012	−0.110
Volunteer FD	−0.033	−0.201
(13) Distance from FD		
Less than 8 km	0	0
More than 8 km	−0.043	−0.280

Table 15.3 (continued)

Factor i	$(\Delta P_{NF})_i/(P_{NS})_r$	$(\Delta P_{NS})_i/(P_{NS})_r$
(14) Access by FD		
From one side of building	0	0
From more than one side of building	0.022	0.184
(15) Average temperature in January		
Higher than 0 °C	0	0
Lower than 0 °C	−0.029	−0.140

Notes (1) The reference conditions are characterized by 0 increments
(2) Positive values indicate improved safety; negative values indicate decreased safety.

The values of the fractional probability increments, $(\Delta P_{NF})_i/(P_{NF})_r$ and $(\Delta P_{NS})_i/(P_{NS})_r$, were determined by means of the Delphi technique. They are listed in Table 15.3. By definition, for the reference conditions these fractional probabilities are equal to zero.

Tables 15.2 and 15.3 contain all the information needed for the use of Eq. 15.4. The calculation procedure is illustrated through a worked example in Section 15.6.

Whether the calculated values are realistic or not is difficult to judge. Although the fractional probability increments developed by the Delphi group seem to be of the right magnitude, there is no doubt that the values listed in Table 15.3 could be improved and the list refined by the addition of further items. Perhaps the greatest weakness of the list is that it leaves the personal quality of the building's occupants, especially their average level of education, undefined. However, with the present societal attitude, which is founded on the political slogan of perfect equality among people, it may be difficult to make the quality of the building's occupants accepted as a legitimate input variable.

Even though the outlined method of calculating fire losses may not be sufficiently accurate for the calculation of fire insurance premiums, it is accurate enough as guidance in design decisions. A question that often arises in the design for fire safety is whether one set of fire safety measures can be replaced by another set. Since tradeoff decisions are usually concerned with only two or three of the items listed in Table 15.3, the possibility of committing a major error in the assessment of P_{NF} and P_{NS} is greatly reduced.

Table 15.4 gives the results of a series of calculations, aimed mainly at comparing the merits of various popular passive and active fire protection measures. The following conclusions may be drawn.

- With basic types of buildings, optimum protection seems to be achieved by the use of smoke detectors and sprinklers (Case E in Table 15.4).[12] The

---

12 The loss values shown in Table 15.4 for Case E would be roughly halved if the upper limit for P_{NF} was not restricted to 0.9.

Table 15.4 Results of calculations aimed at comparing the merits of various passive and active fire safety measures (Harmathy et al. 1989)

Case	Fire loss (human + property) expectation [US$ (1985) per year per square meter floor area][a]					
	Dwellings	Apartments	Hotels, motels	Health-care buildings	Office buildings	Educational buildings
(A) 'Basic' building	0.440	0.545	0.899	0.585	0.208	0.188
(B) As (A) but with self-closing doors	0.337	0.418	0.687	0.439	0.158	0.141
(C) As (A) but with smoke detectors	0.266	0.329	0.563	0.429 (0.297)	0.138 (0.113)	0.131 (0.098)
(D) As (A) but with smoke detectors and self-closing doors	0.202 (0.198)	0.249 (0.246)	0.420	0.364 (0.219)	0.118 (0.083)	0.111 (0.072)
(E) As (A) but with smoke detectors and sprinkler system	0.206 (0.087)	0.253 (0.108)	0.389 (0.171)	0.370 (0.169)	0.120 (0.053)	0.113 (0.051)
(F) As (A) but with increased fire resistance[b]	0.438	0.542	0.894	0.582	0.207	0.187

[a] Values in parentheses are obtained if P_{NF} is allowed to assume values between 0.9 and 1 (see Eq. 15.5).

[b] The fire resistance requirement is increased by reducing the allowable failure probability (i.e. the probability of spread by destruction, P_{SD}) from 0.10 to 0.05.

combined use of these two devices would practically prevent the fire from reaching the postflashover stage. However, the expenses are usually prohibitive,[13] and therefore the combined use of smoke detectors and self-closing doors (Case D) may be considered instead.

● Case F illustrates the futility of trying to improve fire safety by increasing the fire resistance of building elements (by reducing the allowable failure probability, P_{SD}, from 0.1 to 0.05).

● If the equality of fire loss expectations is regarded as the criterion of equivalence of fire safety under different sets of conditions, then the building code requirements may be relaxed for office and educational buildings, but they should be strengthened for hotels and motels.

15.6   Worked Example

Fire Loss Expectation

Calculate the fire loss expectation for a hypothetical building.

Input Information
It is assumed that the compartment boundaries were designed for a 5 per cent failure probability, given flashover (see Section 11.7.1 in Chapter 11), i.e. that $P_f \equiv P_{SD} = 0.05$. With respect to items (5), (6), (7), (9), (10), and (12) (see Table 15.3), the building corresponds to a 'basic' (reference) building. However, it is nonbasic with respect to the following items:

	$(\Delta P_{NF})_i/(P_{NF})_r$	$(\Delta P_{NS})_i/(P_{NS})_r$
(1) Occupancy type: health-care	0.090	0.246
(2) Building height: three-story	−0.016	−0.109
(3) Average floor area: 30 m^{-2}	0.064	−0.124
(4) Furniture arrangement; sparse	0.107	0.123
(8) Smoke detectors: used	0.118	0.268
(11) Self-closing doors: used	0.030	0.530
(13) Distance from FD: 9 km	−0.043	−0.280
(14) Access by FD: from two sides	0.022	0.184
(15) Av. temperature in January: −5 °C	−0.029	−0.140
	0.343	0.698

Probability that, given ignition, no flashover will occur (Eq. 15.7, with $(P_{NF})_r$ from Table 15.2):

$$P_{NF} = 0.787 \times (1 + 0.343) = 1.06 \rightarrow 0.9 \text{ (by virtue of Eq. 15.5)}$$

Probability that, given flashover, no spread will occur (Eq. 15.8, with $(P_{NS})_r$ from Table 15.2):

$$P_{NF} = 0.350 \times (1 + 0.698) = 0.59$$

13 See, for example, Ruegg and Fuller's report (1984) and a paper by Harmathy (1988a).

Thus, fire loss expectation, per year per square meter floor area (Eq. 15.4, with information from Table 2, and $P_{SD} = 0.05$):

$$\mathcal{L} = 78.1 \times 10^{-6} \times [0.9 \times 2170 + (1 - 0.9) \times (0.59 \times 13\,780 + \\ 0.05 \times 63\,700 + (1 - 0.59 - 0.05) \times 58\,700)]$$

$$= 0.406 \text{ US (1985) dollars}$$

Nomenclature

ℓ	average loss (human + property), dollars per incident
$\mathcal{L}$	fire loss (human + property) expectation, dollars per year and per square meter floor area
L	specific fire load, kg m^{-1}
N	expected number of fire incidents, per year and per square meter floor area
P	probability, dimensionless
ΔP	probability increment, dimensionless
U	flow rate, kg m^{-1}
σ	standard deviation, kg m^{-2}
Φ	ventilation parameter, kg s^{-1}

Subscripts

a	of air
f	of failure, conditional on flashover
L	for specific fire load
m	mean
NF	of/in fires with no flashover
NS	of/in fully developed fires with no spread
r	reference value
SD	of/in fully developed fires with spread by destruction
SC	of/in fully developed fires with spread by convection

16 Assessment and Repair of Fire Damage to Concrete

Based on data extracted from the records of the National Fire Incident Reporting System (US), it has been estimated (Harmathy *et al.* 1989) that of 100 fires occurring in low-rise, nonindustrial, nonmercantile buildings 79 will not grow to flashover, seven will reach flashover but will not spread beyond the confines of the compartment of fire origin, 12 will spread by convection (i.e. by the advance of flames and hot gases), and two will spread by the destruction of the compartment boundaries.[1] Similar estimates are not as yet available for industrial or mercantile occupancies.

The destructive potential (severity) of a fully developed fire in the compartment of fire origin is an important factor in the extent of damage. The fire's severity, as discussed in Chapters 10 and 11, depends mainly on two variables: specific fire load and ventilation. In nonindustrial, nonmercantile buildings, the fire load is usually moderate or low (Table 10.1), and the relatively high ratio of window area to floor area for individual compartments (higher than 0.1) ensures good ventilation. Consequently, a fire will probably be fuel-surface-controlled; in other words, short (Eq. 10.11b) and of low destructive potential (Eq. 11.3). If, however, the compartment is poorly ventilated, the fire will be ventilation-controlled and, therefore, long (Eq. 10.11a) and destructive, sometimes even if the fire load is moderate. Thus, poorly ventilated fires are dangerous not only on account of their higher destructive potential, but also on account of their longer duration which makes spread more likely.

The role played by the μ-factor (see Chapter 10) in the total fire damage is perhaps even more important than the fire's destructive potential in the individual compartments. The serious structural damages occur mostly in spreading fires, where some key elements of the building become exposed to fire from more than one direction.

16.1 Assessment of Damage

To assess the severity of a fire incident, it has been customary to examine the

---

1 The probability of failure by heat transmission, assumed in the philosophy of fire-resistant compartmentation, is so small that it is not worth considering.

debris found on the scene. Pieces of melted metals, softened glass, charred wood, etc., have been used for estimating the maximum temperature reached by the fire. Yet, even if it could be assessed reasonably accurately, the maximum temperature of the fire does not provide sufficient basis for evaluating the fire's destructive potential. As discussed in Chapter 10, the destructive potential is measured by the normalized heat load, which is proportional to the total heat absorbed (per unit surface area) by the compartment boundaries, and is essentially independent of the temperature history of the fire.

If estimating the severity of fire happens to be the purpose of investigation, the most expedient procedure to follow consists of determining the temperatures reached at some depths below one or more plane surfaces of the compartment's boundaries. Harmathy (1980a) showed that the maximum temperature rise can be expressed in terms of the normalized heat load as[2]

$$T_{\mathrm{m}} - T_{\mathrm{o}} \simeq 0.435 \, \frac{\kappa^{1/2}}{a} \, H \qquad\qquad [16.1]$$

provided that

$$0.8\tau^{1/2} < \frac{a}{\kappa^{1/2}} < 1.2\tau^{1/2} \qquad\qquad [16.2]$$

where a = depth below the surface of the building element, m;
T_{m} = maximum temperature reached (after the cessation of fire exposure) at a depth a, K;
T_{o} = temperature before fire exposure, K;
H = normalized heat load (defined by Eq. 10.23 or Eq. 10.24), $s^{1/2}$ K;
κ = thermal diffusivity of the material, $m^2\,s^{-1}$;
τ = duration of fire exposure (defined as shown in Fig. 9.5), s.

Substituting for H its expression from Eq. 11.2, and solving the resulting equation for τ, one obtains

$$\tau = \left(1.57 \times 10^{-3} \, \frac{a}{\kappa^{1/2}} \, (T_{\mathrm{m}} - T_{\mathrm{o}}) + 42.1\right)^2 - 1380 \qquad [16.3]$$

where τ (s) now means the duration of a standard test fire that would cause, at a depth a from the surface, the temperature to rise from T_{o} to T_{m}. In Eq. 16.3, 1.57×10^{-3}, 42.1, and 1380 are dimensional constants. To satisfy Eq. 16.2, the depth at which the maximum temperature rise should be determined is about 50 mm for normal-weight concretes and about 30 mm for lightweight concretes.[3]

---

2   This equation is a recast version of Eq. 20 in Harmathy's paper.
3   Owing to the presence of moisture, assessing the 'effective value' of the thermal diffusivity may present some problems. It is estimated that κ is of the order of $0.6 \times 10^{-6}\,m^2\,s^{-1}$ for normal-weight concretes, and $0.3 \times 10^{-6}\,m^2\,s^{-1}$ for lightweight concretes. The problem of estimating the effective value of the thermal diffusivity may be circumvented by using Figs 13.14 and 13.15 to find the equivalent standard exposure times. It should be kept in mind, however, that the maximum temperatures reached (after the cessation of fire exposure) are higher by about 30–50 °C than the temperatures attained during the fire exposure.

The penetration of heat can be conveniently studied on core samples 60—80 mm long taken from the walls and ceilings of the fire-exposed compartments. Sometimes simple visual inspection of the samples is sufficient for assessing the damage. Bessey (1950) noted that the color of concrete changes on heating from gray to light pink at about 300 °C. The pink darkens with further rise of temperature, attains maximum intensity around 600 °C, then begins to lighten and turn to whitish gray by 800 °C. Tovey (1986) warned, however, that a concrete that has not turned pink on heating is not necessarily sound. The color change is due to the presence of ferrous compounds in the aggregates, and is prominent mainly with siliceous-aggregate concretes.

More reliable conclusions on the extent of heat penetration can be drawn by subjecting sections of the core samples to thermogravimetric or thermoluminescence analysis. The thermogravimetric tests make use of the fact that the cement paste undergoes a continuous series of dehydration reactions in the 100—850 °C temperature interval (Fig. 6.2). Thus, a specimen of fire-exposed concrete will exhibit in the test little or no loss of mass up to the maximum temperature attained in the fire (Harmathy 1968). By testing several small samples (about 500 mg each) taken from various sections of the cores, the distribution of maximum temperatures reached at various depths in the compartment's boundary elements during the fire can be determined.

Since the presence of inert aggregates in the test sample tends to obscure the clarity of the conclusion, it is advisable to remove from the sample most of the aggregates. Furthermore, since the dehydration reactions are not all irreversible, it is recommended that the core samples be taken as soon following the fire incident as possible, and the samples be kept in a desiccator until tested.

Apparatus for performing thermogravimetric tests is widely available. Figure 16.1 shows the results of thermogravimetric tests performed on portland cement paste samples exposed to various temperatures. For best results, it is recommended that the tests be conducted at rates of heating no higher than 0.5 °C per minute.

The thermoluminescence analysis utilizes the characteristic of some minerals (quartz, feldspar, calcite, dolomite, calcium fluoride, etc.) to emit very faint light (detectable only by suitable instruments) when heated (Placido 1980; Smith and Placido 1983; Chew 1988). The light intensity versus temperature curve obtained at constant heating rate depends on the sample's thermal and radiation history. Once exposed to elevated temperatures, the emission of light by the material is reduced up to the temperature of exposure. By comparing the thermoluminescence curves for aggregate samples obtained from various sections of the core with the curve of a sample taken from the unexposed material, the temperature history of the concrete element during the fire can be roughly assessed. The material needed for each test is about 500 mg. Unfortunately, the test results are not unambiguous, as the intensity of light emission depends not only on the temperature but also on the duration of previous exposure.

The technique for preparing the samples for thermoluminescence tests and the components of the test equipment have been described by Placido (1980) and Chew (1988).

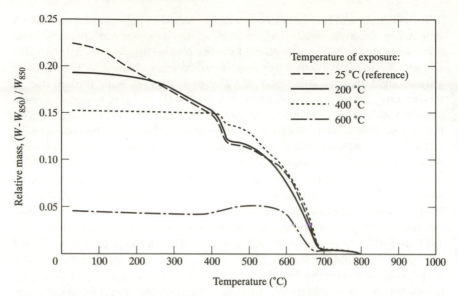

Fig. 16.1 Results of thermogravimetric tests performed on a portland cement paste exposed to various temperatures: W = mass at temperature T, W_{850} = mass at 850 °C; rate of heating 0.5 °C per min (Harmathy 1968)

Since 300 °C is regarded as the limit up to which the strength of concrete is not significantly affected by exposure to fire, it is generally recommended that the repair begin with the removal of the whitish gray and pink layers of concrete. With flexural elements, the fire usually occurs on the side where the strength of concrete is not counted on in the design and, therefore, it does not always make sense to remove indiscriminately all the concrete of which the temperature has exceeded a certain limit.

Decisions on the concrete layers to be removed should be based on a careful examination of the condition of the entire fire-exposed surface. Sometimes it may be sufficient to take 'soundings' on the damaged sections, using a hammer and a chisel (Concrete Society Working Party 1978). The 'ring' of sound concrete and the 'dull thud' of the weak material can be easily distinguished. A Schmidt rebound hammer may be particularly useful, as it delivers a blow of constant force. Low readings on the impact hammer scale, compared with those obtained with undamaged concrete, may reveal the degree of damage.

More accurate information on the extent of damage can be procured by the use of the pulse—echo technique (Carino and Sansalone 1984; Muenow and Abrams 1986). It is analogous to the method used by ships for sounding ocean depths. Mechanical energy produced by the activation of a Schmidt impact hammer is introduced into the concrete. The resulting stress waves, as they propagate through the material, are reflected by cracks or by interfaces between phases of different densities or elastic moduli. The reflected waves (echoes) are monitored by a transducer located in the vicinity of the impactor. The transducer output is displayed on an oscilloscope. If the wave velocity is known, the location of a defect or interface can be determined from the round-trip travel time for the echo.

Some other techniques of estimating the soundness of concrete following fire exposure, namely the ultrasonic pulse velocity test, the Windsor probe test, and the BRE internal fracture test, have been reviewed by Schneider and Nägele (1987). Two methods for determining the thickness of concrete cover over the steel reinforcement have been briefly described by Muenow and Abrams (1986).

If the compressive strength and/or modulus of elasticity of concrete was counted on in the design, it is essential that the layers weakened by the fire be replaced. Figure 6.15 shows that the modulus of elasticity of concrete declines more rapidly with the rise of temperature than does its strength. It has been claimed in a report by The Concrete Society (Concrete Society Working Party 1978) that the permanent loss in the elastic modulus may amount to 40 per cent after exposure to a temperature of 300 °C, and to 85 per cent after exposure to 600 °C. It has also been claimed that the compressive strength of concrete is not severely reduced if its temperature remains below 300 °C.

Unfortunately, because of the heterogeneous nature of concrete, it is next to impossible to make general statements concerning the effect of elevated temperatures on its residual mechanical properties. In contrast, the effect of heating on the residual properties of steel is predictable. As pointed out in Chapter 7, the modulus of elasticity is a structure-insensitive property of steel, therefore its value after fire exposure is not expected to be much different from its value before exposure. Furthermore, unless heated to temperatures above 600 °C, low-carbon, hot-rolled steels will not suffer substantial losses in their yield strength and tensile strength. However, temperatures in excess of 600 °C will cause permanent changes in the grain structure (spheroidization, grain growth), which will result in loss of strength. The finer the grain structure of the steel before the fire, the more severe the loss.

The high-carbon, cold-worked steels used in prestressed concrete flexural elements owe their high strengths partly to their elongated grain structures. Above 450 °C, the grains tend to assume equiaxed shapes, and thereby the steel loses the excess strength imparted to it by cold work. Clearly, the temperature reached by prestressing tendons is a very important aspect of the question of repairability.

The Concrete Society (Concrete Society Working Party 1978) developed a formalized procedure for assessing the fire damage and designing the repair. The classification of damage is based mainly on visual examination. Damages belonging in Class 3 are serious but still repairable. Structures that have suffered Class 4 damages are usually relegated to demolition. The design of repair starts with assigning fire-damage factors to the concrete and to the steel components. The whitish gray and pink concrete is written down to zero strength, and the material that has attained temperatures between 100 and 300 °C is assumed to possess 85 per cent of its original strength. Steel reinforcing bars are usually assumed to have lost 30 per cent of their effectiveness, mainly on account of possible loss of bond between concrete and steel. Bars exposed to fire because of concrete spalling are regarded as having attained a temperature of about 800 °C.

16.2   Techniques of Repair

The repair of a fire-damaged concrete structure consists of the following stages:

(1) removal of unsound concrete,
(2) cleaning of reinforcing bars and (possibly) installation of supplementary bars,
(3) cleaning and roughening of the exposed concrete surface,
(4) replacement of removed concrete.

The removal of unsound concrete and the roughening of the surface is usually done by hand tools and light pneumatic tools, or by sand blasting. According to Ingvarsson and Eriksson (1988) and Silfwerbrand (1990), the removal of concrete, the cleaning and roughening of the new surface, and the cleaning of the reinforcing bars can be done in one step by a technique called water jetting or hydrodemolition, which makes use of the kinetic energy of water at extremely high pressure (up to 400 MPa). Silfwerbrand showed that the surfaces created by water jetting ensure more reliable bonds between the old and new concretes than those created by jack-hammering or sandblasting.

Whenever the amount of concrete to be replaced is substantial, a technique referred to as *shotcreting* or *guniting* is most frequently used. In the dry-mix process, which is generally employed in repairing fire damage, the cement–aggregate mixture is fed by compressed air (at 0.2–0.4 MPa pressure) into a nozzle, where it is intimately mixed with water entering through a ring at a pressure somewhat higher than nozzle pressure. The nozzleman directs the spray more or less perpendicularly to the surface to be coated, from a distance of about 1 m. In overhead work, the shotcrete is applied in 25–50 mm layers at a time. Vertical surfaces may be gunned in layers or a single thickness.

The dry mix is normally proportioned as 3–4 parts of aggregate to 1 part of cement. Because of aggregate rebound from the surface, a dry mix of 3:1 will yield a final proportion of between 2:1 and 2.5:1. The aggregate usually consists of pieces less than 5–10 mm in size. The amount of water is controlled by the nozzleman so as to produce a stiff concrete which adheres to the surface without sagging. It is estimated that the water–cement ratio is about 0.4. With such a low water–cement ratio, the resulting concrete will be of relatively high density (usually between 2240 and 2320 kg m^{-3}) and strength (between 30 and 40 MPa), and of low permeability.

Further information on the practice of shotcreting is available in a report by The Concrete Society (Concrete Society Working Party 1978) and in an ACI Guide (ACI Committee 506 1985).

With prestressed flexural elements, the possibility of shotcrete repair is not very high. Ashton and Bate (1961) claimed that if the temperature of steel in a prestressed beam exceeds 180 °C, the reduction in the stiffness of the beam is likely to be severe, and will normally necessitate replacement of the element. Repair, if practicable, may consist of converting the element to reinforced concrete design, using the tendons as reinforcement, and adding extra bars or mesh within a shotcrete enclosure. Alternatively, external, shotcrete-protected prestressing tendons may be introduced.

Epoxy resins are extensively used in concrete repair. On account of their low molecular masses, all of them are liquid at or slightly above room temperature. Their exceptional molecular mobility during processing permits them to wet thoroughly almost any surface. After the addition of the curing agent, the epoxy resin converts into a highly crosslinked material, without substantial shrinkage or evolution of gases. The high degree of crosslinking produces hardness, strength, and broad chemical resistance (Deanin 1972). The compressive strength of the product is 80–115 MPa, and its tensile strength is 45–85 MPa (Plecnik et al. 1980, 1982).

When used for repairing cracked concrete, the epoxy liquid is injected into the cracks at pressures between 0.7 and 1.4 MPa, or simply poured whenever possible.

Unfortunately, the use of epoxy-repaired concrete elements is limited by the fact that the epoxy plastic rapidly loses its strength if heated above 100 °C, and starts decomposing at a temperature of about 120 °C. Consequently, unless protected at some critical locations by other materials, an epoxy-repaired concrete element may not be regarded as fire-resistant.

Small damages, up to a depth of about 30 mm, may be repaired by hand-application of mortar or polymer-modified mortar. To ensure adequate adhesion, the areas to be repaired should be carefully prepared. Plaster or spayed mineral preparations may also be used to restore fire resistance, but not strength and durability (Tovey 1986).

Nomenclature

a	depth below the surface, m
H	normalized heat load, $s^{1/2}$ K
T_m	maximum temperature reached at a depth a, K (°C)
T_o	original temperature, K (°C)
W	mass at a temperature T, kg
W_{850}	mass at 850 °C, kg
κ	thermal diffusivity, $m^2\,s^{-1}$
τ	duration of fire exposure, s

References

Chapter numbers in parentheses are the chapter(s) of this book to which the reference relates.

Abramowitz M, Stegun I A 1970 *Handbook of Mathematical Functions with Formulas, Graphs, and Mathematical Tables*, Applied Mathematics Series No. 55, National Bureau of Standards, Washington DC, p. 295 (Chap. 13)

Abrams D A 1918 Design of concrete mixtures. *Bulletin No. 1*, Structural Materials Search Laboratory, Lewis Institute, Chicago IL (Chap. 6)

Abrams M S 1971 Compressive strength of concrete at temperatures to 1600 °F. In *Temperature and Concrete*, Publication SP-25, Proceedings of the Symposium on the Effect of Temperature on Concrete, November 1968, Memphis TN, American Concrete Institute, Detroit MI, p. 33 (Chap. 6)

Abrams M S 1979 Behavior of inorganic materials in fire. In *Design of Buildings for Fire Safety*, ASTM STP 685, Smith E E, Harmathy T Z (Eds), American Society for Testing and Materials, Philadelphia PA, p. 14 (Chap. 13)

Abrams M S, Erlin B 1967 Estimating post-fire strength and exposure temperature of prestressing steel by a metallographic method. *Journal of the PCA Research and Development Laboratories* 9: 23 (Chap. 7)

Abrams M S, Gustaferro A H 1968 Fire endurance of concrete slabs as influenced by thickness, aggregate type, and moisture. *Journal of PCA Research and Development Laboratories* 10: 9 (Chaps 12, 13)

Abrams M S, Gustaferro A H 1969 Fire endurance of two-course floors and roofs. *ACI Journal, Proceedings* 66: 92 (Chap. 12)

Abrams M S, Gustaferro A H, Salse E A B 1971 Fire tests of concrete joist floors and roofs. *Research and Development Bulletin RD006.01B*, Portland Cement Association, Skokie IL (Chap. 14)

Abrams M S, Orals D L 1965 Concrete drying methods and their effect on fire resistance. In *Moisture in Materials in Relation to Fire Tests*, ASTM STP 385, American Society for Testing and Materials, Philadelphia PA, p. 52 (Chap. 4)

ACI Committee 216 1989 *Guide for Determining the Fire Endurance of Concrete Elements* ACI 216R-89, American Concrete Institute, Detroit MI (Chaps 6, 12, 13, 14)

ACI Committee 318 1989 Building code requirements for reinforced concrete (ACI 318-89) and commentary. In *ACI Manual of Concrete Practice, 1990 — Part 3: Use of Concrete in Buildings — Design, Specifications, and Related Topics*, American Concrete Institute, Detroit MI (Chaps 6, 14)

ACI Committee 506 1985 Guide to shotcrete. ACI 506R-85. In *ACI Manual of Concrete Practice 1990*, Part 5, American Concrete Institute, Detroit MI (Chap. 16)

Alexeeff G V, Packham S C 1984 Evaluation of smoke toxicity using concentration−time products. *Journal of Fire Sciences* **2**: 362 (Chap. 9)

Allen D E, Lie T T 1974 Further studies of the fire resistance of reinforced concrete columns. *Technical Paper No. 416, NRCC 14047*, Division of Building Research, National Research Council of Canada, Ottawa (Chap. 13)

American Concrete Institute 1989 *ACI Manual of Concrete Practice, 1989 — Part 1: Materials and General Properties of Concrete*, ACI, Detroit MI (Chap. 6)

American Society for Testing and Materials 1982 ASTM E05.11 Task Group, Repeatability and reproducibility of results of ASTM E 119 fire tests. *Research Report No. RR: E5-1003*, ASTM, Philadelphia PA (Chaps 11, 13)

American Society for Testing and Materials 1983a C 88—83 Standard test for soundness of aggregates by use of sodium sulfate or magnesium sulfate. *1990 Annual Book of ASTM Standards*, Vol. 04.02, ASTM, Philadelphia PA (Chap. 6)

American Society for Testing and Materials 1983b E 662-83 Standard test method for specific optical density of smoke generated by solid materials. In *Fire Test Standards*, 3rd Edition, 1990, ASTM, Philadelphia PA (Chap. 9)

American Society for Testing and Materials 1986 C 135-86, Test method for true specific gravity of refractory materials by water immersion. *1990 Annual Book of ASTM Standards*, Vol. 15.01, ASTM, Philadelphia PA (Chap. 4)

American Society for Testing and Materials 1987a D 2863-87 Standard method for the minimum oxygen concentration to support candle-like combustion of plastics (oxygen index). In *Fire Test Standards*, 3rd Edition, 1990, ASTM, Philadelphia PA (Chap. 9)

American Society for Testing and Materials 1987b E 162-87 Standard test method for surface flammability of materials using a radiant heat energy source. In *Fire Test Standards*, 3rd Edition, 1990, ASTM, Philadelphia PA (Chap. 9)

American Society for Testing and Materials 1988a E 119−88, Standard methods of fire tests of building construction and materials. In *ASTM Fire Test Standards*, 3rd Edition, 1990, Philadelphia PA (Chaps 3, 4, 12)

American Society for Testing and Materials 1988b D 2843-88 Standard test method for density of smoke from the burning or decomposition of plastics. In *Fire Test Standards*, 3rd Edition, 1990, ASTM, Philadelphia PA (Chap. 9)

American Society for Testing and Materials 1988c D 3894-88 Standard test method for evaluation of fire response of rigid cellular plastics using a small corner configuration. In *Fire Test Standards*, 3rd Edition, 1990, ASTM, Philadelphia PA (Chap. 9)

American Society for Testing and Materials 1989a E 84−89a, Standard test method for surface burning characteristics of building materials. In *ASTM Fire Test Standards*, 3rd Edition, 1990, Philadelphia PA (Chaps 3, 9)

American Society for Testing and Materials 1989b E 176-89a, Standard terminology of fire standards. *1989 Annual Book of ASTM Standards*, Vol. 04.07, ASTM, Philadelphia PA (Chap. 15)

American Society for Testing and Materials 1990a E 1321-90 Standard test method for determining material ignition and flame spread properties. In *Fire Test Standards*, 3rd Edition, 1990, ASTM, Philadelphia PA (Chap. 9)

American Society for Testing and Materials 1990b E 1352-90 Standard test method for cigarette ignition resistance of mock-up upholstered furniture assemblies. In *Fire Test Standards*, 3rd Edition, 1990, ASTM, Philadelphia PA (Chap. 9)

American Society for Testing and Materials 1990c E 1353-90 Standard test methods for cigarette ignition resistance of components of upholstered furniture. In *Fire Test Standards*, 3rd Edition, 1990, ASTM, Philadelphia PA (Chap. 9)

American Society for Testing and Materials 1990d E 1354-90 Standard test method for heat and visible smoke release rates for materials and products using an oxygen consumption calorimeter. In *Fire Test Standards*, 3rd Edition, 1990, ASTM, Philadelphia PA (Chap. 9)

American Welding Society 1979 *Structural Welding*, AWS D1. 4-79, AWS, Miami FL (Chap. 7)

Anderberg Y 1978 Mechanical properties of reinforcing steel at elevated temperatures. *Tekniska meddelande nr 36*, Halmstad Järnverk AB, Lund, Sweden (Chap. 7)

Anderberg Y 1983 *Properties of Materials at High Temperatures, Steel*, published by Division of Building Fire Safety and Technology, Lund Institute of Technology, Sweden, on behalf of RILEM F-75015, Paris (Chap. 7)

Anderberg Y, Forsén N E 1982 Fire resistance of concrete structures. *Report LUTVDG/(TVBB-3009)*, Department of Fire Safety Engineering, Lund University, Lund, Sweden (Chap. 14)

Anderberg Y, Thelandersson S 1973 Stress and deformation characteristics of concrete at high temperatures, 1. General discussion and critical review of literature. *Bulletin 34*, Lund Institute of Technology, Lund, Sweden (Chap. 6)

Anderberg Y, Thelandersson S 1976 Stress and deformation characteristics of concrete at high temperatures, 2. Experimental investigation and material behaviour model. *Bulletin 54*, Lund Institute of Technology, Lund, Sweden (Chaps 6, 14)

Anonymous 1971 Horizontal projections in the prevention of spread of fire from storey to storey. *Report No. TR52/75/397*, Commonwealth Experimental Building Station, Australia (Chap. 11)

Anonymous 1976 Fire protection. *Bulletin No. 6*, Canadian Steel Industries Construction Council, Willowdale ON, Canada (Chap. 12)

Anonymous 1987 Early detection of fires in ventilation and air conditioning systems. *Cerberus Alarm*, No. 102, December, p. 1 (Chap. 9)

Ashton L A, Bate S C C 1961 The fire resistance of prestressed concrete beams. *Proceedings of the Institution of Civil Engineers* 17: 15 (Chap. 16)

Associate Committee on the National Building Code 1973 Measures for fire safety in high buildings. *NRCC 13366*, National Research Council of Canada, Ottawa (Chap. 9)

Associate Committee on the National Building Code 1990 *National Building Code of Canada*, 10th Edition, NRCC 30619, National Research Council of Canada, Ottawa (Chap. 11)

Associate Committee on the National Building code 1990 *The Supplement to the National Building Code of Canada*, 3rd Edition, National Research Council of Canada, Ottawa (Chaps 6, 12, 14)

Atkinson G A 1971 The building law in Western Europe: how responsibility for safety and good performance is shared. *Current Paper No. 6/71*, Building Research Station, UK (Chap. 2)

Atkinson G A 1974 Building regulations — the international scene. *Current Paper No. 16/74*, Building Research Establishment, UK (Chap. 2)

Attwood P C 1980 Penetration of fire partitions by plastic pipe. *Fire Technology* 16: 37 (Chap. 11)

Babrauskas V 1979 Full scale burning behavior of upholstered chairs. *NBS Technical Note No. 1103*, National Bureau of Standards, Center for Fire Research, Washington DC (Chap. 9)

Babrauskas V 1980a Estimating room flashover potential. *Fire Technology* **16**: 94 (Chap. 9)

Babrauskas V 1980b Flame length under ceilings. *Fire and Materials* **4**: 119 (Chap. 9)

Babrauskas V 1982a Development of the cone calorimeter — a bench-scale heat release rate apparatus based on oxygen consumption. *NBSIR 82-2611*, National Bureau of Standards, Center for Fire Research, Washington DC (Chap. 9)

Babrauskas V 1982b Will the second item ignite? *Fire Safety Journal* **4**: 281 (Chap. 9)

Babrauskas V 1988 Burning rates. In *SFPE Handbook of Fire Protection Engineering*, DiNenno P J (Ed.), 1st Edition, National Fire Protection Association—Society of Fire Protection Engineers, Quincy/Boston MA, p. 2-1 (Chap. 9)

Babrauskas V, Wickström U G 1979 Thermoplastic pool compartment fires. *Combustion and Flame* **34**: 195 (Chap. 10)

Babrauskas V, Williamson R B 1975 Post-flashover compartment fires. *Report No. UBC FRG 75-1*, Fire Research Group, University of California, Berkeley CA (Chap. 10)

Babrauskas V, Williamson R B 1978 The historical basis of fire testing — Parts I and II. *Fire Technology* **14**: 184, 304 (Chap. 3)

Babrauskas V, Lawson J R, Walton W D, Twilley W H 1982 Upholstered furniture heat release rates measured with a furniture calorimeter. *NBSIR 82-2604*, National Bureau of Standards, Center for Fire Research, Washington DC (Chap. 9)

Babuschkin W I, Mtschedlow-Petrossian O P 1959 Zur Thermodynamik der Reaktionen in den Systemen $Ca(OH)_2-SiO_2-H_2O$, $\beta-C_2S-H_2O$ und C_3S-H_2O unter normalen und hydrothermalen Bedingungen. *Silikattechnik* **10**: 605 (Chap. 6)

Baldwin R, Law M, Allen G, Griffiths L G 1970 Survey of fire-loads in modern office buildings — some preliminary results. *Fire Research Note No. 808*, Joint Fire Research Organization, Borehamwood UK (Chap. 10)

Banks J, Rardin R L 1982 International comparison of fire losses. *Fire Technology* **18**: 268 (Chap. 1)

Barnett C R 1988 Fire separation between external walls of buildings. In *Fire Safety Science*, Proceedings of the Second Symposium of the International Association of Fire Safety Science, Wakamutsu T, Hasemi Y, Sekizawa A, Seeger P G, Pagni P J, Grant C E (Eds), Hemisphere Publishing Corp., New York, p. 841 (Chap. 11)

Beck V, Eaton C, Jarman M, Johnson P, Merewether T, Reddaway L, Tweeddale M 1989 *Analysis of Fire Risk, Part 2*. Task Group 2 Report, The Warren Centre for Advanced Engineering, University of Sydney, Australia (Chap. 1)

Becker J, Bizri H, Bresler B 1974 Fires-T — a computer program for the fire response of structures — thermal. *Report No. UCB FRG 74-1*, Fire Research Group, University of California, Berkeley CA (Chap. 13)

Becker J, Bresler B 1974 FIRES-RC — a computer program for the fire response of structures — reinforced concrete frames. *Report No. UCB FRG 74-3*, Fire Research Group, University of California, Berkeley CA (Chap. 13)

Becker J, Bresler B, Bizri H 1974 Reinforced concrete frames in fire environments. *Meeting Preprint 2250*, ASCE National Structural Engineering Meeting, Cincinnati OH, 22—26 April (Chap. 14)

Becker W, Stanke J 1970 Brandversuche an Stahlbetonfertigstützen (2. Teil). *Deutscher Ausschuss für Stahlbeton*, Heft 215, Wilhelm Ernst & Sohn, Berlin, p. 1 (Chap. 14)

Belles D W, Fisher F L, Williamson R B 1988 How well does ASTM E-84 predict fire performance of textile wallcoverings? *Fire Journal* **82**(1): 24 (Chap. 9)

Benjamin I A 1977 Development of a room fire test. In *Fire Standards and Safety*, ASTM STP 614, Robertson A F (Ed.), American Society for Testing and Materials, Philadelphia

PA, p. 300 (Chap. 9)

Bennetts I D 1981 Elevated temperature behaviour of concrete and reinforcing steel. *Report No. MRL/PS23/81/001*, BHP Melbourne Research Laboratories, Clayton, Victoria, Australia (Chaps 6, 7)

Bennetts I D 1982 Behaviour of concrete elements in fire — Part 2. *Report MRL/PS23/82/005*, Melbourne Research Laboratories, Clayton, Victoria, Australia (Chap. 14)

Berl W G, Halpin B M 1980 Human fatalities from unwanted fires. *John Hopkins APL Technical Digest* **1**(2): 129 (Chap. 1)

Bessey G E 1950 Investigations on building fires: Part 2. The visible changes in concrete or mortar exposed to high temperatures. *National Building Studies Technical Paper No. 4*, HMSO, London (Chap. 16)

Beyler C 1988 Thermal decomposition of polymers. In *SFPE Handbook of Fire Protection Engineering*, DiNenno P J (Ed.), 1st Edition, National Fire Protection Association–Society of Fire Protection Engineers, Quincy/Boston MA, p. 1-165 (Chap 9)

Billmeyer F W 1971 *Textbook of Polymer Science*, John Wiley & Sons, New York (Chaps 8, 9)

Birch F, Clark H 1940 The thermal conductivity of rocks and its dependence upon temperature and composition. *American Journal of Science* **238**: 529 (Chap. 6)

Blinov V I, Khudiakov G N 1957 The burning of liquid pools. *Doklady Akademi Nauk SSSR* **113**: 1094 (Chap. 9)

Bobrowski J (Ed.) 1975 *Fire Resistance of Concrete Structures*, Report of a Joint Committee of The Institution of Structural Engineers and The Concrete Society, The Institution of Structural Engineers, London (Chap. 14)

Bobrowski J (Ed.) 1978 *Design and Detailing of Concrete Structures for Fire Resistance*, Interim Guidance by a Joint Committee of The Institution of Structural Engineers and The Concrete Society, The Institution of Structural Engineers, London (Chaps 6, 7, 12, 13, 14)

Böge A (Ed.) 1983 *Das Techniker Handbuch*, 7. Auflage, Braunschweig, Germany (Chap. 7)

Bogue R H 1955 *The Chemistry of Portland Cement*, 2nd Edition, Reinhold Publishing Corporation, New York (Chap. 6)

Boutin J P 1983 *Practique du Calcul de la Résistance au Feu des Structures en Béton*, Eyrolles, Paris, France (Chap. 14)

Boyer H E, Gall T L (Eds) 1985 *Metals Handbook, Desk Edition*, American Society for Metals, Metals Park OH (Chap. 7)

Brenden J J 1967 Calorific values of the volatile pyrolysis products of wood. *Combustion and Flame* **11**: 437 (Chap. 9)

Brenden J J 1970 Determining the utility of a new optical test procedure for measuring smoke from various wood products. *Research Paper FPL 137*, US Department of Agriculture, Forest Service, Madison WI (Chap. 9)

Bresler B 1976a Response of reinforced concrete frames to fire. Reprint from the preliminary report, *Tenth Congress, International Association for Bridge and Structural Engineering*, Tokyo, 6–11 September (Chaps 13, 14)

Bresler B 1976b Response of reinforced concrete frames to fire. *Report No. UCB FRG 76-12*, Fire Research Group, University of California, Berkeley CA (Chap. 14)

Bresler B 1977 Fire protection of modern buildings: Engineering response to new problems. *Report No. UBC FRG WP 77-3*, Department of Civil Engineering, University of California, Berkeley CA (Chap. 14)

Bresler B, Iding R H 1983 Fire response of prestressed concrete members. In *Fire Safety of Concrete Structures*, Publication SP-80, Abrams M S (Ed.), American Concrete Institute, Detroit MI, p. 69 (Chaps 6, 13, 14)

Bresler B, Thielen G, Nizamuddin Z, Iding R 1976 Limit state behavior of reinforced concrete frames in fire environment. *Proceedings of the ASCE-IABSE 2nd Conference on Tall Buildings*, Hong Kong, September, p. 310 (Chap. 14)

British Standards Institution 1981 BS 476: Part 7 — *Surface spread of flame test for materials*, BSI, London (Chap. 3)

British Standards Institution 1983 BS 970: Part 1. *General inspection and testing procedures for carbon, carbon manganese, alloy and stainless steels*, BSI, London (Chap. 7)

British Standards Institution 1985 BS 8110:Part 1 — *Code of Practice for Design and Construction*; Part 2 — *Code of Practice for Special Circumstances*; Part 3 — *Design Charts for Simply Reinforced Beams, Doubly Reinforced Beams and Rectangular Columns*, BSI, London (Chap. 14)

British Standards Institution 1987 BS 476 : Part 7 *Method for classification of surface spread of flame of products*, BSI, London (Chap. 9)

Brockenbrough R L, Johnston B G 1968 *USS Steel Design*, United States Steel Corporation, Pittsburgh PA (Chap. 7)

Bruce H D 1959 Experimental dwelling-room fires. *Report No. 1941*, Forest Products Laboratory, US Department of Agriculture, Madison WI (Chap. 9)

Bruggeman D A G 1936 Über die Geltungsbereiche und die Konstantenwerte der verschiedenen Mischkörperformeln Lichteneckers. *Physikalische Zeitschrift* **37**: 906 (Chap. 5)

Brunauer S 1945 *The Adsorption of Gases and Vapors*, Vol. I, *Physical Adsorption*, Princeton University Press, NJ, p. 153 (Chap. 4)

Brunauer S, Greenberg S A 1962 The hydration of tricalcium silicate and β-dicalcium silicate at room temperature. In *Chemistry of Cement*, Vol. I, Proceedings of the Fourth International Symposium, 2–7 October, 1960, Washington DC, National Bureau of Standards Monograph 43 — Vol. I, US Government Printing Office, p. 135 (Chap. 6)

Brunauer S, Emmett P H, Teller E 1938 Adsorption of gases in multimolecular layers. *Journal of the American Chemical Society* **60**: 309 (Chap. 4)

Brunauer S, Hayes J C, Hass W E 1954 The heats of hydration of tricalcium silicate and β-dicalcium silicate. *Journal of Physical Chemistry* **58**: 279 (Chap. 6)

Bryan J L 1979 Design to cope with incipient fires. In *Design of Buildings for Fire Safety*, ASTM STP 685, Smith E E, Harmathy T Z (Eds), American Society for Testing and Materials, Philadelphia PA, p. 169 (Chap. 1)

Brydson J A 1975 *Plastics Materials*, Butterworths, London (Chap. 8)

Bullen M L 1977 A combined overall and surface energy balance for fully developed ventilation controlled liquid fuel fires in compartments. *Fire Research* **1**: 171 (Chap. 10)

Bullen M L, Thomas P H 1979 Compartment fires with noncellulosic fuels. *17th Symposium (International) on Combustion*, The Combustion Institute, Pittsburgh PA, p. 1139 (Chap. 10)

Burgess D S, Strasser A, Grumer J 1961 Diffusive burning of liquids in open trays. *Fire*

Research Abstracts and Reviews **3**: 177 (Chap. 9)

Burnett D S 1987 *Finite Element Method from Concept to Applications*, Addison—Wesley Publishing Company, Reading MA (Chap. 13)

Bush B, Anno G, McCoy R, Gaj R, Small R D 1991 Fuel loads in U.S. cities. *Fire Technology* **27**: 5 (Chap. 10)

Butcher E G, Parnell A C 1983 *Designing for Fire Safety*, John Wiley & Sons, Chichester UK (Chap. 11)

Butcher E G, Chitty T B, Ashton L A 1966 The temperature attained by steel in building fires. *Fire Research Technical Paper No. 15*, HMSO, London (Chap. 10)

Butcher E G, Bedford G K, Fardell P J 1968a Further experiments on temperatures reached by steel in building fires. Paper No. 1, *Joint Fire Research Organization Symposium No. 2*, HMSO, London, p. 2 (Chap. 10)

Butcher E G, Clark J J, Bedford G K 1968b A fire test in which furniture was fuel. *Fire Research Note No. 695*, Joint Fire Research Organization, Borehamwood UK (Chap. 10)

Butler C P 1971 Notes on charring rates in wood. *Fire Research Note No. 896*, Joint Fire Research Organization, Borehamwood, UK (Chap. 9)

Cammerer J S 1942 Die Wärmeleitzahl von Lichtbeton, insbesondere von Gas- und Schaumbeton. *Fortschritte und Forschungen im Bauwesen*, Series B, No. 2, p. 74 (Chap. 6)

Canadian Prestressed Concrete Institute 1978 *Fire Resistance Ratings for Prestressed and Precast Concrete*, CPCI, Canada (Chaps 6, 14)

Canadian Standards Association 1984 *CAN3-A23.3-M84, Design of Concrete Structures for Buildings*, CSA, Rexdale ON (Chap. 14)

Carino N J, Sansalone M 1984 Pulse—echo method for flaw detection in concrete. *NBS Technical Note 1199*, US Department of Commerce, National Bureau of Standards, Gaithersburg MD (Chap. 16)

Carlson C C, Selvaggio S L, Gustaferro A H 1966 A review of studies of the effects of restraint on the fire resistance of prestressed concrete. In *Feuerwiderstandsfähigkeit von Spannbeton — Fire Resistance of Prestressed Concrete*, Kordina K (Ed.), Fédération Internationale de la Précontrainte, Bauverlag GmbH, Wiesbaden, Germany (Chap. 14)

Carman A P, Nelson R A 1921 The thermal conductivity and diffusivity of concrete. *Bulletin 122*, Engineering Experiment Station, University of Illinois (Chap. 6)

Carnelley T 1879 Über die Beziehung zwischen den Schmetzpunkten der Elemente und ihren Ausdehnungscoëfficienten durch Wärme. *Berichte der deutschen chemischen Gesellschaft* **12**: 439 (Chap. 5)

Carslaw H S, Jaeger J C 1959 *Conduction of Heat in Solids*, 2nd Edition, Oxford University Press, UK (Chap. 13)

Castillo C, Durrani A J 1990 Effect of transient high temperature on high-strength concrete. *ACI Materials Journal* **87**: 47 (Chap. 6)

Castino G T, Beyreis J R, Metes W S 1975 Flammability studies of cellular plastic and other building materials used for interior finish. *File Subject 723*, Underwriters Laboratories Inc., Northbrook IL (Chap. 9)

Cement and Concrete Association 1975 *Guides to Good Practice — FIB/CEB Recommendations for the Design of Reinforced and Prestressed Concrete Structural Members for Fire Resistance*, CCA, Wexham Springs UK (Chap. 12)

Chew M Y L 1988 Assessing heated concrete and masonry with thermoluminescence. *ACI Materials Journal* **85**: 537 (Chap. 16)

Choi K K 1987 Fire stops for plastic pipe. *Fire Technology* **23**: 267 (Chap. 11)

Cibula E 1971 The structure of building control — an international comparison. *Current Paper No. 28/71*, Building Research Station, UK (Chap. 2)

Clarke F B, Ottoson J 1976 Fire death scenarios and firesafety planning. *Fire Journal* **70**(3): 20 (Chap. 1)

Collins R E 1961 *Flow of Fluids through Porous Materials*. Reinhold Publishing Corp. New York (Chap. 4)

Comité Euro-International du Béton 1978a *International System of Unified Standard Codes of Practice for Structures*. Vol. I — *Common Unified Rules for Different Types of Construction and Material*; Vol. II — *CEB—FIP Model Code for Concrete Structures*. CEB—FIP international recommendations, *Bulletin d'Information No. 124—125*, CEB, Paris, France (Chap. 14)

Comité Euro-International du Béton 1978b Trial and comparison calculations based on the CEB/FIP Model Code for Concrete Structures. *Bulletin d'Information No. 129*, CEB, Paris, France(Chap. 14)

Comité Euro-International du Béton 1982 Design of concrete structures for fire resistance. Preliminary draft of an Appendix to the CEB—FIP Model Code, *Bulletin d'Information No. 145*, CEB, Paris, France (Chaps 11, 12, 13, 14)

Concrete Society Working Party 1978 Assessment of fire-damaged concrete structures and repair by gunite. *Concrete Society Technical Report No. 15*, The Concrete Society, London (Chap. 7, 16)

Conley J E, Wilson H, Klinefelter T A 1948 Production of lightweight concrete aggregates from clays, shales, slates, and other materials. *R. I. 4401*, Bureau of Mines, US Department of the Interior, Washington DC (Chap. 6)

Copeland L E, Bragg R H 1955 Self-desiccation in portland cement pastes. *ASTM Bulletin*, No. 204 (February) p. 34 (Chap. 6)

Copeland L E, Kantro D L 1969 Hydration of portland cement. In *Hydration of Cements*, Vol. II, Proceedings of The Fifth International Symposium on the Chemistry of Cement, 7—11 October 1968, Tokyo, p. 387 (Chap. 6)

Copeland L E, Kantro D L, Verbeck G 1962 Chemistry of hydration of portland cement. In *Chemistry of Cement*, Vol. I, Proceedings of the Fourth International Symposium, 2—7 October 1960, Washington DC, National Bureau of Standards Monograph 43 — Volume I, US Government Printing Office, p. 429 (Chap. 6)

Copier W J 1979 The spalling of normal weight and lightweight concrete on exposure to fire. *Heron* **24**(2) (Chap. 4)

Corben R W, Newitt D M 1955 The mechanism of the drying of solids, Part VI — The drying characteristics of porous granular material. *Transactions of the Institution of Chemical Engineers* **33**: 52 (Chap. 4)

Cornell C A 1969 A probability-based structural code. *ACI Journal, Proceedings* **66**: 974 (Chap. 11)

Cruz C R 1968 Apparatus for measuring creep of concrete at high temperatures. *Journal, PCA Research and Development Laboratories* **10**: 36; also *Research Department Bulletin*, No. 225, Portland Cement Association, Skokie IL (Chap. 6)

Culver C G 1978 Characteristics of fire loads in office buildings. *Fire Technology* **14**: 51 (Chap. 10)

Dalkey N C 1969 The Delphi method: an experimental study of group opinion. *Report RM-5888-PR*, Rand Corporation, Santa Monica CA (Chap. 15)

Day M F, Jenkinson E A, Smith A I 1961 Effect of elevated temperatures on high-tensile steel wires for prestressed concrete. *Proceedings of the Institution of Civil Engineers* **16**: 55 (Chap. 7)

Dayan A 1982 Self-similar temperature, pressure and moisture distributions within an intensely heated porous half space. *International Journal of Heat and Mass Transfer* **25**: 1469 (Chap. 4)

Dayan A, Gluekler E L 1982 Heat and mass transfer within an intensely heated concrete slab. *International Journal of Heat and Mass Transfer* **25**: 1461 (Chap. 4)

Deanin R D 1972 *Polymer Structure, Properties and Applications*, Cahners Publishing Co., Boston MA (Chaps 8, 16)

de Boer J H 1958 The shapes of capillaries. In *The Structure and Properties of Porous Materials*, Everett D H, Stone F S (Eds), Proceedings of the Tenth Symposium of the Colston Research Society, held in the University of Bristol, 24–27 March, p. 68 (Chap. 4)

Debye P 1914 *Vorträge über die kinetische Theorie der Materie und Elektrizität*. Gehalten in Göttingen auf Einladung der Kommission der Wolfskehlstiftung, B G Teubner, Leipzig, p. 46 (Chap. 5)

Delichatsios M A 1988 Air entrainment into buoyant jet flames and pool fires. In *SFPE Handbook of Fire Protection Engineering*, DiNenno P J (Ed.), 1st Edition, National Fire Protection Association–Society of Fire Protection Engineers, Quincy/Boston MA, p. 1-306 (Chap. 9)

deRis J N 1969 Spread of a laminar diffusion flame. *12th Symposium (International) on Combustion*, The Combustion Institute, Pittsburgh PA, p. 241 (Chap. 9)

deRis J 1985 Flammability testing state-of-the-art. *Fire and Materials* **9**: 75 (Chap. 9)

Desai C S 1979 *Elementary Finite Element Method*, Prentice-Hall, Englewood Cliffs NJ (Chap. 13)

Dettling H 1964 Die Wärmedehnung des Zementsteines, des Gesteine und der Betone. *Bulletin No. 164*, Deutscher Ausschuss für Stahlbeton, Berlin (Chap. 6)

Deutsches Institut für Normung 1961 DIN 17007 Part 2. *Material numbers; System of the Principal Group 1; Steel* (Chap. 7)

Deutsches Institut für Normung eV 1978 DIN 18230 *Structural fire protection in industrial building construction*: Part 1: Required fire resistance period; Part 2: Determination of the burning factor *m*; Appendix 1 to Part 1: Calorific value and *m* factors. Draft in manuscript form, Berlin (Chap. 11)

Deutsches Institut für Normung 1980 DIN 17100 *Steels for General Structural Purposes; Quality Specifications* (Chap. 7)

Deutsches Institut für Normung 1981 DIN 4102 Part 1 — *Fire behaviour of building materials and building components — building materials, concepts, requirements and tests* (Chap. 3)

De Vries D A 1952 The thermal conductivity of granular materials. In *Problems Relating to Thermal Conductivity, Bulletin de l'Institut International du Froid, Annexe 1952–1*, Louvain (Belgique), p. 115 (Chap. 5)

Dias W P S, Khoury G A, Sullivan P J 1990 Mechanical properties of hardened cement paste exposed to temperatures up to 700 °C (1292 °F). *ACI Materials Journal* **87**: 160 (Chap. 6)

Diederichs U, Schneider U 1981 Bond strength at high temperatures. *Magazine of Concrete Research*, **33**(115): 75 (Chap. 6)

DiNenno P J, Beyler C L, Custer R L P, Walton W D, Watts J M (Eds) 1988 *SFPE*

Handbook of Fire Protection Engineering, 1st Edition, National Fire Protection Association—Society of Fire Protection Engineers, Quincy/Boston MA (Chap. 9)

Dorn J E 1954 Some fundamental experiments on high-temperature creep. *Journal of the Mechanics and Physics of Solids* 3: 85 (Chap. 5)

Dougill J W 1983 Materials dominated aspects of design for structural fire resistance of concrete structures. In *Fire Safety of Concrete Structures*, Abrams M S (Ed.), Publication SP-80, American Concrete Institute, Detroit MI, p. 151 (Chap. 4)

Douglas J 1961 A survey of numerical methods for parabolic differential equations. In *Advances in Computers*, Alt F L (Ed.), Academic Press, New York, p. 1 (Chap. 13)

Drysdale D 1985 *An Introduction to Fire Dynamics*, John Wiley & Sons, Chichester UK (Chap. 9)

Drysdale D D 1988 Thermochemistry. In *SFPE Handbook of Fire Protection Engineering*, DiNenno P J (Ed.), 1st Edition, National Fire Protection Association—Society of Fire Protection Engineers, Quincy/Boston MA, p. 1-146 (Chap. 9)

DTU (Document Technique Unifié) 1980 Méthode de prévision par le calcul du comportement au feu des structures en béton. *Cahiers du Centre Scientifique et Technique du Bâtiment No. 208*, Paris, France (Chap. 14)

Dunham J W, O'Connor W J, Ingberg S H, Thorud B M, Diener C N 1942 Fire resistance classification of building materials. *BMS Report No. 92*, National Bureau of Standards, Washington DC (Chap. 12)

Dusinberre G M 1949 *Numerical Analysis of Heat Flow*, McGraw-Hill, New York (Chap. 13)

Eitel W 1952 *Thermochemical Methods in Silicate Investigation*, Rutgers University Press, New Brunswick NJ (Chaps 5, 6)

El-Shayeb M 1986 Fire resistance of concrete-filled and reinforced concrete columns. PhD Dissertation, University of New Hampshire, Durham NH (Chap. 13)

Emery A F, Carson W W 1969 Evaluation of use of the finite element method in computation of temperature. *ASME Publication, No. ZUX ASME 69-WA/HT-38*, paper presented at the American Society of Mechanical Engineers Winter Annual Meeting, 16—20 November, Los Angeles CA (Chap. 13)

Emmons H W 1943 The numerical solution of heat-conduction problems. *Transactions of the American Society of Mechanical Engineers* 65: 607 (Chap. 13)

Emmons H W 1981 The calculation of a fire in a large building. *20th Joint ASME/AIChE National Heat Transfer Conference*, August, Milwaukee (Chap. 9)

Eshbach O W (Ed.) 1952 *Handbook of Engineering Fundamentals*, 2nd Edition, John Wiley & Sons, New York (Chap. 14)

Everett D H 1958 Some problems in the investigation of porosity by adsorption methods. In *The Structure and Properties of Porous Materials*, Everett D H, Stone F S (Eds.), Proceedings of the Tenth Symposium of the Colston Research Society, held in the University of Bristol, 24—27 March, p. 95 (Chap.14)

Fang J B 1975 Measurement of the behavior of incidental fires in a compartment. *NBSIR 75-679*, National Bureau of Standards, Washington, DC (Chap. 9)

Fédération Internationale de la Précontrainte 1984 *Practical Design of Reinforced and Prestressed Concrete Structures*, based on the CEB—FIP Model Code (MC78). FIP Recommendations, Thomas Telford Limited, London (Chap. 14)

Feldman R F, Sereda P J 1969 Written discussion of 'Structures and physical properties of cement paste' by Verbeck G J and Helmuth R H. In *Properties of Cement Paste and Concrete*, Vol. III, Proceedings of The Fifth International Symposium on the Chemistry of Cement, 7–11 October 1968, Tokyo, p. 36 (Chaps 4, 6)

Fenimore C P, Martin F J 1966 Flammability of polymers. *Combustion and Flame* **10**: 135) (Chap. 9)

Feret R 1897 Studies on the internal constitution of hydraulic mortars. *Bulletin de la Société d'Encouragement pour l'Industrie Nationale*, Volume II (Chap. 6)

Fernandez-Pello A C 1977 Fire spread over vertical fuel surfaces under the influence of externally applied thermal radiation. *Home Fire Project Technical Report No. 19*, Harvard University, MA (Chap. 9)

Fernandez-Pello A C 1978 A theoretical model for the upward laminar spread of flames over vertical fuel surfaces. *Combustion and Flame* **31**: 135 (Chap. 9)

Ferran J 1956 Contribution minéralogique à l'étude de l'adhérence entre les constituants hydratés des ciments et les matériaux enrobés. *Revue des Matériaux de Construction*, No. 490-491, p. 155; No. 492, p. 191 (Chap. 6)

Fire Defense Agency 1988 *White Book on Fire Service in Japan*, Ministry of Home Affairs, Tokyo, Japan (Chap. 1)

Fisher E A 1923, 1927 Some factors affecting the evaporation of water from soil, Parts I and II. *Journal of Agricultural Science* **13**: 121; **17**: 407 (Chap. 4)

Fisher R 1985 The building regulations 1985. *Fire Prevention* No. 184 (November), p. 16 (Chap. 2)

Fleming R P 1988 Automatic sprinkler system calculations. In *SFPE Handbook of Fire Protection Engineering*, DiNenno P J (Ed.), 1st Edition, National Fire Protection Association–Society of Fire Protection Engineers, Quincy/Boston MA, p. **3**-22 (Chap. 9)

Forest Products Laboratory 1974 *Wood Handbook: Wood as an Engineering Material*, Agriculture Handbook No. 72, Forest Service, US Department of Agriculture, US Government Printing Office, Washington DC (Chap. 8)

Forsythe G E, Wasow W R 1960 *Finite-Difference Methods for Partial Differential Equations*, John Wiley & Sons, New York (Chap. 13)

Francis A O, Burdette E G, Deatherage J H 1991 Elastic modulus, Poisson's ratio, and compressive strength relationships at early ages. *ACI Materials Journal* **88**: 3 (Chap. 6)

Gamble W L 1989 Predicting protected steel member fire endurance using spread-sheet programs. *Fire Technology* **25**: 256 (Chap. 13)

Garnham Wright J H 1983 *Building Control by Legislation: The UK Experience*, John Wiley & Sons, New York (Chap. 2)

Gaskill J R 1970 Smoke development in polymers during pyrolysis or combustion. *Journal of Fire and Flammability* **1**: 183 (Chap. 9)

Gebhart B 1961 *Heat Transfer*, McGraw-Hill, New York, pp. 73, 111 (Chap. 13)

Geller L B, Buchanan R M, Reeves J E, Larocque G E, Svikis V D, Hanes F E, Soles J A, Gray W M, Twidale M A 1962 Jet-piercing research project. *Mines Branch Investigation Report IR 62-27*, Department of Mines and Technical Surveys, Ottawa, Canada (Chap. 6)

Geymayer H G 1972 Effect of temperature on creep of concrete; a literature review. In *Concrete for Nuclear Reactors*, Vol. I, Publication SP-34, American Concrete Institute, Detroit MI, p. 565 (Chap. 6)

Gomberg A, Buchbinder B, Offensend F F 1982 Evaluating alternative strategies for reducing residential fire loss — the fire loss model. *NBSIR 82-2551*, National Bureau of Standards, Washington, DC (Chap. 9)

Green C H, Brown R 1978 Life safety: What is it and how much is it worth? *Current Paper No. 52/78*, Building Research Establishment, UK (Chap. 1)

Griffith A 1937 Physical properties of typical American rocks. *Bulletin No. 131*, Engineering Experiment Station, Iowa State College, IA (Chap. 6)

Gross D 1962 Experiments on the burning of cross piles of wood. *Journal of Research, National Bureau of Standards* **66C**: 99 (Chap. 9)

Gross D, Robertson A F 1965 Experimental fires in enclosures. *10th Symposium (International) on Combustion*, The Combustion Institute, Pittsburgh PA, p. 931 (Chap. 10)

Gross D, Loftus J J, Robertson A F 1967 Method for measuring smoke from burning materials. In *Symposium on Fire Test Methods — Restraint and Smoke 1966*, ASTM STP 422, American Society for Testing and Materials, Philadelphia PA, p. 166 (Chap. 9)

Gross H 1975 High-temperature creep of concrete. *Nuclear Engineering and Design* **32**: 129 (Chap. 6)

Grubits S J 1970 Smoke produced by burning building materials in oxygen-deficient atmospheres. *Technical Record 44/153/391*, Commonwealth Experimental Building Station, Australia (Chap. 9)

Gustaferro A H 1970 Temperature criteria at failure. In *Fire Test Performance*, ASTM STP 464, American Society for Testing and Materials, Philadelphia PA, p. 68 (Chap. 14)

Gustaferro A H 1986 Design implementation — concrete structures. In *Design of Structures against Fire*, Anchor R D, Malhotra H L, Purkiss J A (Eds), Elsevier Applied Science Publishers, London UK, p. 189 (Chap. 14)

Gustaferro A H, Abrams M S, Litvin A 1971a Fire resistance of lightweight insulating concretes. In *Lightweight Concrete*, Publication SP-29, American Concrete Institute, Detroit MI, p. 161 (Chap. 13)

Gustaferro A H, Abrams M S, Salse E A B 1971b Fire resistance of prestressed concrete beams — Study C: Structural behavior during fire tests. *Research and Development Bulletin RD009.01B*, Portland Cement Association, Skokie IL (Chap. 14)

Gustaferro A H, Martin L D 1989 *Design for Fire Resistance of Precast Prestressed Concrete*. 2nd Edition, Prepared for the PCI Fire Committee, Prestressed Concrete Institute, Chicago IL (Chap. 14)

Gustafson D P (Ed.) 1980 *Reinforced Concrete Fire Resistance*. Prepared under the direction of the CRSI Engineering Practice Committee by the CRSI Committee on Fire Ratings, Concrete Reinforcing Steel Institute, Chicago IL (Chaps 11, 12, 13, 14)

Hägglund B 1983 A room fire simulation model. *FOA Report C 20501-D6*, National Defense Research Institute, Stockholm, Sweden (Chap. 9)

Hägglund B, Jansson R, Onnermark B 1974 Fire development in residential rooms after ignition by nuclear explosions. *FOA Report C 20016-D6 (A3)*, Forsvarets Forskningsanstalt, Stockholm, Sweden (Chap. 9)

Haksever A 1977a *Zur Frage des Trag- und Verformungsverhaltens ebener Stahlbetonrahmen im Brandfall*, Heft 35, Institut für Baustoffkunde und Stahlbetonbau der Technischen Universität Braunschweig, Germany (Chap. 14)

Haksever A 1977b Rechnerische Untersuchung zum Brandverhalten statisch unbestimmter

Stahlbetonstabtragwerke. In *Arbeitsbericht 1975—7*, Vol. I *Sonderforschungsbericht 148, 'Brandverhalten von Bauteilen'*, Technische Universität Braunschweig, Germany (Chap. 14)

Haksever A 1982 *Stahlbetonstützen mit Rechteckquerschnitten bei natürlichen Bränden*. Heft 52, Institut für Baustoffe, Massivbau und Brandschutz der Technischen Universität Braunschweig, Germany (Chap. 14)

Haksever A, Anderberg Y 1982 Comparison between measured and computed structural response of some reinforced concrete columns in fire. *Fire Safety Journal*, **4**: 293 (Chap. 14)

Hall J R, Sekizawa A 1991 Fire risk analysis: general conceptual framework for describing models. *Fire Technology* **27**: 33 (Chap. 15)

Hamilton R L, Crosser O K 1962 Thermal conductivity of heterogeneous two-component systems. *Industrial & Engineering Chemistry Fundamentals* (American Chemical Society) 411: 187 (Chap. 5)

Hansen T C 1968 Theories of multi-phase materials applied to concrete, cement mortar and cement paste. In *The Structure of Concrete*, Brooks A E, Newman K (Eds), Proceedings of an International Conference, London, September 1965, Cement and Concrete Association/Williams Clowes and Sons, London, p. 16 (Chap. 6)

Harada T, Takeda J, Yamane S, Furumura F 1972 Strength, elasticity and thermal properties of concrete subjected to elevated temperatures. In *Concrete for Nuclear Reactors*, Vol. I, Publication SP-34, American Concrete Institute, Detroit MI, p. 377 (Chap. 6)

Harmathy T Z 1961 A treatise on theoretical fire resistance rating. In *Symposium on Fire Test Methods*, ASTM STP 301, American Society for Testing and Materials, Philadelphia PA, p. 10 (Chaps 8, 13)

Harmathy T Z 1963 Temperature distribution in homogeneous slabs during fire tests. *Transactions of the Engineering Institute of Canada* **6** (B-6), Paper No. EIC-63-MECH 6 (Chap. 13)

Harmathy T Z 1964 Variable-state methods of measuring the thermal properties of solids. *Journal of Applied Physics* **35**: 1190 (Chaps 5, 6, 8)

Harmathy T Z 1965a Effect of moisture on the fire endurance of building elements. In *Moisture in Materials in Relation to Fire Tests*, ASTM STP 385, American Society for Testing and Materials, Philadelphia PA, p. 74 (Chaps 4, 13)

Harmathy T Z 1965b Ten rules of fire endurance rating. *Fire Technology* **1**: 93 (Chap. 12)

Harmathy T Z 1967a Moisture sorption of building materials, *Technical Paper No. 242*, Division of Building Research, National Research Council of Canada, Ottawa (Chaps 4, 6)

Harmathy T Z 1967b *Simultaner Feuchtigkeits- und Wärmetransport in porigen Systemen mit besonderem Hinweis auf Trocknung*. Doctoral dissertation, Technische Hochschule, Wien (Chap. 4)

Harmathy T Z 1967c Experimental study on moisture and fire endurance. *Fire Technology* **2**: 52 (Chap. 4)

Harmathy T Z 1967d Deflection and failure of steel-supported floors and beams in fire. In *Symposium on Fire Test Methods — Restraint and Smoke*, ASTM STP 422, American Society for Testing and Materials, Philadelphia PA, p. 40 (Chaps 5, 7)

Harmathy T Z 1967e A comprehensive creep model. *Transactions of the ASME, Series D, Journal of Basic Engineering* **89**(Series D): 496 (Chap. 5)

Harmathy T Z 1968 Determining the temperature history of concrete constructions following fire exposure. *Journal of the American Concrete Institute*, **65**(11): 959 (Chap. 16)

Harmathy T Z 1969 Simultaneous moisture and heat transfer in porous systems with particular reference to drying. *Industrial & Engineering Chemistry Fundamentals* **8**: 92 (Chap. 4)

Harmathy T Z 1970a Thermal properties of concrete at elevated temperatures. *ASTM Journal of Materials* **5**: 47 (Chaps 5, 6, 13)

Harmathy T Z 1970b Thermal performance of concrete masonry walls in fire. In *Fire Test Performance*, ASTM STP 464, American Society for Testing and Materials, Philadelphia PA, p. 209 (Chaps 12, 13)

Harmathy T Z 1971 Peak-time method for measuring thermal diffusivity of small solid specimens. *AIChE Journal* **17**: 198 (Chap. 6)

Harmathy T Z 1972 A new look at compartment fires, Parts I and II. *Fire Technology* **8**: 196, 326 (Chaps 9, 10)

Harmathy T Z 1976 Design of buildings for fire safety. *Fire Technology* **12**: 95, 219 (Chap. 11)

Harmathy T Z 1977 Building design and the fire hazard. *Wood and Fiber* **9**: 127 (Chaps 9, 15)

Harmathy T Z 1977/1978 Performance of building elements in spreading fires. *Fire Research* **1**: 119 (Chaps 12, 13)

Harmathy T Z 1978a Experimental study on the effect of ventilation on the burning of piles of solid fuels. *Combustion and Flame* **31**: 259 (Chap. 9)

Harmathy T Z 1978b Mechanism of burning of fully developed compartment fires. *Combustion and Flame* **31**: 265 (Chap. 10)

Harmathy T Z 1979a Design to cope with fully developed fires. In *Design of Buildings for Fire Safety*, ASTM STP 685, Smith E E, Harmathy T Z (Eds), American Society for Testing and Materials, Philadelphia PA, p. 198 (Chaps 10, 12)

Harmathy T Z 1979b Effect of the nature of fuel on the characteristics of fully developed compartment fires. *Fire and Materials* **3**: 49 (Chap. 10)

Harmathy T Z 1980a The possibility of characterizing the severity of fires by a single parameter. *Fire and Materials* **4**: 71 (Chaps 3, 10, 16)

Harmathy T Z 1980b Ventilation of fully developed compartment fires. *Combustion and Flame* **37**: 25 (Chap. 10)

Harmathy T Z 1980/1981 Some overlooked aspects of the severity of compartment fires. *Fire Safety Journal* **3**: 261 (Chap. 10)

Harmathy T Z 1981 The fire resistance test and its relation to real-world fires. *Fire and Materials* **5**: 93 (Chap. 3)

Harmathy T Z 1983a Fire severity: basis of fire safety design. In *Fire Safety of Concrete Structures*, Abrams M S (Ed.), Publication SP-80, American Concrete Institute, Detroit MI, p. 115 (Chaps 3, 10)

Harmathy T Z 1983b Properties of building materials at elevated temperatures. *DBR Paper No. 1080*, Division of Building Research, NRCC 20956, National Research Council of Canada, Ottawa (Chaps 6, 8)

Harmathy T Z 1984 Burning, pyrolysis, combustion and char-oxidation. *Fire and Materials* **8**: 224 (Chap. 9)

Harmathy T Z 1985a Correction of the results of standard fire resistance tests. *Journal of Testing and Evaluation* **13**: 303 (Chap. 3)

Harmathy T Z 1985b $k\rho c$ or $\sqrt{k\rho c}$? — Thermal inertia or thermal absorptivity? *Fire Technology* **21**: 146 (Chap. 3)

Harmathy T Z 1986 A suggested logic for trading between fire safety measures. *Fire and*

Materials **10**: 141 (Chap. 15)

Harmathy T Z 1987 On the equivalent fire exposure. *Fire and Materials* **11**: 95 (Chap. 11)

Harmathy T Z 1988a On the economics of mandatory sprinklering of dwellings. *Fire Technology* **24**: 245 (Chaps 9, 15)

Harmathy T Z 1988b How much fire resistance is really needed? *Concrete International* **10**(12): 40 (Chap. 11)

Harmathy T Z 1990/1991 Design of buildings against fire spread (a review). *Journal of Applied Fire Science* **1**: 65 (Chap. 10)

Harmathy T Z, Allen L W 1973 Thermal properties of selected masonry unit aggregates. *Journal of the American Concrete Institute, Proceedings* **70**: 132 (Chap. 6)

Harmathy T Z, Berndt J E 1966 Hydrated portland cement and lightweight concrete at elevated temperatures. *Journal of the American Concrete Institute, Proceedings* **63**: 93 (Chap. 6)

Harmathy T Z, Lie T T 1971 Experimental verification of the rule of moisture moment. *Fire Technology* **7**: 17 (Chap. 4)

Harmathy T Z, Mehaffey J R 1982 Normalized heat load: a key parameter in fire safety design. *Fire and Materials* **6**: 27 (Chaps 3, 10)

Harmathy T Z, Mehaffey J R 1987 The normalized heat load concept and its use. *Fire Safety Journal* **12**: 75 (Chap. 10)

Harmathy T Z, Oleszkiewicz I 1987 Fire drainage system. *Fire Technology* **23**: 26 (Chap. 11)

Harmathy T Z, Oracheski E W 1970 Equivalent thickness of concrete masonry units. *Building Research Note No. 71*, Division of Building Research, National Research Council of Canada, Ottawa (Chap. 12)

Harmathy T Z, Stanzak W W 1970 Elevated-temperature tensile and creep properties of some structural and prestressing steels. In *Fire Test Performance*, ASTM STP 464, American Society for Testing and Materials, Philadelphia PA, p. 186 (Chap. 7)

Harmathy T Z, Sultan M A 1988 Correlation between the severities of the ASTM E 119 and ISO 834 fire exposures. *Fire Safety Journal* **13**: 163 (Chap. 3)

Harmathy T Z, Sultan M A, MacLaurin J W 1987 Comparison of severity of exposure in ASTM E 119 and ISO 834 fire resistance tests. *Journal of Testing and Evaluation* **15**: 371 (Chap. 3)

Harmathy T Z *et al.* 1989 A decision logic for trading between fire safety measures. *Fire and Materials* **14**: 1 (Chaps 2, 11, 15, 16)

Hartree D R 1958 *Numerical Analysis*, 2nd Edition, Oxford University Press, UK (Chap. 13)

Hartzell G E, Priest D N, Switzer W G 1985 modeling of intoxication of rats by carbon monoxide and hydrogen cyanide. *Journal of Fire Science* **3**: 115 (Chap. 9)

Hasemi Y 1986 Thermal modelling of upward flame spread. In *Fire Safety Science*, Proceedings of the First International Symposium, Grant C E, Pagni P J (Eds), International Association of Fire Safety Science, Hemisphere Publishing Corp., Washington, DC, p. 87 (Chap. 9)

Hashin Z 1962 Elastic moduli of heterogeneous materials. *Journal of Applied Mechanics* **29**: 143 (Chap. 5)

Helmuth R A, Turk D H 1966 Elastic moduli of hardened portland cement and tricalcium silicate pastes. In *Effect of Porosity*, Proceedings of the Symposium on Structure of Portland Cement Paste and Concrete, Washington 1965, Highway Research Board, Special Report 90, p. 139 (Chap. 6)

Hertz K 1980 Bond between concrete and deformed bars exposed to high temperatures. Paper presented at the *Meeting of CIB W 14*, Athens, 19–23 May (Chap.6)

Heselden A J M 1961 Some fires in a single compartment with independent variation of fuel surface area and thickness. *Fire Research Note No. 469*, Joint Fire Research Organization, Borehamwood UK (Chap. 10)

Heskestad G 1983 Luminous heights of turbulent diffusion flames. *Fire Safety Journal* **5**: 103 (Chap. 9)

Heskestad G 1988 Fire plumes. In *SFPE Handbook of Fire Protection Engineering*, DiNenno P J (Ed.), 1st Edition, National Fire Protection Association–Society of Fire Protection Engineers, Quincy/Boston MA, p. **1**-107 (Chap. 9)

Hidnert P, Souder W 1950 *Thermal Expansion of Solids*, US Department of Commerce, National Bureau of Standards Circular 486 (Chap. 5)

Hilado C J 1970 The effect of chemical and physical factors on smoke evolution from polymers. *Journal of Fire and Flammability* **1**: 217 (Chap. 9)

Hilado C J 1982 *Flammability Handbook for Plastics*, Technomic Publication, Westport CT (Chap. 9)

Hirano T, Noreikis S E, Waterman T E 1974 Postulations of flame spread mechanisms. *Combustion and Flame* **22**: 353 (Chap. 9)

Hirsch T J 1962 Modulus of elasticity of concrete affected by elastic moduli of cement paste matrix and aggregate. *Journal of the American Concrete Institute, Proceedings* **59**: 427 (Chap. 6)

Holmes M, Anchor R D, Cook G M E, Crook R N 1982 The effects of elevated temperatures on the strength properties of reinforcing and prestressing steels. *The Structural Engineer* **60B**(1): 7 (Chap. 7)

Home Office 1990 *Fire Statistics — United Kingdom, 1988* Research and Statistics Department, London, UK (Chap. 1)

Honda K 1917 On the thermal expansion of different kinds of steel at high temperatures. *The Science Reports of the Tohoku Imperial University*, First Series, Sendai, Japan, **6**: 203 (Chap. 7)

Hottel H C 1959 Certain laws governing diffusive burning of liquids. *Fire Research Abstracts and Reviews* **1**: 41 (Chap. 9)

Hsu T T, Slate F O, Sturman G M, Winter G 1963 Microcracking of plain concrete and the shape of the stress–strain curve. *Journal of the American Concrete Institute, Proceedings* **60**: 209 (Chap. 6)

Huang C L D 1979 Multi-phase moisture transfer in porous media subject to temperature gradient. *International Journal of Heat and Mass Transfer* **22**: 1295 (Chap. 4)

Huang C L D, Siang H H, Best C H 1979 Heat and moisture transfer in concrete slabs. *International Journal of Heat and Mass Transfer* **22**: 252 (Chap. 4)

Huebner K H, Thornton E A 1982 *The Finite Element Method for Engineers*, 3rd Edition, John Wiley & Sons, New York (Chap. 13)

Huggett C 1980 Estimation of heat release rate by means of oxygen consumption measurements. *Fire and Materials* **4**: 61 (Chap. 9)

Hull W A, Ingberg S H 1926 Fire resistance of concrete columns. *Technologic Paper No. 272*, National Bureau of Standards, US Department of Commerce, Washington DC (Chap. 14)

Hurlbut C S 1959 *Mineralogy*, John Wiley & Sons, New York (Chap. 6)

Iding R, Bresler B 1984 Prediction of fire response of buildings using finite element method.

In *Computing in Civil Engineering, Proceedings of Third Conference*, sponsored by ASCE, CEPA, ICES, NICE — San Diego CA, 2–6 April, p. 213 (Chaps 13, 14)

Iding R, Lee J, Bresler B 1975 Behavior of reinforced concrete under variable elevated temperatures. *Report No. UCB FRG 75-8*, Fire Research Group, University of California, Berkeley CA (Chap. 13)

Ingberg S H 1928 Tests of severity of building fires. *NFPA Quarterly* **22**: 43 (Chaps 10, 11)

Ingvarsson H, Eriksson B 1988 Hydrodemolition for bridge repairs. *Nordisk Betong*, No. 2-3, p. 49 (Chap. 16)

Institution of Engineers of Australia 1989 *Fire Engineering for Building Structures*, Barton, Australia, p. 27 (Chap. 1)

International Standards Organization 1990 ISO/CD 834-1, ISO/CD 834-2, ISO/CD 834-3 — *Fire resistance tests — Elements of building construction, Parts 1, 2, and 3*. Doc. Refs. ISO TC92/SC2/WG1 N322, N321, N321, 10 May (Chaps 3, 4)

International Standards Organization 1991 ISO 5658 (DP) — *Fire test — Reaction to fire — Spread of flame tests of building materials* (Chap. 9)

International Technical Information Institute 1976 *User's Practical Selection Handbook for Optimum Plastics, Rubbers and Adhesives*, ITII, Tokyo, Japan (Chap. 8)

Issen L A 1980 Single-family residential fire and live loads survey. *Report FAA-H-37-72, Task 200, NBSIR 80-2155*, National Bureau of Standards, Washington DC (Chap. 10)

Issen L A, Gustaferro A H, Carlson C C 1970 Fire tests of concrete members: an improved method for estimating thermal restraint forces. In *Fire Test Performance*, ASTM STP 464, American Society for Testing and Materials, Philadelphia PA, p. 153 (Chap. 14)

Johnson K O 1961 Moisture and temperature distributions during drying of insulation boards. *Tappi* **44**: 599 (Chap. 4)

Jones C N 1962 The approximate determination of the fire endurances of structural elements. *Report UP No. 147*, Commonwealth Experimental Building Station, Australia (Chap. 12)

Jones W W 1983 A review of compartment fire models. *IR 83-2684*, National Bureau of Standards, Washington DC (Chap. 9)

Kalousek G L, Prebus A F 1958 Crystal chemistry of hydrous calcium silicates III. Morphology and other properties of tobermorite and related phases. *Journal of the American Ceramic Society* **41**: 124 (Chap. 6)

Kamei S 1937 Untersuchung über die Trocknung fester Stoffe. *Journal of the Society of Chemical Industry, Japan* **40**: 251, 374 (Chap. 4)

Kamei S, Shiomi S 1937 Untersuchung über die Trocknung fester Stoffe. *Journal of the Society of Chemical Industry, Japan* **40**: 257, 325, 366 (Chap. 4)

Kaplan S, Garrick B J 1981 On the quantitative definition of risk. *Risk Analysis* **1**: 11 (Chap. 15)

Kaplan H L, Grand A F, Hartzell G E 1983 *Combustion Toxicology: Principles and Test Methods*, Technomic Publishing Co., Lancaster PA (Chap. 9)

Kardestuncer H, Norrie D H 1987 *Finite Element Handbook*, McGraw-Hill, New York, p. 4.259 (Chap. 13)

Karter M J 1986 Fire losses in the United States during 1985. *Fire Journal*, September, p. 26 (Chap. 15)

Karter M J 1988 U.S. fire losses in 1987. *Fire Journal* **82**(5): 33 (Chap. 1)

Kashiwagi T 1974 Experimental observation of flame spread characteristics over selected carpets. *JFF/Consumer Product Flammability* **1**: 367 (Chap. 9)

Kawagoe K 1958 Fire behaviour in rooms. *Report No. 27*, Building research Institute, Japan (Chap. 10)

Kawagoe K, Sekine T 1963 Estimation of fire temperature–time curve in rooms. *B.R.I. Occasional Report No. 11*, Building Research Institute, Japan (Chap. 10)

Keen B A 1914 The evaporation of water from soil. *Journal of Agricultural Science* **6**: 456 (Chap. 4)

Kelley K K 1934 Contributions to the data on theoretical metallurgy — II. High-temperature specific heat equations for inorganic substances. *Bulletin 371*, Bureau of Mines, US Government Printing Office, Washington DC (Chaps 5, 6)

Kelley K K 1936 Contributions to the data on theoretical metallurgy — V. Heats of fusion of inorganic substances. *Bulletin 393*, Bureau of Mines, US Government Printing Office, Washington DC (Chaps 5, 6)

Kelley K K 1949 Contributions to the data on theoretical metallurgy — X. High-temperature heat-content, heat capacity, and entropy data for inorganic compounds. *Bulletin 476*, Bureau of Mines, US Government Printing Office, Washington DC (Chaps 5, 6, 7)

Kelley K K 1960 Contributions to the data on theoretical metallurgy — XIII. High-temperature heat-content, heat capacity, and entropy data for the elements and inorganic compounds. *Bulletin 584*, Bureau of Mines, US Government Printing Office, Washington DC (Chaps 5, 6)

Kelley K K, Southard J C, Anderson C T 1941 Thermodynamic properties of gypsum and its dehydration products. *Technical Paper 625*, US Department of Interior, Bureau of Mines, Washington DC (Chap. 8)

Kersken-Bradley M 1982 A safety concept for structural fire design. In the proceedings of *VFDB (Vereinigung zur Förderung des Deutschen Brandschutzes eV) 6th International Seminar*, Karlsruhe, 21–24 September 1982, p. 155 (Chap. 11)

Keyser C A 1974 *Materials in Science and Engineering*, 2nd Edition, Charles E Merrill Publ. Co., Columbus OH (Chap. 5)

Kingery W D 1959 *Property Measurements at High Temperatures*, John Wiley & Sons, New York (Chap. 5)

Kingery W D, McQuarrie M C, Adams M, Loeb A L, Francl J, Coble R L, Vasilos T 1954 Thermal conductivity: I to X. *Journal of the American Ceramic Society* **37**: 67 (Chap. 5)

Kingery W D, Bowen H K, Uhlmann D R 1976 *Introduction to Ceramics*, John Wiley & Sons, New York (Chaps 5, 8)

Kletz T A 1971 Hazard analysis — a quantitative approach to safety. *Symposium Series No. 34*, Institution of Chemical Engineers, London, UK, p. 75 (Chap. 1)

Klingsch W 1976 *Traglastberechnung instationär thermisch belasteter schlanker Stahlbetondruckglieder mittels zwei- und dreidimensionaler Diskretisierung*. Heft 33, Institut für Baustoffkunde und Stahlbetonbau der Technischen Universität Braunschweig, Germany (Chap. 14)

Klote J H 1988 Smoke Control. In *SFPE Handbook of Fire Protection Engineering*, DiNenno P J (Ed.), 1st Edition, National Fire Protection Association–Society of Fire Protection Engineers, Quincy/Boston MA, p. 3-143 (Chap. 9)

Kordina K, Meter-Ottens C 1981 *Beton Brandschutz Handbuch*. Beton-Verlag GmbH, Düsseldorf, Germany (Chaps 4, 13)

Korn G A, Korn T M 1961 *Mathematical Handbook for Scientists and Engineers*, McGraw-Hill, New York, p. 20.5-2 (Chap. 13)

Kozu S, Masuda M, Ueda J 1929 Changes in axial ratio, in interfacial angle and in volume of calcite, caused by heating. *The Science Reports of the Tohoku Imperial University*, 3rd Series, Sendai, Japan, 3(3): 247 (Chap. 6)

Krause R F, Gann R G 1980 Rate of heat release measurements using oxygen consumption.

Journal of Fire and Flammability **12**: 117 (Chap. 9)

Kroschwitz J I, Mark H F, Bikales N M, Overberger C G, Menges G (Eds) 1985–1989 *Encyclopedia of Polymer Science and Engineering*, Vols 1 to 17, John Wiley & Sons, New York ((Chaps 3, 8)

Labour Canada 1988 *Annual Report 1986 — Fire Losses in Canada*. Catalogue No. L012-0428/86B, Ottawa, Canada (Chap. 1)

Lastrina F A, Magee R S, McAlevy R F 1971 Flame spread over fuel beds: solid-phase energy considerations. *13th Symposium (International) on Combustion*, The Combustion Institute, Pittsburgh PA, p. 935 (Chap. 9)

Law M 1971 A relationship between fire grading and design and contents. *Fire Research Note, No. 877*, Joint Fire Research Organization, Borehamwood UK (Chap. 11)

Law M, Arnault P 1972 Fire loads, natural fires and standard fires. In *Tall Building — Criteria and Loading*, Proceedings of the International Conference on Planning and Design of Tall Buildings, Lehigh University, Bethlehem PA, August, Vol. 1b, p. 475 (Chap. 10)

Lawson D I, Simms D L 1952 The ignition of wood by radiation. *British Journal of Applied Physics* **3**: 288 (Chap. 9)

Lea F M 1970 *The Chemistry of Cement and Concrete*, 3rd Edition, Edward Arnold, Glasgow (Chap. 6)

Lémeray M 1900 Sur une relation entre la dilation et la température de fusion des métaux simples. *Comptes Rendus des Scéances de l'Académie des Sciences* **131**: 1291 (Chap. 5)

Lerch W, Bogue R H 1934 Heat of hydration of portland cement paste. *Bureau of Standards Journal of Research* **12**: 645 (Chap. 6)

Levin B C, Paabo M, Gurman J L, Harris S E, Braun E 1987 Toxicological interaction between carbon monoxide and carbon dioxide. *Toxicology* **47**: 135 (Chap. 9)

Lie T T 1972 *Fire and Buildings*, Architectural Science Series, Applied Science Publishers Ltd, London (Chap. 3)

Lie T T 1974 Characteristic temperature curves for various fire severities. *Fire Technology* **10**: 315 (Chap. 10)

Lie T T 1977 Temperature distributions in fire-exposed building columns. *Journal of Heat Transfer* **99** Series C: 113 (Chap. 13)

Lie T T 1978 Calculation of the fire resistance of composite concrete floor and roof slabs. *Fire Technology* **14**: 28 (Chap. 12)

Lie T T 1984 A procedure to calculate fire resistance of structural members. *Fire and Materials* **8**: 40 (Chap. 13)

Lie T T 1988 Fire temperature–time relations. In *SFPE Handbook of Fire Protection Engineering*, DiNenno P J (Ed.), 1st Edition, National Fire Protection Association–Society of Fire Protection Engineers, Quincy/Boston MA, p. 3-81 (Chap. 11)

Lie T T 1989 Fire resistance of reinforced concrete columns: a parametric study. *Journal of Fire Protection Engineering*, **1**: 121 (Chaps 4, 12, 13, 14)

Lie T T (Ed.) 1992 *Structural Fire Protection: Design Manual of Practice*, American Society of Civil Engineers, New York (Chap. 12)

Lie T T, Allen D E 1972 Calculation of the fire resistance of reinforced concrete columns. *Technical Paper No. 378, NRCC 12797*, Division of Building Research, National Research Council of Canada, Ottawa (Chaps 12, 14)

Lie T T, Celikkol B 1991 A method to calculate the fire resistance of circular reinforced concrete columns. *ACI Materials Journal* **88**: 84 (Chap. 13)

Lie T T, Harmathy T Z 1972 A numerical procedure to calculate the temperature of protected steel columns exposed to fire. *Fire Study No. 28*, Division of Building Research, National Research Council of Canada, Ottawa (Chap. 13)

Lie T T, Harmathy T Z 1974 Fire endurance of concrete-protected steel columns. *ACI Journal, Proceedings* **71**: 29 (Chap. 12)

Lie T T, Irwin R J 1990 Evaluation of the fire resistance of reinforced concrete columns with rectangular crosssection. *Internal Report No. 601*, Institute for Research in Construction, National Research Council of Canada, Ottawa (Chap. 14)

Lie T T, Lin T D 1985 Fire performance of reinforced concrete columns. In *Fire Safety: Science and Engineering*, ASTM STP 882, Harmathy T Z (Ed.), American Society for Testing and Materials, Philadelphia PA, p. 176 (Chaps 13, 14)

Lie T T, Woollerton J L 1988 Fire resistance of reinforced concrete columns — test results. *Internal Report No. 569*, Institute for Research in Construction, National Research Council of Canada, Ottawa (Chap. 14)

Lie T T, Lin T D, Allen D E, Abrams M S 1984 Fire resistance of reinforced concrete columns. *DBR Paper No. 1167, NRCC 23065*, Division of Building Research, National Research Council of Canada, Ottawa (Chaps 13, 14)

Lin T D, Abrams M S 1983 Simulation of realistic thermal restraint during fire tests of floors and roofs. In *Fire Safety of Concrete Structures*, Publication SP-80, Abrams M S (Ed.), American Concrete Institute, Detroit MI, p. 1, (Chaps 12, 13, 14)

Lin T D, Ellingwood B, Piet O 1988 Flexural and shear behavior of reinforced concrete beams during fire tests. *Research and Development Bulletin RD091 T*, PCA R & D Serial No. 1849, Portland Cement Association, Skokie IL (Chap. 14)

Ling W C T, Williamson R B 1982 Using fire tests for quantitative risk analysis. In *Fire Risk Assessment*, ASTM STP 762, Casino G T, Harmathy T Z (Eds), American Society for Testing and Materials, Philadelphia PA, p. 38 (Chap. 15)

Linstone H A, Turoff M 1975 *The Delphi Method: Techniques and Applications*, Addison Wesley, London (Chap. 15)

Ljunggren P 1960 Determination of mineralogical transformations of gypsum by differential thermal analysis. *Journal of the American Ceramic Society* **43**: 227 (Chap. 8)

Loh H T, Fernandez-Pello A C 1986 Flow assisted flame spread over thermally thin fuels. In *Fire Safety Science*, Proceedings of the First International Symposium, Grant C E, Pagni P J (Eds), International Association of Fire Safety Science, Hemisphere Publishing Corp., Washington DC, p. 65 (Chap. 9)

Loudon A G, Stacy E F 1966 The thermal and acoustic properties of lightweight concretes. *Structural Concrete (London)* **3**: 58 (Chap. 6)

Lykov A V, Mykhaylov Y A 1961 *Theory of Energy and Mass Transfer*, Prentice-Hall, Englewood Cliffs NJ (Chap. 13)

Lynam G C 1934 *Growth and Movement in Portland Cement Paste*, 1st Edition, Oxford University Press, London (Chap. 6)

Lynch J F, Ruderer C G, Duckworth W H 1966 *Engineering Properties of Selected Ceramic Materials*. Compiled by Battelle Memorial Institute Columbus Laboratories, published by The American Ceramic Society, Columbus, OH (Chap. 5)

Lyubimova T Yu, Pinus E R 1962 Crystallization structure in the contact zone between aggregate and cement in concrete. *Kolloidnyi Zhurnal* **24**: 578 (in Russian) (Chap. 6)

Madorsky S L 1964 *Thermal Degradation of Organic Polymers*, John Wiley & Sons, New York (Chap. 9)

Magnusson S E 1978 Reducing life hazards due to fire — a governmental investigation. *FoU-brand* No. 1, p. 1 (Chap. 1)

Magnusson S E, Pettersson, O 1980/1981 Rational design methodology for fire-exposed load bearing structures. *Fire Safety Journal*, **3**: 227 (Chap. 11)

Magnusson S E, Thelandersson S 1970 Temperature—time curves of complete process of fire development. *Acta Polytechnica Scandinavica, Civil Engineering and Building Construction Series No. 65*, Stockholm (Chaps 9, 10, 11)

Malhotra H L 1956 The effect of temperature on the compressive strength of concrete. *Magazine of Concrete Research* **8**: 85 (Chap. 6)

Malhotra H L 1969 Fire resistance of structural concrete beams. *Fire Research Note No. 741*, Fire Research Station, Joint Fire Research Organization, Borehamwood UK (Chap. 14)

Malhotra H L 1977 The behaviour of polymers used in building construction. *Fire Research Note No. 1071*, Building Research Establishment, Fire Research Station, Borehamwood UK (Chap. 8)

Malhotra H L 1980 Current thinking and developments in fire testing in Europe. *Fire and Materials* **4**: 177 (Chap. 3)

Malhotra H L 1982 *Design of Fire-Resisting Structures*, Surrey University Press, Glasgow UK (Chap. 12)

Mantell C L (Ed.) 1958 *Engineering Materials Handbook*, McGraw-Hill, New York (Chap. 7)

Maréchal J C 1969 Le fluage du béton en fonction de la température. *Matériaux et Constructions — Materials and Structures* **2**(8): 111 (Chap. 6)

Maréchal J C 1972a Variations in the modulus of elasticity and Poisson's ratio with temperature. In *Concrete for Nuclear Reactors*, Vol. I, Publication SP-34, American Concrete Institute, Detroit MI, p. 495 (Chap. 6)

Maréchal J C 1972b Creep of concrete as a function of temperature. In *Concrete for Nuclear Reactors*, Vol. I, Publication SP-34, American Concrete Institute, Detroit MI, p. 547 (Chap. 6)

Mark H 1942 Intermolecular forces and mechanical behavior of high polymers, *Industrial and Engineering Chemistry* **34**: 1343 (Chap. 8)

Markstein G H, deRis J N 1972 Upward fire spread over textiles. *14th Symposium (International) on Combustion*, The Combustion Institute, Pittsburgh PA, p. 1085 (Chap. 9)

Mason B, Berry L G 1968 *Elements of Mineralogy*, W H Freeman and Company, San Francisco (Chap. 6)

Maxwell J C 1904 *A Treatise on Electricity and Magnetism*, Vol. 1, 3rd Edition, Clarendon Press, Oxford, UK, p. 440 (Chap. 5)

McAdams W H 1954 *Heat Transmission*, 3rd Edition, McGraw-Hill, New York (Chap. 13)

McBurney J W, Lovewell C E 1933 Strength, water absorption and weather resistance of building bricks produced in the United States. American Society for Testing Materials, *Proceedings of the Thirty-sixth Annual Meeting* **33** (II): 636 (Chap. 8)

McCaffrey B J 1979 Purely buoyant diffusion flames: some experimental results. *NBSIR 79-1910*, National Bureau of Standards, Center for Fire Research, Washington DC (Chap. 9)

McCaffrey B 1988 Flame height. In *SFPE Handbook of Fire Protection Engineering*, DiNenno P J (Ed.), 1st Edition, National Fire Protection Association—Society of Fire Protection Engineers, Quincy/Boston MA, p. 1-298 (Chap. 9)

McCaffrey B J, Quintiere J G, Harkleroad M F 1981 Estimating room temperatures and the likelihood of flashover using fire test data correlations. *Fire Technology* **17**: 98 (Chap. 9)

McCarter R J, Broido A 1965 Radiative and convective energy from wood crib fires. *Pyrodynamics* **2**: 65 (Chap. 9)

McGuire J H 1958 The estimation of fire resistance. *Fire Research Note No. 348*, Joint Fire Research Organization, Borehamwood UK (Chap. 12)

McGuire J H 1973 Penetration of fire partitions by plastic DWV pipes. *Fire Technology* **9**: 5 (Chap. 11)

McGuire J H 1975 Small-scale fire tests of walls penetrated by telephone cables. *Fire Technology* **11**: 73 (Chap. 11)

McGuire J H, Tamura G T 1975 Simple analysis of smoke-flow problems in high buildings. *Fire Technology* **11**: 15 (Chap. 9)

McGuire J H, Tamura G T, Wilson A G 1970 Factors in controlling smoke in high buildings. In *Fire Hazards in Buildings, ASHRAE Symposium Bulletin*, San Francisco CA, January, p. 8 (Chap. 9)

McGuire J H, Stanzak W W, Law M 1975 The scaling of fire resistance problems. *Fire Technology* **11**: 191 (Chap. 12)

Mchedlov-Petrosyan O P, Babushkin W I 1962 Thermodynamics of the hardening process of cement. In *Chemistry of Cement*, Vol. I, Proceedings of the Fourth International Symposium, 2–7 October 1960, Washington DC, National Bureau of Standards Monograph 43 — Vol. I, US Government Printing Office, Washington DC, p. 430 (Chap. 6)

McKinney P E, Clark C L 1938 *Compilation of Available High-Temperature Creep Characteristics of Metals and Alloys*, Joint Research Committee on Effect of Temperature on Properties of Metals, American Society for Testing Materials and The American Society of Mechanical Engineers, Philadelphia/New York (Chap. 7)

Mehaffey J R, Harmathy T Z 1981 Assessment of fire resistance requirements. *Fire Technology* **17**: 221 (Chap. 11)

Mehaffey J R, Harmathy T Z 1984 Failure probabilities of constructions designed for fire resistance. *Fire and Materials* **8**: 96 (Chap. 11)

Mehaffey J R, Harmathy T Z 1986 Thermal response of compartment boundaries to fire. In *Fire Safety Science, Proceedings of the First International Symposium*, Grant C E, Pagni P J (Eds), Hemisphere Publ. Co., Washington DC, p. 111 (Chap. 11)

Menzel C A 1955 A method for determining the moisture condition of hardened concrete in terms of relative humidity. *American Society for Testing Materials, Proceedings* **55**: 1085 (Chap. 4)

Meyer-Ottens C 1974 Abplatzversuche an Prüfkörpern aus Beton, Stahlbeton und Spannbeton bei verschiedenen Temperaturbeanspruchungen. *Schriftenreihe des Deutschen Ausschusses für Stahlbeton*, Heft 241, Berlin (Chap. 4)

Mitler H E 1985 The Harvard fire model. *Fire Safety Journal* **9**: 7 (Chap. 9)

Modak A T, Croce P A 1977 Plastic pool fires. *Combustion and Flame* **30**: 251 (Chap. 9)

Mori M 1983 *The Finite Element Method and Its Applications*, Macmillan, New York ((Chap. 13)

Muenow R A, Abrams M S 1986 Nondestructive testing methods for evaluating damage and repair of concrete exposed to fire. In *Evaluation and Repair of Fire Damage to Concrete*, Publication SP-92, Harmathy T Z (Ed.), American Concrete Institute, Detroit MI, p. 63 (Chap. 16)

Muskat M, Wyckoff R D 1946 *The Flow of Homogeneous Fluids through Porous Media*, 1st Edition, McGraw-Hill, New York (Chap. 4)

Nasser K W 1971 Creep of concrete at low stress–strength ratios and elevated temperatures. In *Temperature and Concrete*, Publication SP-25, Proceedings of the Symposium on the Effect of Temperature on Concrete, November 1968, Memphis TN, American Concrete Institute, Detroit MI, p. 137 (Chap. 6)

Nasser K W, Neville A M 1967 Creep of old concrete at normal and elevated temperatures. *Journal of the American Concrete Institute, Proceedings* **64**: 97 (Chap. 6)

National Fire Protection Association 1986 *NFPA 550, Firesafety Concepts Tree*. NFPA Technical Committee on Systems Concepts for Fire Protection in Structures, NFPA, Quincy MA (Chap. 15)

Neville A M 1971 *Hardened Concrete: Physical and Mechanical Aspects*, American Concrete Institute, Detroit MI (Chaps 5, 6)

Neville A M, Brooks J J 1987 *Concrete Technology*, Longman Scientific & Technical, Harlow, UK (Chap. 6)

Nissan A H, Kaye W G, Bell J R 1959 Mechanism of drying thick porous bodies during the falling rate period, I. The pseudo-wet-bulb temperature. *A.I.Ch.E. Journal* **5**: 103 (Chap. 4)

Nizamuddin Z, Bresler B 1979 Fire response of reinforced concrete slabs. *Journal of the Structural Division, Proceedings of the American Society of Civil Engineers*, **105** (No. ST8): p. 1653 (Chap. 14)

Norme française 1985 *NF P 92.501 — Fire protection — Building materials — Reaction to fire* (Chap. 3)

Nyame B K, Illston J M 1980 Capillary pore structure and permeability of hardened cement paste. In *7th International Congress on the Chemistry of Cement*, Vol. III, *Communications*, Paris 1980, p. VI-181 (Chap. 6)

Ödeen K 1963 Theoretical determination of temperature development in a number of constructions subjected to fire. *Bulletin No. 9*, Division of Building Construction, Royal Institute of Technology, Stockholm, Sweden (Chap. 10)

Oleszkiewicz I 1990 Fire exposure of exterior walls and flame spread on combustible cladding. *Fire Technology* **26**: 357 (Chaps 9, 11)

Orals D L, Quigg P S 1976 *Communication cable 'Poke Through' Floor Fire Test*, United States Gypsum Company, Des Plaines IL (Chap. 11)

Pape R, Waterman T E 1979 Understanding and modeling preflashover compartment fires. In *Design of Buildings for Fire Safety*, ASTM STP 685, Smith E E, Harmathy T Z (Eds), American Society for Testing and Materials, Philadelphia PA, p. 106 (Chap. 9)

Parker W J 1972 Flame spread model for cellulosic materials. *Journal of Fire and Flammability* **3**: 254 (Chap. 9)

Parker W J 1982 Calculations of the heat release rate by oxygen consumption for various applications. *NBSIR 81-2427-1*, National Bureau of Standards, Center of Fire Research, Washington DC (Chap. 9)

Parker W J, Long M E 1972 Development of a heat release rate calorimeter at NBS. In *Ignition, Heat Release, and Noncombustibility of Materials*, ASTM STP 502, American Society for Testing and Materials, Philadelphia PA, p. 135 (Chap. 9)

Pauls J 1988 Movement of people. In *SFPE Handbook of Fire Protection Engineering*, DiNenno P J (Ed.), 1st Edition, National Fire Protection Association–Society of Fire Protection Engineers, Quincy/Boston MA, p. 1-246 (Chap. 9)

Perry B 1986 International fire losses 1981–1984. *Fire Prevention* No. 189, p. 26 (Chap. 1)

Perry J H (Ed.) 1950 *Chemical Engineers' Handbook*, 3rd Edition, Section 3: Physical and Chemical Data, McGraw-Hill, New York, p. 107 (Chaps 5, 6, 8)

Petrella R V 1979 The mass burning rate and mass transfer number of selected polymers, wood, and organic liquids. *Polymer–Plastics Technology and Engineering* **13**: 83 (Chap. 9)

Pettersson O 1975 The connection between a real fire exposure and the heating conditions according to standard fire tests — with special application to steel structures. *Bulletin No. 39*, Division of Structural Mechanics and Concrete Construction, Lund Institute of Technology, Lund, Sweden (Chap. 11)

Pettersson O 1986 Structural fire behaviour — development trends. In *Fire Safety Science*, Proceedings of the 1st International Symposium, Grant C E, Pagni P J (Eds), International Association of Fire Safety Science, Hemisphere Publishing Corp., Washington DC, p. 229 (Chap. 11)

Pettersson O, Magnusson S E, Thor J 1976 Fire engineering design of steel structures. *Bulletin No. 50*, Swedish Institute of Steel Construction, Stockholm, Sweden (Chaps 10, 13)

Piasta J, Sawicz Z, Rudzinski L 1984 Changes in the structure of hardened cement paste due to high temperature. *Matériaux et Constructions — Materials and Structures* **17**(100): 291 (Chap. 6)

Placido F 1980 Thermoluminescence test for fire-damaged concrete. *Magazine of Concrete Research* **32**(111): 112 (Chap.16)

Platt D G, Elms D G, Buchanan A H 1989 Modelling fire spread — a time based probability approach. *Research Report 89/7*, Department of Civil Engineering, University of Canterbury, Christchurch, New Zealand (Chap. 11)

Plecnik J, Bresler B, Cunningham J D, Iding R 1980 Temperature effects on epoxy adhesives. *Journal of the Structural Division, Proceedings of the American Society of Civil Engineers*, **106**(ST1): 99 (Chap. 16)

Plecnik J, Bresler B, Chan H M, Pham M, Chao J 1982 Epoxy-repaired concrete walls under fire exposure. *Journal of the Structural Division, Proceedings of the American Society of Civil Engineers*, **108**(ST8): 1894 (Chap. 16)

Popovics S 1990 Analysis of the concrete strength versus water–cement ratio relationship. *ACI Materials Journal* **87**: 517 (Chap. 6)

Portland Cement Association 1968 *Design and Control of Concrete Mixtures*, 11th Edition, PCA, Skokie IL (Chap. 4)

Powers T C 1949 The non-evaporable water content of hardened portland cement paste — its significance for concrete research and its method of determination. *ASTM Bulletin*, No. 158. p. 68; *Bulletin 29*, Research and Development Laboratories of the Portland Cement Association, Skokie IL (Chap. 6)

Powers T C 1958 Structure and physical properties of hardened portland cement paste. *Journal of the American Ceramic Society* **41**: 1; *Bulletin 94*, Research and Development Laboratories of the Portland Cement Association, Skokie IL (Chap. 6)

Powers T C 1959 The physical structure of and engineering properties of concrete. Lecture at the Institution of Civil Engineers, London, March 1956, *Bulletin 90*, Research and Development Laboratories of the Portland Cement Association, Skokie IL (Chap. 6)

Powers T C 1962 Physical properties of cement paste. In *Chemistry of Cement*, Vol. II, Proceedings of the Fourth International Symposium, 2–7 October, 1960, Washington DC, National Bureau of Standards Monograph 43 — Vol. II, US Government Printing Office, p. 577 (Chap. 6)

Powers T C 1964 The physical structure of portland cement paste. In *The Chemistry of Cements*, Vol. I, W Taylor (Ed.), Academic Press, London, p. 391 (Chap. 6)

Powers T C 1968 Mechanisms of shrinkage and reversible creep of hardened cement paste. In *The Structure of Concrete*, Brooks A E, Newman K (Eds), Proceedings of an International Conference, London, September 1965, Cement and Concrete Association/Williams Clowes and Sons Ltd, London, p. 319 (Chaps 5, 6)

Powers T C, Brownyard T L 1948 Studies of the physical properties of hardened portland cement paste. *Bulletin 22*, Research Laboratories of the Portland Cement Association, Skokie IL [a collection of papers reprinted from the *Proceedings of the American Concrete Institute* **43** (1946-1947), pp. 101, 249, 469, 549, 669, 845, 933] (Chap. 6)

Powers T C, Copeland L E, Mann H M 1959 Capillary continuity or discontinuity in cement pastes. *Journal of the PCA Research and Development Laboratories* **1**: 38 (Chap. 6)

Prahl J, Emmons H W 1975 Fire induced flow through an opening. *Combustion and Flame* **25**: 369 (Chap. 10)

Przetak L 1977 *Standard Details for Fire-Resistive Building Construction*, McGraw-Hill, New York (Chap. 11)

Purser D A 1988 Toxicity assessment of combustion products. In *SFPE Handbook of Fire Protection Engineering*, DiNenno P J (Ed.), 1st Edition, National Fire Protection Association—Society of Fire Protection Engineers, Quincy/Boston MA, p. **1**-200 (Chap. 9)

Quintiere J 1977 Growth of fire in building compartments. In *Fire Standards and Safety*, ASTM STP 614, Robertson A F (Ed.), American Society for Testing and Materials, Philadelphia PA, p. 131 (Chap. 9)

Quintiere J 1981 A simplified theory for generalizing results from a radiant panel rate of flame spread apparatus. *Fire and Materials* **5**: 52 (Chap. 9)

Quintiere J 1985 Some factors influencing fire spread over room linings and in the ASTM E 84 tunnel test. *Fire and Materials* **9**: 65 (Chap. 9)

Quintiere J 1988a The application of flame spread theory to predict material performance. *Journal of Research of the National Bureau of Standards* **93**: 61 (Chap. 9)

Quintiere J 1988b Analytical methods of fire-safety design. *Fire Technology* **24**: 333 (Chap. 9)

Quintiere J G 1988c Surface flame spread. In *SFPE Handbook of Fire Protection Engineering*, DiNenno P J (Ed.), 1st Edition, National Fire Protection Association—Society of Fire Protection Engineers, Quincy/Boston MA, p. **1**-360 (Chap. 9)

Quintiere J G, Harkleroad M T 1985 New concepts for measuring flame spread properties. In *Fire Safety: Science and Engineering*, ASTM STP 882, T Z Harmathy (Ed.), American Society for Testing and Materials, Philadelphia PA, p. 239 (Chap. 9)

Quintiere J G, McCaffrey B J, Braven K D 1979 Experimental and theoretical analysis of quasi-steady small-scale enclosure fires. *17th Symposium (International) on Combustion*, The Combustion Institute, Pittsburgh PA, p. 1125 (Chapter 10)

Ramachandran G 1988 Value of human life. In *SFPE Handbook of Fire Protection Engineering*. DiNenno P J (Ed.), 1st Edition, National Fire Protection Association—Society of Fire Protection Engineers, Quincy/Boston MA, p. **4**-74 (Chap. 15)

Ramachandran V S, Feldman R F, Beaudoin J J 1981 *Concrete Science*, Heyden, London (Chaps 4, 5, 6)

Rasbash D J 1975 Relevance of firepoint theory to the assessment of fire behaviour of combustible materials. *International Symposium on Fire Safety of Combustible Materials*, Edinburgh University, p. 169 (Chap. 9)

Rasbash D J, Phillips R P 1978 Quantification of smoke produced at fires. Test methods for smoke and methods of expressing smoke evolution. *Fire and Materials* **2**: 102 (Chap. 9)

Rasbash D J, Rogowski Z W, Stark G W V 1956 Properties of fires of liquids. *Fuel* **31**: 94 (Chap. 9)

Razelos P 1973 Methods of obtaining approximate solutions. Section 4 in *Handbook of Heat Transfer*, Rohsenow W M, Hartnett J P (Eds), McGraw-Hill, New York, p. 4-1 (Chap. 13)

Rea M S, Clark F R S, Ouellette M J 1985 Photometric and psychophysical measurement of exit signs through smoke. *DBR Paper No 1291, NRCC 24627*, Division of Building Research, National Research Council of Canada, Ottawa (Chap. 9)

Read R E H, Adams F C, Cooke G M E 1980 *Guidelines for the Construction of Fire Resisting Structural Elements*, Building Research Establishment, London (Chap. 12)

Richter E, Sager H 1980 Erprobung von Hochtemperatur-Dehnmeßstreifen. *Sonderforschungsbereich 148*, Arbeitsbericht 1978–1980, Teil I, A/B-2, Technical University of Braunschweig, p. 355 (Chap. 7)

Richtmyer R D, Morton K W 1967 *Difference Methods for Initial Value Problems*, Interscience Publishers, New York (Chap. 13)

Rihani D N, Doraiswamy L K 1965 Estimation of heat capacity of organic compounds from group contributions. *Industrial & Engineering Chemistry Fundamentals* **4**: 17 (Chap. 5)

Ritter H L, Drake L C 1945 Pore-size distribution in porous materials. *Industrial and Engineering Chemistry, Analytical Edition* **17**: 782 (Chap. 4)

Roberts A F 1964a Calorific values of partially decomposed wood samples. *Combustion and Flame* **8**: 245 (Chap. 9)

Roberts A F 1964b Ultimate analysis of partially decomposed wood samples. *Combustion and Flame* **8**: 345 (Chap. 9)

Roberts A F, Clough G 1963 Thermal decomposition of wood in an inert atmosphere. *9th Symposium (International) on Combustion*, The Combustion Institute, Pittsburgh PA, p. 158 (Chap. 9)

Robertson A F 1973 Tests indicate venting increases smoke from some polymerics. *Fire Engineering*, September, p. 97 (Chap. 9)

Robertson A F 1979 A flammability test based on proposed ISO spread of flame test. Third progress report, *IMCO FP/215*, Intergovernmental Maritime Consultative Organization (Chap. 9)

Robertson A F 1984 Recommendations of fire test procedures for surface flammability of bulkhead and deck finish materials. Resolution A 516(13), *Fire Test Procedures*, International Maritime Organization, London (Chap. 9)

Robertson A F, Gross D 1958 An electrical-analog method for transient heat-flow analysis. *Journal of Research of the National Bureau of Standards* **61**: 105 (Chap. 12)

Rogowski B F W 1976 Plastics in buildings — fire problems and control. *Report No. CP 39/76*, Building Research Establishment, Fire Research Station, Borehamwood, UK (Chap. 8)

Ronov A B, Yaroshevskiy A A 1967 Chemical structure of the earth's crust. *Geochemistry International* **4**: 1041 (Chap. 6)

Roux H J 1982 A discussion on fire risk assessment. In *Fire Risk Assessment*, ASTM STP 762, Casino G T, Harmathy T Z (Eds), American Society for Testing and Materials, Philadelphia PA, p. 16 (Chap. 15)

Roux H J, Berlin G N 1979 Toward a knowledge-based fire safety system. In *Design of Buildings for Fire Safety*, ASTM STP 685, Smith E E, Harmathy T Z (Eds), American Society for Testing and Materials, Philadelphia PA, p. 3 (Chap. 9)

Rudolph K, Richter E, Hass R, Quast U 1986 Principles for calculation of load-bearing and deformation behaviour of composite structural elements under fire. In *Fire Safety Science*, Proceedings of the First International Symposium, Grant C E, Pagni P J (Eds), International Association of Fire Safety Science, Hemisphere Publishing Corp., Washington DC, p. 301 (Chap. 14)

Ruegg R T, Fuller S K 1984 A benefit—cost model of residential fire sprinkler systems. *NBS Technical Note 1203*, National Bureau of Standards, Washington DC (Chaps 1, 9, 15)

Saito H 1966 Explosive spalling of prestressed concrete in fire. In *Feuerwiderstandsfähigkeit von Spannbeton — Fire Resistance of Prestressed Concrete*, Kordina K (Ed.), Fédération Internationale de la Précontrainte, Bauverlag GmbH, Wiesbaden, Germany (Chap. 4)

Saito K, Quinitiere J G, Williams F A 1986 Upward turbulent flame spread. In *Fire Safety Science*, Proceedings of the First International Symposium, Grant C E, Pagni P J (Eds), International Association of Fire Safety Science, Hemisphere Publishing Corp., Washington, p. 75 (Chap. 9)

Schaffer E L 1984 Structural fire design: wood. *Research Paper FPL 450*, US Department of Agriculture, Forest Science, Forest Products Laboratory, Madison WI (Chap. 8)

Scheidegger A E 1960 *The Physics of Flow through Porous Media*, University of Toronto Press, Toronto, pp. 62, 69, 245 (Chap. 4)

Schifiliti R P 1988 Design of detection systems. In *SFPE Handbook of Fire Protection Engineering*, DiNenno P J (Ed.), 1st Edition, National Fire Protection Association—Society of Fire Protection Engineers, Quincy/Boston MA, p. 3-1 (Chap. 9)

Schleich J B 1987 Computer assisted analysis of the fire resistance of steel and composite concrete—steel structures (REFAO—CAFIR). *Report EUR 10828 EN*, Commission of the European Communities, Technical Steel Research, Directorate-General, Science, Research and Development, Luxembourg (Chap. 14)

Schneider U 1976 Behaviour of concrete under thermal steady state and nonsteady state conditions. *Fire and Materials*, **1**: 103 (Chaps 6, 14)

Schneider U 1977 Festigkeits- und Verformungsverhalten von Beton unter stationärer und instationärer Temperaturbeanspruchung. *Die Bautechnik* **54**: 123 (Chap. 6)

Schneider U (Ed.) 1985 *Properties of Materials at High Temperatures, Concrete*, published by Department of Civil Engineering, Kassel University, Germany, on behalf of RILEM, F-75015 Paris (Chap. 6)

Schneider U, Nägele E 1987 *Repairability of Fire Damaged Structures*, Draft Report for CIB W14, Gesamthochschule Kassel Universität, Kassel, Germany (Chap. 16)

Schneider U, Diederichs U, Rosenberger W, Weiss R 1980 Hochtemperaturverhalten von Festbeton (High-temperature behaviour of concrete), *Sonderforschungsbereich 148*, Arbeitsbericht 1978—1980, Teil II, B 3, Technical University of Braunschweig, p. 3 (Chap. 6)

Schulman D 1988 Money is the main reason performance codes will not continue to work. *Fire Journal* **8** (January/February), p. 14 (Chap. 2)

Schwartz K J, Lie T T 1985 Investigating the unexposed temperature criteria of standard ASTM E 119 *Fire Technology* **21**: 169 (Chap. 10)

Seader J D, Chien W P 1975 Mass optical density as a correlating parameter for the NBS smoke chamber. *Journal of Fire and Flammability* **5**: 151 (Chap. 9)

Seader J D, Einhorn I N 1977 Some physical, chemical, toxicological, and physiological aspects of fire smokes. *16th Symposium (International) on Combustion*, The Combustion Institute, Pittsburgh PA, p. 1423 (Chap. 9)

Seekamp H 1959 Brandversuche mit stark bewehrten Stahlbetonsäulen. *Deutscher Ausschuss für Stahlbeton*, Vol. 132, Wilhelm Ernst & Sohn, Berlin, p. 1 (Chap. 14)

Seekamp H, Becker W, Struck W 1964 Verhalten von Stahlbeton und Spannbeton beim Brand. *Deutscher Ausschuss für Stahlbeton*, Vol. 162, Wilhelm Ernst & Sohn, Berlin, p. 1 (Chap.14)

Segerlind L J 1984 *Applied Finite Element Analysis*, 2nd Edition, John Wiley & Sons, New York (Chap. 13)

Shah S P, Slate F O 1968 Internal microcracking, mortar–aggregate bond and the stress–strain curve of concrete. In *The Structure of Concrete*, Brooks A E, Newman K (Eds), Proceedings of an International Conference, London, September 1965, Cement and Concrete Association/Williams Clowes and Sons, Ltd, London, p. 82 (Chap. 6)

Shepherd C B, Hadlock C, Brewer R C 1938 Evaporation of surface moisture. *Industrial and Engineering Chemistry* **30**: 388 (Chap. 4)

Sherwood T K 1929, 1930, 1932 The drying of solids, Parts I to IV. *Industrial and Engineering Chemistry* **21**: 12, 976; **22**: 132; **24**: 307 (Chap. 4)

Sherwood T K, Comings E W 1933 The drying of solids, Part V. *Industrial and Engineering Chemistry* **25**: 311 (Chap. 4)

Shideler J J 1957 Lightweight-aggregate concrete for structural use. *Journal of the American Concrete Institute* **54**: 299 (Chap. 6)

Shields T J, Silcock G W H 1987 *Buildings and Fire*. Longman Scientific & Technical, Harlow, UK (Chap. 9)

Shoub H 1961 Early history of fire endurance testing in the United States. In *Fire Test Methods*, ASTM STP 301, American Society for Testing and Materials, Philadelphia PA, p. 1 (Chap. 3)

Silfwerbrand J 1990 Improving concrete bond in repaired bridge decks. *Concrete International* **12** (September): 61 (Chap. 16)

Simmons W F, Cross H C 1955 *Elevated-temperature properties of carbon steels*. ASTM STP 180, American Society for Testing Materials, Philadelphia PA (Chap. 7)

Simms D L, Hird D, Wraight H G H 1960 The temperature and duration of fires — Part I. Some experiments with models with restricted ventilation. *Fire Research Note No. 412*, Joint Fire Research Organization, Borehamwood UK (Chap. 10)

Smith C I, Kirby B R, Lapwood D G, Cole K J, Cunningham A P, Preston R R 1981 The reinstatement of fire damaged steel framed structures. *Fire Safety Journal* **4**: 21 (Chap. 7)

Smith E E 1972 Heat release rate of building materials. In *Ignition, Heat Release, and Noncombustibility of Materials*, ASTM STP 502, American Society for Testing and Materials, Philadelphia PA, p. 119 (Chap. 9)

Smith E E, Green T J 1987 Release rates for a mathematical model. In *Mathematical Modeling of Fires*, ASTM STP 983, Mehaffey J R (Ed.), American Society for Testing and Materials, Philadelphia PA, p. 7 (Chap. 9)

Smith G V 1972 *Elevated temperature static properties of wrought carbon steel*, ASTM STP 503, American Society for Testing and Materials, Philadelphia PA (Chap. 7)

Smith L, Placido F 1983 Thermoluminescence: A comparison with residual strength of various concretes. In *Fire Safety of Concrete Structures*, Publication SP-80, Abrams M S (Ed.), American Concrete Institute, Detroit MI (Chap. 16)

Smithells C J, Brandes E A (Eds) 1976 *Metals Reference Book*, 5th Edition, Butterworths, London (Chap. 7)

Sosman R B 1927 *The Properties of Silica*, The Chemical Catalog Co., New York (Chap. 6)

Spanier J, Oldham K B 1987 *An Atlas of Functions*, Hemisphere Publishing, Washington DC, p. 40:9 (388) (Chap. 13)

Spiegelhalter F 1977 Guide to design of cavity barriers and fire stops. *BRE Current Paper No CP 7/77*, Building Research Establishment, Fire Research Station, Borehamwood UK (Chap. 11)

Stanke J 1970 Brandversuche an Stahlbetonfertigstützen (3. Teil). *Deutscher Ausschuss für Stahlbeton*, Vol. 215, Wilhelm Ernst & Sohn, Berlin, p. 75 (Chap. 14)

Stanzak W W, Lie T T 1973 Fire resistance of unprotected steel columns. *Journal of the Structural Division, ASCE* **99**, No. ST5, *Proceedings*, Paper 9719, 837 (Chap. 12)

Steckler K D, Quintiere J G, Rinkinen W J 1982 Flow induced by fire in a compartment. *19th Symposium (International) on Combustion*, The Combustion Institute, Pittsburgh PA, p. 913 (Chap. 10)

Steward F R 1970 Prediction of the height of turbulent diffusion buoyant flames. *Combustion Science and Technology* **2**: 203 (Chap. 9)

Sultan M A, Halliwell R E 1990 Optimum location of fire alarms in apartment buildings. *Fire Technology* **26**: 342 (Chap. 9)

Sultan M A, Harmathy T Z, Mehaffey J R 1986a Heat transmission in fire test furnaces. *Fire and Materials* **10**: 47 (Chaps 3, 13)

Sultan M A, Lie T T, Lin J 1991 Heat transfer analysis for fire-exposed concrete slab–beam assemblies. *Internal Report No. 605*, Institute for Research in Construction, National Research Council of Canada, Ottawa (Chap. 13)

Sumi K, D'Souza M V 1982 Fire Safety of soft furnishings — an overview. *Fire and Materials* **6**: 16 (Chap 9)

Sumi K, Tsuchiya Y 1975 Toxicity of decomposition products. *Journal of Combustion Toxicology* **2**: 13 (Chap. 9)

Taday H 1989 International fire losses — Part 2 — Fires by occupancy. *Fire Prevention* No. 225, p. 23 (Chap. 1)

Taday H 1990 International fire losses — Part 3. *Fire Prevention* No. 226, p. 16 (Chap. 1)

Tamura G T, Shaw C Y 1976 Studies on exterior wall air tightness and air infiltration of tall buildings. *ASHRAE Transactions*, **82**, Part I, p. 122 (Chap. 9)

Tamura G T, Wilson A G 1966 Pressure differences for a nine-storey building as a result of chimney effect and ventilation system operation. *ASHRAE Transactions*, **72**, Part I, p. 180 (Chap. 9)

Tamura G T, Wilson A G 1967 Pressure differences caused by chimney effect in three high buildings and building pressures caused by chimney action and mechanical ventilation. *ASHRAE Transactions*, **73**, Part II, p. 2.1 (Chap. 9)

Taplin J H 1959 A method for following the hydration reaction in portland cement paste. *Australian Journal of Applied Science* **10**: 329 (Chap. 6)

Taylor H F W 1969 The calcium silicate hydrates. In *Hydration of Cements*, Vol. II,

Proceedings of The Fifth International Symposium on the Chemistry of Cement, 7—11 October 1968, Tokyo, p. 1 (Chap. 6)

Taylor H F W 1990 *Cement Chemistry*, Academic Press, London (Chap. 6)

Taylor W 1974 Review of design trends and possible fire hazards in plastics furniture. *Report No. 25/74*, Building Research Establishment, Fire Research Station, Borehamwood, UK (Chap. 8)

Taylor W 1981 Spread of fire in vertical concealed spaces. Unpublished progress report, National Research Council of Canada, Ottawa (Chap. 11)

Tewarson A 1980 Heat release rate in fire. *Fire and Materials* **4**: 185 (Chap. 9)

Tewarson A, Pion R F 1976 Flammability of plastics, I. Burning intensity. *Combustion and Flame* **26**: 85 (Chap. 9)

Thomas F G, Webster C T 1953 Investigations on building fires, Part VI. The fire resistance of reinforced concrete columns. *National Building Studies Research Paper No. 18*, Department of Scientific and Industrial Studies, HMSO, London (Chap. 14)

Thomas P H 1963 The size of flames from natural fires. *9th Symposium (International) on Combustion*, The Combustion Institute, Pittsburgh PA, p. 844 (Chap. 10)

Thomas P H, Bullen M L 1979 On the role of $K\rho c$ of room lining materials in the growth of room fires. *Fire and Materials* **3**: 68 (Chap. 9)

Thomas P H, Nilsson L 1973 Fully developed compartment fires — new correlations on burning rates. *Fire Research Note No. 979*, Joint Fire Research Organization, Borehamwood UK (Chap. 10)

Thomas P H, Heselden A J M, Law M 1967 Fully developed compartment fires — two kinds of behaviour. *Fire Research Technical Paper No. 18*, Joint Fire Research Organization, Borehamwood UK (Chap. 10)

Thor J, Sedin G 1979/1980 Fire risk evaluation and cost benefit of fire protective measures in industrial buildings. *Fire Safety Journal* **2**: 153 (Chap. 15)

Thorne J L 1979 *Plastics Process Engineering*, Marcel Dekker, New York (Chap. 8)

Todor D N 1976 *Thermal Analysis of Minerals*, Abacus Press, Tunbridge Wells, UK (Chap. 6)

Touloukian Y S, Judd W R, Roy R F 1981 *Physical Properties of Rocks and Minerals*, McGraw-Hill/CINDAS Data Series on Material Properties, Vol. II-2, McGraw-Hill, New York (Chap. 6)

Tovey A K 1986 Assessment and repair of fire-damaged concrete structures — an update. In *Evaluation and Repair of Fire Damage to Concrete*, Publication SP-92, Harmathy T Z (Ed.), American Concrete Institute, Detroit MI, p. 47 (Chaps 7, 16)

Tsuchiya Y 1981 Dynamic toxicity factor — evaluating fire gas toxicity. *Journal of Combustion Toxicology* **8**: 187 (Chap. 9)

Tsuchiya Y, Sumi K 1967 Thermal decomposition products of polyvinyl chloride. *Journal of Applied Chemistry* **17**: 364 (Chap. 9)

Tsuchiya Y, Sumi K 1971 Computation of the behavior of fire in an enclosure. *Combustion and Flame* **16**: 131 (Chap. 10)

Tsuchiya Y, Sumi K 1973 Combined lethal effect of temperature, CO_2, CO_2, and O_2 of simulated fire gases. *Journal of Fire and Flammability* **44**: 132 (Chap. 9)

Tsuchiya Y, Sumi K 1974 Smoke-producing characteristics of materials. *Journal of Fire and Flammability* **5**: 64 (Chap. 9)

Tudhope M 1989 International fire losses 1985-1987. *Fire Prevention* No. 219, p. 34 (Chap. 1)

Turner P S 1946 Thermal expansion in reinforced plastics. *Journal of Research of the National Bureau of Standards* **37**: 239; RP 1745 (Chap. 5)

Underwriters Laboratories Inc. 1981 *Fire Resistance Directory*, Northbrook, IL (Chap. 11)
Underwriters' Laboratories of Canada 1980 *List of Equipment and Materials*, Vol. II, *Building Construction*, Scarborough ON (Chap. 11)
Underwriters' Laboratories of Canada 1988 *ULC-S101,1M Criteria for Use in Extension of Data from Fire Endurance Tests*, ULC Subject C263(e) — M1988, ULC, Scarborough ON, Canada (Chap. 12)
Underwriters' Laboratories of Canada 1989 *Design of Structures for Fire Resistance*, Proposed 1st edition, Scarborough ON, Canada (Chap. 12)
US Bureau of Census 1990 *Statistical Abstract of the United States*, 110th Edition, Washington DC, Table No. 123, p. 85 (Chap. 1)

Venecanin S D 1990 Thermal incompatibility of concrete components and thermal properties of carbonate rocks. *ACI Materials Journal* **87**: 602 (Chap. 6)
Verbeck G J, Helmuth R H 1969 Structures and physical properties of cement paste. In *Properties of Cement Paste and Concrete*, Vol. III, Proceedings of The Fifth International Symposium on the Chemistry of Cement, 7—11 October 1968, Tokyo, p. 1 (Chap. 6)

Wagman D D, Evans W H, Parker V B, Schumm R H, Halow I, Bailey S M, Churney K L, Nuttall R L 1982 The NBS tables of chemical thermodynamic properties — selected values for inorganic and C_1 and C_2 organic substances in SI units. *Journal of Physical and Chemical Reference Data* **11**, Supplement No. 2 (Chap. 6)
Walton W D, Thomas P H 1988 Estimating temperatures in compartment fires. In *SFPE Handbook of Fire Protection Engineering*, DiNenno P J (Ed.), 1st Edition, National Fire Protection Association—Society of Fire Protection Engineers, Quincy/Boston MA, p. 2-16 (Chap. 9)
Wangaard F F 1958 Wood. In Section **29** of *Engineering Materials Handbook*, 1st Edition, Mantell C L (Ed.), McGraw-Hill, New York (Chap. 8)
Washburn E W 1921 Note on a method of determining the distribution of pore sizes in a porous material. *Proceedings of the National Academy of Sciences of the United States of America* **7**: 115 (Chap. 4)
Waterman T E 1968 Room flashover — criteria and synthesis. *Fire Technology* **4**: 25 (Chap. 9)
Watts J M 1988 Fire risk assessment schedules. In *SFPE Handbook of Fire Protection Engineering*. DiNenno P J (Ed.), 1st Edition, National Fire Protection Association—Society of Fire Protection Engineers, Quincy/Boston MA, p. 4-89 (Chap. 15)
Waubke N V, Schneider U 1974 Tensile stresses in concrete due to fast vapour flow. In *Pore Structure and Properties of Materials*, Proceedings of the International Symposium RILEM/IUPAC, Prague, 18—21 September, Final Report, Part III, Academia, Prague, p. D-213 (Chap. 4)
Webster C T, Raftery M M 1959 The burning of fires in rooms — Part II. Tests with cribs and high ventilation on various scales. *Fire Research Note No. 401*, Joint Fire Research Organization, Borehamwood UK (Chap. 10)
Webster C T, Smith P G 1964 The burning of well ventilated compartment fires — Part IV. Brick compartment, 2.4 m (8 ft) cube. *Fire Research Note No. 578*, Joint Fire Research Organization, UK (Chap. 10)
Webster C T, Wraight H, Thomas P H 1959 The burning of fires in rooms — Part I.

Small scale tests with cribs and high ventilation. *Fire Research Note No. 389*, Joint Fire Research Organization, Borehamwood UK (Chap. 10)

Webster C T, Raftery M M, Smith P G 1961 The burning of well ventilated compartment fires — Part III. The effect of wood thickness. *Fire Research Note No. 474*, Joint Fire Research Organization, Borehamwood UK (Chap. 10)

Weigler H, Fischer R 1972 Influence of high temperatures on strength and deformations of concrete. In *Concrete for Nuclear Reactors*, Vol. I, Publication SP-34, American Concrete Institute, Detroit MI, p. 481 (Chap. 6)

Weil N A 1964 Parametric effects governing the mechanics of ceramic materials. In *International Symposium on High Temperature Technology*, organized by Stanford Research Institute, held in Asilomar Pacific Grove CA, published by Butterworth, London (Chap. 5)

West R R, Sutton W J 1954 Thermography of gypsum. *Journal of the American Ceramic Society* 37: 221 (Chap 8)

Wickström U 1985 Application of the standard fire curve for expressing natural fires for design purposes. In *Fire Safety: Science and Engineering*, ASTM STP 882, Harmathy T Z (Ed.), American Society for Testing and Materials, Philadelphia PA, p. 145 (Chaps 10, 11)

Wiebe H F 1906 Über die Beziehung des Schmelzpunktes zum Ausdehnungskoeffizienten der starren Elemente. *Verhandlunge der deutschen physikalischen Gesellschaft* 8: 91 (Chap. 5)

Williams F A 1976 Mechanisms of fire spread. In *16th Symposium (International) on Combustion*, The Combustion Institute, Pittsburgh PA, p. 1281 (Chap. 9)

Williams-Leir G 1966 Approximations for spatial separation. *Fire Technology* 2: 136 (Chap. 11)

Williams-Leir G 1970 Another approximation for spatial separation. *Fire Technology* 6: 189 (Chap. 11)

Wilmot T 1979 European fire costs — the wasteful statistical gap. *The Geneva Papers on Risk and Insurance*, Geneva, Switzerland (Chap. 1)

Wilmot T 1989 *Bulletin No. 7*, World Fire Statistics Centre, London, UK (Chap. 1)

Wilson A G 1973 Design of large buildings for safety and health. *The Engineering Journal* 56: 33 (Chap. 2)

Wilson E L, Nickell 1966 Application of the finite element method to heat conduction analysis. *Nuclear Engineering and Design* 4: 276 (Chap. 13)

Witteveen J 1982 A systematic approach towards improved methods of structural fire engineering design. In the proceedings of *VFDB (Vereinigung zur Förderung des Deutschen Brandschutzes eV) 6th International Seminar*, Karlsruhe, 21−24 September 1982, p. 203 (Chap. 11)

Wittmann F H (Ed.) 1982 *Fundamental Research on Creep and Shrinkage of Concrete*, Martinus Nijhoff, The Hague, The Netherlands (Chap. 6)

Workshop CIB W14 1986 Design guide — structural fire safety. *Fire Safety Journal* 10: 77 (Chaps 10, 11)

World Fire Statistics Centre 1983 *Fire International* No. 82 (August/September), p. 45 (Chap. 1)

Yung D, Beck V R 1989 A risk−cost assessment model for evaluating fire risks and protection costs in apartment buildings. In *International Symposium on Fire Engineering for Building Structures and Safety*, Melbourne 14−15 November, The Institution of Engineers, Barton, Australia (Chaps 2, 15)

Zahn J J 1977 Reliability-based design procedures for wood structures. *Forest Products Journal* **27**: 21 (Chap. 11)

Zener C, Hollomon J H 1944 Effect of strain rate on the plastic flow of steel. *Journal of Applied Physics* **15**: 22 (Chap. 5)

Zienkiewicz O C, Cheung Y K 1967 *The Finite Element Method in Structural and Continuum Mechanics*, McGraw-Hill, London (Chap. 13)

Zoldners N 1960 Effect of high temperatures on concretes incorporating different aggregates. *American Society for Testing Materials, Proceedings* **60**: 1087 (Chap. 6)

Zoldners N 1971 Thermal properties of concrete under sustained elevated temperatures. In *Temperature and Concrete*, Publication SP-25, American Concrete Institute, Detroit MI, p. 1 (Chap. 6)

Zukoski E E 1986 Fluid dynamic aspects of room fires. In *Fire Safety Science*, Proceedings of the First International Symposium, Grant C E, Pagni P J (Eds), International Association of Fire Safety Science, Hemisphere Publishing Corp., Washington DC, p. 1 (Chap. 9)

Zukoski E E, Cetegen B M, Kubota T 1985 Visible structure of buoyant diffusion flames. *20th Symposium (International) on Combustion*, The Combustion Institute, Pittsburgh PA, p. 361 (Chap. 9)

Index